THE ENCYCLOPAEDIA OF IGNORANCE

Other titles of interest from Pergamon Press

A. G. BELL
The Machine Plays Chess

M. BOTVINNIK
Anatoly Karpov — His Road to the World Championship

H. B. COLE
The British Labour Party

M. COOK and R. McHENRY
Sexual Attraction

L. R. B. ELTON and H. MESSEL
Time and Man

F. H. GEORGE
Machine Takeover

INSTITUTE OF MARXISM–LENINISM
Leonid Ilyich Brezhnev: A Short Biography

VIKTOR KORCHNOI
Viktor Korchnoi's Best Games

H. LUCAS
Pensions and Industrial Relations

D. A. REAY
The History of Man-powered Flight

THE ENCYCLOPAEDIA OF IGNORANCE

Everything you ever wanted to know about the unknown

Edited by

RONALD DUNCAN

and

MIRANDA WESTON-SMITH

PERGAMON PRESS

OXFORD · NEW YORK · TORONTO · SYDNEY · PARIS · FRANKFURT

U.K.	Pergamon Press Ltd., Headington Hill Hall, Oxford OX3 0BW, England
U.S.A.	Pergamon Press Inc., Maxwell House, Fairview Park, Elmsford, New York 10523, U.S.A.
CANADA	Pergamon of Canada Ltd., 75 The East Mall, Toronto, Ontario, Canada
AUSTRALIA	Pergamon Press (Aust.) Pty. Ltd., 19a Boundary Street, Rushcutters Bay, N.S.W. 2011, Australia
FRANCE	Pergamon Press SARL, 24 rue des Ecoles, 75240 Paris, Cedex 05, France
FEDERAL REPUBLIC OF GERMANY	Pergamon Press GmbH, 6242 Kronberg-Taunus, Pferdstrasse 1, Federal Republic of Germany

---

First edition 1977
Reprinted 1978

British Library Cataloguing in Publication Data

The encyclopaedia of ignorance.
1. Science
I. Duncan, Ronald II. Weston-Smith, Miranda
500 Q158.5 77-30376

ISBN 0-08-021238-7 (Hardcover)
ISBN 0-08-022426-1 (Flexicover)

Printed in Great Britain by William Clowes & Sons, Limited
London, Beccles and Colchester

CONTENTS

EDITORIAL PREFACE

Compared to the pond of knowledge, our ignorance remains atlantic. Indeed the horizon of the unknown recedes as we approach it. The usual encyclopaedia states what we know. This one contains papers on what we do not know, on matters which lie on the edge of knowledge.

In editing this work we have invited scientists to state what it is they would most like to know, that is, where their curiosity is presently focused. We found that this approach appealed to them. The more eminent they were, the more ready to run to us with their ignorance.

As the various disciplines have become increasingly specialised, they have tended to invent a language, or as we found in the computer field, a jargon almost incomprehensible to anybody outside that subject. We have tried to curtail this parochialism and have aimed this book at the informed layman, though possibly at university level, in the hope that he will be encouraged to read papers outside his own subject.

Clearly, before any problem can be solved, it has to be articulated. It is possible that one or two of our papers might direct research or stimulate it. In so far as it succeeds in stating what is unknown the volume will be of use to science historians. A decade hence many of the problems mentioned in these pages will have been solved.

It could be said that science has to date advanced largely on the elbows and knees of technology. Even the concept of relativity depended on technology to prove its validity. In some disciplines we have already reached the point when the Heisenberg principle applies and the observer alters the object observed. And it may well be in cosmology especially, in our attitudes to space and time, that our concepts are our limiting factor. Perhaps imagination is a part of our technology? Perhaps some answers depend only on asking the correct question?

Ronald Duncan
Miranda Weston-Smith

ACKNOWLEDGEMENT

The Editors wish to acknowledge with gratitude the considerable help and the advice they have received from Professor O. R. Frisch, F.R.S. while compiling this book.

Photograph Bertl Gaye, Cambridge, U.K.

Otto R. Frisch, O.B.E., F.R.S.

Jacksonian Professor of Natural Philosophy, University of Cambridge, 1947-72, now Professor Emeritus.

He is best known as the physicist who in 1939 collaborated with Lise Meitner to produce the first definite identification and explanation of the phenomenon of "nuclear fission", a phrase which he himself coined.

Carried out research in many parts of the world, including Berlin, Copenhagen, Oxford, Los Alamos, Harwell and Cambridge, and is the author of many publications on atomic and nuclear physics.

WHY

Some 15 years ago, WHY was a magic word, used by a small boy to keep Daddy talking.

"Daddy, why does the sun go down in the West?"

"Because West is what we call the place where the sun goes down."

"But why does it go down?"

"It doesn't really; it's the Earth that turns round."

"Why?"

"Because there is no friction to stop it."

"Why . . ."

But what do we mean when we say WHY? We expect some answer; what kind of answer? "Why did Jones break his leg?"

"Because his tibia hit the kerb" says the surgeon.

"Because some fool dropped a banana skin" says Mrs Jones.

"Because he never looks where he goes" says a colleague.

"Because he subconsciously wanted a holiday" says a psychiatrist.

For any event there are several styles of answering the question why it happened. (For further confusion, see Schopenhauer's essay "Die vierfache Wurzel des Satzes vom Grunde".) But when we ask why something is so, then we are on different ground. It would be defeat for a scientist to accept the answer "because God made it so". But often there is another answer which perhaps comes to the same thing. To the question "why do intelligent beings exist?" it seems legitimate to reply "because otherwise there would be nobody to ask that question". Many popular WHYs can be answered in that manner.

Let us try another question. In my car, why does the spark plug ignite the mixture at a particular instant? There are two answers:

(1) because the cam shaft causes a spark at just that instant;

(2) because a spark at that instant gives good engine efficiency.

Answer (1) is what a physicist expects. Still, he may ask why the cam shaft has been made to cause a spark at just that time; then (2) is the answer required. It introduces a new character: the designer, with intelligence and a purpose (in this case the design of an efficient car engine).

The teleological explanation, (2), is here certainly the more telling one (except to that mythical personage, the pure scientist). To the question "why is John running?" the reply "in order to catch the bus" is satisfactory; the reply "because his brain is sending the appropriate messages to his leg muscles" (though basically correct) would be regarded as a leg-pull.

A couple of centuries ago, physical laws were often formulated in teleological language; that this was possible appeared to show that the laws had been designed to fulfil some divine purpose. For instance, a beam of light passing through refracting media (as in a telescope) was shown to travel along the path that requires the minimum time, according to the wave theory of light. But the basic law of refraction can be deduced without reference to that parsimonious principle and, moreover, allows light to travel equally on a path that requires not the least but the most time (at least compared to all neighbouring pathways).

Today such minimum (or maximum) principles are regarded merely as pretty (and sometimes useful) consequences of more basic laws, like the law of refraction; it was anyhow always obscure what divine purpose was served by making light go the quickest (or sometimes the slowest) way. Teleological explanations are not accepted nowadays; not in physics.

In biology it is otherwise; nobody doubts that many features of animals serve a purpose; claws serve to kill, legs to run, wings to fly. But is it a divine purpose? The great debate has not altogether ceased, but the large majority of scientists are agreed that natural selection can account for the appearance of purposeful design, even though some of them find it hard to imagine how such a marvellous instrument as the human eye (let alone the human brain) could have developed under the pressure of natural selection alone.

The power of artificial selection is, of course, well known to any plant or animal breeder. Admittedly, natural selection lacks the breeder's guiding hand. But it has acted through millions of years and on uncounted billions of individuals, and its power of favouring any improved adaptation to the life a species has to live is inexorable. The dug-up skeletons of horses show the development, over some millions of generations, of a rabbit-size creature to the powerful runner of today whose ancestors survived pursuit by ever faster predators. Natural selection still works today: of two varieties of moths of the same species, the darker variety predominates in smoky cities where it is well camouflaged, as his lighter cousins are in birchwoods where they in turn predominate.

As to the human eye, any light-sensitive organ, however primitive, is useful, and any improvement in sensitivity, resolution and mobility is strongly favoured by natural selection. But what about feathers? Even if a very unlikely mutation caused a reptile to have offspring with feathers instead of scales, what good would that do, without muscles to move them and a brain rebuilt to control those muscles? We can only guess. But let me mention the electric eel. It used to be a puzzle how his electric organ could have grown to its present size when in its early stages it would have been quite useless as a weapon. We now have an answer: even a feeble electric organ helps with navigation in muddy waters, and its gradual improvement has led, as it were, from a radar to a death ray.

Much about the process of evolution is still unknown; but I have no doubt that natural selection provides the justification for teleological answers.

Finally, let us go back to physics and ask a question to which, it seems, there is no answer: Why did a particular radium nucleus break up at a particular time? When the theory of atomic nuclei was young it was suggested that their complexity provided the answer: an alpha particle could escape only when all the others were in a particular configuration, as unlikely as twenty successive zeros in a game of roulette. Even with the configurations changing about 10^{20} times a second, it could take years before the right one turned up. That theory has been given up; for one thing there are much less complex nuclei with similar long lives.

Probability theory started as a theory of gambling. The apparent caprice of Lady Luck was attributed to our unavoidable ignorance of the exact way a dice was thrown; if we knew the exact way we could predict the outcome. Sure, we would have to know exactly how the dice was thrown and every detail of the surface on which it fell much more accurately than we could conceivably hope to achieve; but "in principle" it would be possible.

From those humble beginnings in a gamblers' den, the theory of probability grew in power until it took over large parts of physics. For instance, the observable behaviour of gases was accounted for by the innumerable random collisions of its molecules. Just like computing the profitability of a gambling house or an insurance company, this could be done without predicting the behaviour of single molecules. It might still be possible in principle to predict where a given molecule would be one second later; but to do that we would have to know the positions and velocities of millions of other molecules with such precision that to write down, not those numbers themselves but merely the number of decimals required, would be more than a man's life work!

With that in mind, you might find it easier to accept that quantum theory uses the concept of probability without justifying it by ignorance. Today most physicists believe that it is impossible even "in principle" to predict when a given radioactive nucleus will break up. Indeed it is only a few properties (such as the wavelength of light sent out by a given type of excited atom) for which the quantum theory allows us to calculate accurate values; in most other cases all we get is a probability that a particular event will take place in a given time.

To some people this idea of probability as a physical attribute of, say, an unstable atom seems distasteful; the idea of inexorable laws, even if we can never follow their work in detail, has not lost its appeal. Einstein felt it was essential; "God does not play dice with the world" he said. Could not the seeming randomness of atomic events result from the activities of smaller, still unknown entities? The random movements of small particles (pollen grains, etc.) in a fluid, observed in 1827 through the microscope of a botanist, Robert Brown, were later understood as resulting from the impact of millions of molecules, whose existence was merely a matter for speculation in 1827. Perhaps we shall similarly explain the random behaviour of atoms, in 40 years or so?

Such entities, under the non-committal name of "hidden variables", have been speculated on; so far they have remained hidden. Should they come out of hiding they would probably do no more than restore the illusion that the behaviour of atomic particles can be predicted "in principle". On the other hand, they may possibly predict new and unexpected physical phenomena, and that would be very exciting. I have no serious hope of that, but I can't foretell the future.

Sir Hermann Bondi, F.R.S.

Chief Scientific Adviser, Ministry of Defence and Professor of Mathematics at King's College, University of London.

Formulated the Steady State Theory of the origin of the universe with Thomas Gold.

Director-General of the European Space Research Organization 1967-71. Former Chairman of the U.K. Ministry of Defence Space Committee and the U.K. National Committee for Astronomy, and Secretary of the Royal Astronomical Society. Published widely on many aspects of Cosmology and Astrophysics.

THE LURE OF COMPLETENESS

Theories are an essential part of science. Following Karl Popper it is clear that theories are necessary for scientific progress and equally clear that they must be sufficiently definite in their forecasts to be empirically falsifiable. However, it does not follow that it is desirable, let alone necessary, for a theory to be comprehensive in the sense of leaving no room for the unknown or at least the undefined. In many instances it is the very task of the theory to describe the common features of a large group of phenomena, their range of variety necessarily stemming from something outside the theory. Consider, for example, the Galilean theory that in the gravitational field all bodies suffer the same acceleration and that in a suitably limited volume of space (e.g. a golf course) this acceleration is everywhere the same. Yet golf balls fly about in a vast variety of motions (even if one abstracts from air resistance) according as to how, when and where they have been hit by golf clubs.

The theory is explicitly designed to omit any statement of how the bodies were set in motion and indeed gains its importance from this, for otherwise it would not have its vital universality. By concentration on *accelerations*, dynamical theories allow for an input from arbitrary initial conditions of position and velocity. Any dynamical theory not doing so would be condemned to have an absurdly limited *applicability*. It is not that a dynamicist would regard initial conditions in any sense as inexplicable, but he would not view it as *his* business to explain them.

This is a very widespread characteristic of many scientific theories. Thus in Maxwellian electrodynamics the forces making charges (or current-carrying conductors) are explicitly outside the theory (provided they are electrically neutral), in hydrodynamics the position and motion of the boundaries are viewed as an external input to the theory; in the theory of the excited levels of atoms the exciting agency is taken as externally given, in General Relativity the equation of state of matter is viewed as outside the purview of the theory, etc.

In all these fields a theory that had no room for something outside itself as an essential input would be uselessly narrow. Is this a universal characteristic of scientific theories?

It may be worthwhile trying to classify the exceptions. On the one hand we have *historical* theories. Any theory of the origin of the solar system, of the origin of life on Earth, of the origin of the Universe is of an exceptional nature in the sense stated above in that it tries to describe an event in some sense *unique*.

Looking first at the problem of the origin of the solar system we do not, as yet, know of any other planetary systems though many astronomers suspect that they may be fairly or very common. Up to now the challenge is therefore to devise a sequence of occurrences by which the event of the origin of our solar system *could* have happened. This has turned out to be an extremely severe test, and it has been possible to disprove a variety of theories by demonstrating that no planetary system could have formed thus. We have reached the stage of having theories of how *a* planetary system could have formed, but not one with the actual properties of *our* planetary system, nor the stage of having several theories, each accounting satisfactorily for the features of our solar system, so that each *could* be an adequate description, but leaving us in ignorance of which it was in fact. However, it is reasonable to expect that before so very long we will have significant empirical evidence on the frequency of occurrence of planetary systems and perhaps on what are common features of such systems. Such discoveries will add statistical arguments to be considered and may do much to reassure one that one is not dealing with a truly singular event. In the case of the origin of life it may take much longer before there is any evidence on its frequency of occurrence, and we must recall Monod's warning on the indications in favour of its uniqueness, or at least its extreme rarity. Lastly the origin and evolution of the Universe are almost by definition without peers, and thus of intrinsic uniqueness.

The applicability of scientific argumentation to unique historical events is debatable. A *description* of what happened is surely the most ambitious that could be aimed for. A theory wider than this (e.g. one allowing for a whole range for the time dependence of Hubble's constant) simply does not serve the purpose of accounting for the properties of *the* Universe. For what can be the meaning of the set of unrealised universes? What was it that selected the model with the actually occurring time dependence of Hubble's constant from all the others?

But even in cosmology this demand for completeness that looks so sensible for a global feature like Hubble's constant dissolves into nonsense for characteristics on a lesser scale. The bewildering variety of our Universe is surely one of its most striking features. Even very large subunits, like galaxies, have a taxonomy of amazing complexity, varying amongst each other widely in size, in shape, in constituents, in clustering. Would one really ask of a theory of the origin and evolution of our Universe that it should result in a catalogue of galaxies with their individual properties, arranged cluster by cluster? This is surely driving the demand for completeness to absurdity. The best one can reasonably aim for is that one's theory of the Universe should provide a background against which galaxies of the kinds we know of can form. We could not wish for and could not imagine a theory of the Universe telling us why the Virgo cluster formed in one area and our local system in another near to it. We need a separate input for this, and since something external to the Universe is meaningless, our only alibi can be randomness, which fortunately is an intrinsic property of matter.

Thus we see that even the theory of a necessarily unique system like the Universe not only cannot, but must not, be complete. Similarly we would view with the utmost suspicion a theory of the origin of the solar system that necessarily led to just the set of planetary orbits, masses and satellites that we actually find.

Another case, different from the historical one, where completeness of description looks attractive at first sight, is the case of the study of overall systems, as in thermodynamics. In a certain sense a system in thermodynamic equilibrium is fully described by a small set of parameters (volume, temperature, entropy, etc.), a set we like to think of as complete. However, the very power and elegance of the thermodynamic appraisal lies in its *essential incompleteness.* Whatever the interactions between the constituent particles, whatever their character, the system's parameters give a valid and most useful description of its state. It is true that this is a description of the overall state rather than of all the detail that goes on in the micro-scale, but this detail is generally not required. The fact that we can say a great deal about such a system without knowing about it in detail is a source of pride rather than of regret at the incompleteness of our knowledge.

Similarly the existence of systems parameters such as linear momentum and angular momentum derives its value precisely because completeness of knowledge of the interior of the system is not required. No understanding whatever is needed of anatomy, physiology, or the properties of leather to establish that one cannot pull oneself up by one's bootstraps. Indeed one can argue that science is only possible because one can say *something* without knowing *everything.* To aim for completeness of knowledge can thus be essentially unsound. It is far more productive to make the best of what one knows, adding to it as means become available.

Yet in some sense the lure of completeness seems to have got hold of some of the greatest minds in physics; Einstein, Eddington, Schrödinger and most recently Heisenberg have aimed for "world equations" giving a complete description of all forces in the form perhaps of a "unified field theory". A vast number of hours and indeed years of the time of these towering intellects have been spent on this enterprise, with the end result (measured as one should measure science, by the lasting influence on others) of precisely zero.

In my view it is by no means fortuitous that all this endeavour was in vain, for I think that to aim at such completeness of description is mistaken in principle.

Science is by its nature inexhaustible. Whenever new technologies become available for experiment and observation, the possibility, indeed the probability exists that something previously not dreamt of is discovered. To look only at extraterrestrial research in the last quarter century the van Allen radiation belts, the solar wind, Mars craters, radio sources, quasars, pulsars, X-ray stars are all in this class, owing their discovery to space probes and satellites of various kinds, to radio telescopes and to new instrumentation for optical telescopes. Most of these discoveries were totally unforeseen in the antecedent picture though some of their aspects later turned out to be compatible with it. To suggest that at any stage of technical progress in experimentation and observation we have reached such a level of completeness that it is worth a major effort to encapsulate this imagined (not to say imaginary) completeness in an accordingly supposedly harmonious mathematical formulation makes little sense to me. Where there are empirical reasons to join together previously separate branches then this is a worthwhile enterprise likely to lead to important insights but where there are no such indications one is probably only indulging in a mathematical game rather than in science, for one is hardly likely to find testable observable consequences of such a purposeless unification. Of course, the fashion was started by General Relativity which unified inertia and gravitation with great success. But this was based on Galileo's observation that all bodies fall equally fast. Theory followed experiment by 300 years, for his experiment established the equality of inertia and passive gravitational mass. What is there to guide us in attempting to unify gravitation with electromagnetism and perhaps with weak or strong nuclear interactions? There are no experiments beyond those involved in General Relativity that are joining any of these fields. Until a new technology enables us to perform such experiments, the unification is virtually bound to be sterile.

The counter-argument to my scepticism has generally been that one should rely for guidance on a supposed concept of "mathematical beauty". Experience indicates that while an individual theoretician may perhaps find such a concept heuristically helpful, it is not one on which different people can agree, in stark contrast to the unanimity with which the yardstick of experimental disproof is accepted. Hence the failure of the work of Einstein and others on unified field theories to be followed up, hence the total waste of all this effort. To my mind, which is perhaps not very appreciative of the significance of mathematical beauty, the whole concept looks meaningless and arbitrary depending as it does on whether somebody invents a concise notation or whether a similarity with a previously established mathematical field can be adduced.

The aim of this article has been to show that our most successful theories in physics are those that explicitly leave room for the *unknown*, while confining this room sufficiently to make the theory empirically disprovable. It does not matter whether this room is created by allowing for arbitrary forces as Newtonian dynamics does, or by allowing for arbitrary equations of state for matter, as General Relativity does, or for arbitrary motions of charges and dipoles, as Maxwell's electrodynamics does. To exclude the unknown wholly as a "unified field theory" or a "world equation" purports to do is pointless and of no scientific significance.

R. A. Lyttleton, F.R.S.

Professor of Theoretical Astronomy, and Fellow of St John's College, University of Cambridge.

Gold Medallist of the Royal Astronomical Society 1959, and Royal Medallist of the Royal Society 1965. Member of the Institute of Astronomy, Cambridge.

Research interests include astrophysics, cosmogony, physics, dynamics, and geophysics.

THE NATURE OF KNOWLEDGE

If most of us are ashamed of shabby clothes and shoddy furniture, let us be more ashamed of shabby ideas and shoddy philosophies.

Albert Einstein

When asked where one would like to see scientific research directed, there springs to mind the question whether truly scientific research can be directed at all. It is true that once the principles of a subject have been laid down for us by geniuses such as Newton and Maxwell,† then research could be directed to studying their consequences, though even here great ingenuity requiring a high order of mathematical skill and imagination to overcome difficulties may be needed to pass from the principles to accounting for known observations and also making future predictions. The theory of the motion of the Moon provides a case in point: deep problems of mathematics have arisen all along and have had to be surmounted with the outcome of increasing accuracy of prediction, though even today, three centuries after Newton, there still remain puzzling features of the lunar motion. Are these difficulties of importance, and how much effort should be devoted to understanding them sufficiently to tackle them with success? Who is to decide such questions, and how can those doing so come to a proper assessment of such matters? With the early work to achieve more accurate measures of atomic weights, considered to be necessarily in integer-ratios, one can imagine proposals for such research turned down as of unlikely value, yet hidden there was the superlative power of nuclear energy.

Before stating my view as to where I consider effort might best be put, let us notice that our educational systems provide instruction from infancy up to doctoral level on everything appertaining to science, mathematics pure and applied, chemistry, physics, and so on, in order that pupils may achieve familiarity and facility with a subject as it exists. But the ideal objective, if a subject is not to become moribund, should be to learn it so well as to be scientifically critical of it with the object of bringing about advances in it. Sums of money vaster than ever before are enabling huge numbers of people to be drafted into scientific research, with all concerned inspired by an earnest hope that important contributions in the shape of new discoveries and fundamental principles will be made thereby. But when it comes to novel ideas, to the proposal of new hypotheses for realms in which there may be insufficient or even no established principles at all to serve as a guide, then controversy can arise and acute differences of opinion emerge, not always expressed in terms that pay due regard to the amenities of proper scientific discussion. Indeed, so strongly may some believe in their ideas, and so desirable may they consider it to promulgate them widely, that resort may be made to inadmissible ways and means to shield and defend them. (If you, dear reader, have never had any experience of this sort of thing, then your case has been more fortunate than most.)

This having briefly been said, my answer would be that the prime need of science today, so urgent that the body-scientific may choke to death in some fields of endeavour if nothing is done, is for steps to be taken to inform all those working in science what Science is really about, what is its true objective. Of course every scientist thinks he knows this, and this is right – he thinks he does, yet it has to be admitted that many very serious-minded, solid, and knowledgeable people work hard in science all their lives and produce nothing of the smallest importance, while others, few by comparison and perhaps seemingly carefree and not highly erudite, exhibit a serendipity of mind that enables them to have valuable ideas in any subject they may choose to take up. Few of the former have ever stopped off from their industrious

† My examples have to be from fields I have studied, and are not meant to suggest that there have not been equal men in other fields.

habits to consider the question of How does one know when one knows?! It is true that much has been written discussing the subject, but mainly by writers handicapped by lack of experienced appreciation of the technicalities of mathematics and science, and even lack of acquaintance with the very things they are talking about. It is reported that a somewhat pretentious hostess once asked Einstein what were the philosophical and religious implications of his theory of relativity, to which he replied, of course correctly, "None as far as I know". Occasionally a Karl Pearson may come along, leaving aside his regular work, to set down his considered views of the matter for the intended benefit of others, but praiseworthy as such excursions may be, they may do more harm than good if they misconceive the matter, as indeed Pearson did. The few that properly understand the nature and object of science are usually as a result so busy with the work created by their attitude, which leads them to unending fruitful research, that they have little time for anything but actually *doing* science. Just as, contrariwise, those that do not understand the real object of science gradually come to rest in it and perhaps take up administration wherein they may be a ready prey to non-scientific approaches, or they may go in for popular exposition and thereby attain great repute as scientists of eminence by writing for the nursery. The late H. F. Baker, a pure mathematician for whom no point was too fine, when asked his opinion of one of Sir James Jeans's famous popularisations of astronomy, said, "I wish I really knew what he puts on one page".

As best one can apprehend something essentially difficult to discover, the attitude of most of those attempting scientific research seems to be that they believe themselves to be trying to find out the properties of some real material world that actually has independent existence and works in some discoverable structured mechanical way that can be ascertained if only it is studied carefully enough, "Observe, and observe, and observe", and that the "explanation" will emerge of its own accord if only the matter is studied sufficiently long and laboriously. But it will seldom if ever do so: indeed too much in the way of observations can reveal so many seeming complications and contradictions that almost any hypothesis may appear to be ruled out. A new idea may be likened to a new-born babe: it is to be carefully nurtured and given every consideration rather than attacked with the choking diet of a multitude of so-called facts because it cannot prove at once that it will one day grow into a Samson. In this connection, it has been conjectured that had Newton known all the complex details of the lunar motion now known he might never have believed that so simple a law as the inverse-square could possibly explain them. Indeed, even after his death, the observed advance of the perigee (which Newton had in fact already solved) seemed so puzzling that it led to proposals that an inverse-cube term should be added to resolve the problem. It must also be recognised that it is often only when a theory of some phenomenon exists can new discoveries be made and new or improved theories be invented. The advance of the perihelion of Mercury could not even have been conceived or discovered till Newtonian theory was available, while the problem posed by what seems so simple a phenomenon resisted solution for over half a century awaiting the right interpretation, as many consider the general-relativity explanation to be.

But, important as it is to examine and if possible measure every possible aspect of the phenomenon of interest, it may come as a surprise to some and perhaps even be received coldly if it is stated that such activity is only a preliminary to science and not science itself, any more than the manufacture of golf-courses and equipment is golf itself, to give a trifling parallel. The true purpose of science is to invent hypotheses upon which can be developed mathematical theories and formalisms that enable predictions to be made in response to recognised objectives. But the statement raises further questions. First as to where the objectives come from, which is a subjective matter that need not be discussed here if it is admitted that men do find themselves with desired objectives, and second, which will be discussed, is where do the hypotheses come from and how are they to be invented. Newtonian dynamics enables predictions to be made of the motions of the planets and in realms far beyond that, while Maxwell's equations do this for the domain of electricity and magnetism, and quantum-theory for the infinitesimal realm of atoms and molecules.

Now although it is an essential precursor for the formulation of scientific theories that the phenomena of interest are first of all observed and if possible experimented with to at least some extent, it is of the utmost importance to recognise that no secure meaning or interpretation can be given to any observations until they are understood theoretically, or at best in terms of some hypothesis and theory based on it whether right or wrong. And it is here that one comes up against the strangely paradoxical nature of science, for the observations of phenomena are first needed to inspire someone to imagine an appropriate theory, yet they (the observations) cannot be claimed to be properly understood until a formal theory of them is available, and especially is this so where new phenomena are concerned when there are no (theoretical) means whereby the relevance of observations or experiments can be safely assessed. There can be no "facts", no reliable "evidence", until there are hypotheses and theories to test out. Before the advent of gravitational theory, comets were firmly believed to be by far the most important of all heavenly objects, and the official (verbal) theory of the phenomenon was that they were immaterial aethereal portents sent by the gods to warn man of coming violence and pestilence. The railway-lines can be observed definitely to meet in the distance, and interpretation of the observation requires some theory of space, and the lines might even actually meet if some impish engineer decided to make them do so a mile or so away! But the prediction on either hypothesis, parallel or meeting, could be tested out, for example by walking along the track to find out, or making accurate local measurements suggested by either hypothesis and a theory of space (Euclidean in this case would be adequate). The great Newton himself fell into error in this way, for when he found that the theoretical extrapolation of gravity to the distance of the Moon, which necessarily involved the "known" size of the Earth, gave an acceleration inadequate to account for the lunar motion, he abandoned his theory as contradicted by "facts". Six whole years were to go by, with the world the poorer, before it was discovered that the measured size of the Earth was considerably in error and not the gravitational theory. Observations are by no means always to be trusted as reliable guides.

But many "theories" have been constructed and proffered as scientific theories that do not really qualify as such at all. Some amount to no more than a more or less ingenious narration redescribing, sometimes imperfectly and all too often selectively, the data as known, without adding anything to these data, themselves describable in terms of some established theory: the shape of the Earth, for instance, requires for its description the theory of the Euclidean geometry of space, though the hypothesis of its near-sphericity was not always examined by some with due scientific consideration. But unless such a theory can make some prediction of its own, or suggest some crucial experiment, itself a kind of prediction, and few if any verbal theories can do this, it is not a scientific theory in a proper sense. Yet much science-literature today abounds with verbal "theories" wrapped as established stories round data susceptible to more than one interpretation than that proposed and omitting to include equally established data as irrelevant when this is not yet known. Chaucer knew about such verbal rationalisations when he wrote warningly, "You can by argument make a place A mile broad of twenty foot of space". Such theories can add nothing to the data and may even degrade them, and unless the story makes some verifiable prediction unique to the theory itself and does not conflict with other available data, one might just as well accept the data in their entirety. By means of verbal theory, the Moon could be held to the Earth with a piece of string if gravitation were to cease: the force is in the right direction, it would be possible to make a quarter of a million miles of string, and an astronaut could attach one end to the Moon, and so on. But when numbers (theory) are put in, the verbal theory collapses at once, and it is found that a cable of stoutest steel several hundred kilometres thick would be needed. Verbal descriptive theories are analogous to putting a curve through points (representing, say, a series of observations): given a number of points in a plane, say, a curve can be put through them with as great precision as one wishes mathematically, but unless the resulting curve can predict the next point and the next (not necessarily absolutely accurately), the curve is not a theory of the phenomenon and is valueless. The correct theory may not quite go through any of the points.

Before discussing how new ideas and new hypotheses, on which new theories may be built and tested, come to be invented, let us leave the theme for a moment and consider what attitude a scientist should adopt towards such novelties, or indeed towards existing ideas and theories. In a recent lecture Medawar dealt with this briefly by the piece of advice, "Never fall in love with your hypothesis". But controlled energy and enthusiasm are needed to work upon and examine a hypothesis sufficiently carefully, and these are qualities turned on as it were by emotional drive, and if one succeeds in not actually falling in love with one's ideas, which state notoriously weakens if not altogether disables a person's judgement and critical faculty, then how far should one go in relation to a new idea, whether one's own or someone else's? This is obviously a subjective question, but knowably or not, if an idea comes to the awareness of a scientist, he will begin to adopt some attitude to it. This will result from interaction of the idea with all his previous experience, remembered or not, and his character and temperament and so on, and these will combine of their own accord to determine an attitude.

The scientific attitude to adopt in regard to any hypothesis in my view (and we are talking of subjective things) can be represented schematically by means of a simple model of a bead that can be moved on a short length of horizontal wire (see diagram on next page). Suppose the left-hand end denoted by 0 (zero) and the right-hand end by 1 (unity), and let 0 correspond to complete disbelief unqualified, and the right-hand end 1 to absolute certain belief in the hypothesis. Now the principle of practice that I would urge on all intending scientists in regard to any and every hypothesis is:

> *Never let your bead ever quite reach the position 0 or 1.*

This is quite possible, for however close to the end one may have set it, there are still an infinite number of points to move the bead to in either direction in the light of new data or new arguments or whatever. If genuine scientific data reach your attention that increase your confidence in the hypothesis, then move your bead suitably towards 1, but never let it quite get there. If decreasing confidence is engendered by genuine data, then let your bead move towards 0, but again never let it quite reach there. Your changing confidence must be the result of your own independent scientific judgement of the data or arguments or proofs and so on, and not be allowed to result from arguments based on reputation of others, nor upon such things as numerical strength of believers or disbelievers. When Einstein heard that a book was being brought out entitled "A Hundred Against Einstein", he merely said "One would be enough!" My own beads for Newtonian dynamics and Maxwell's equations are very near to 1, and for flying-saucers and the Loch Ness monster very near to 0. But these it must be emphasised are my own subjective beads, and it seems there exist people whose beads for UFOs are near to 1 or even at

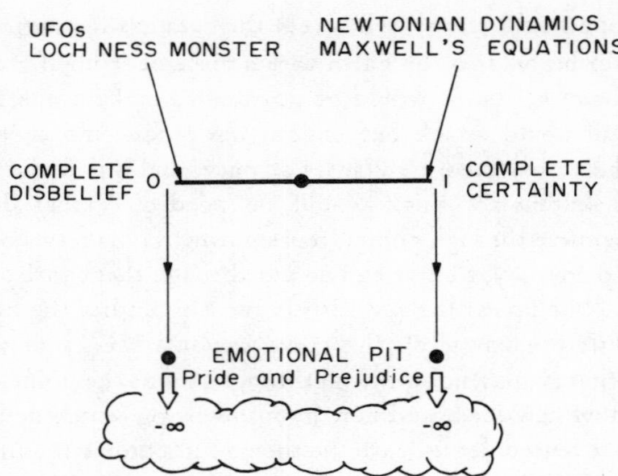

it and beyond, the consequences of which we proceed to discuss.

It seems to be a common defect of human minds that they tend to crave for complete certainty of belief or disbelief in anything. Not only is this undesirable scientifically, but it must be recognised that no such state is attainable in science. However successful and reliable a theory may be up to any point of time, further data may come along and show a need for adjustment of the theory, while at the other extreme, however little confidence one has in a hypothesis, new data may change the situation. We come now to the reason why one should never allow a bead ever to get right to 0 or 1: it is that, if one does so, the bead will fall into a deep potential-well associated with every facet of non-scientific or even anti-scientific emotion. In some cases the depth may tend to infinity, especially with advancing years, and no amount of data conflicting with the certain belief or disbelief will ever get the bead out of the well back onto the even tenor of the wire. Any attempt to bring about the uplifting of a bead so situated, by means of data or reason, can sometimes lead the owner of the bead to manifest further attitudes unworthy scientifically. In some cases it may be useless to discuss the hypothesis or theory to which the bead relates. On the other hand, if the bead is kept somewhere on the wire *between* 0 and 1 always, it can if necessary be moved quite readily in response to new data with the owner remaining calmly tranquil rather than undergoing an emotional upset. With such reaction to hypotheses and theories, one can get genuine scientific pleasure from adjusting one's beads to take account of new data and new arguments. From the small sample that my experience has limited me to, it seems regrettably to be the case that few even among scientists are always capable of keeping their beads on the wire, and much tact may be needed if one wishes to help to restore them to a rational level on the wire, if indeed in some cases it is possible at all. In Nazi Germany, it would have been dangerous indeed to have one's bead on the wire even near to 1 as an attitude to the theory that theirs was a super-race destined to rule the world; 99.9 per cent of the beads were deep down the well and only violent efforts proved sufficient to move some of them. So one of the things I would like to see scientists directed to do is always to keep their beads safely on the wire, in order that their minds may be receptive to new ideas and advances. In the words of one Chan, "Human mind like parachute: work best when open", and *open* means on the wire somewhere between 0 and 1.

When it comes to the question of how new ideas are to be arrived at, we meet up with the little-recognised fact that there is no such thing as "*the* scientific method": there is no formal procedure, no fixed set of rules, whereby new problems can be tackled or the correct interpretation of data in a new area rigidly attained, or whereby the necessary ideas to establish new principles can be reached by logical induction. However, the history of science shows that certain types of mind can see into a problem more deeply than others by some inner light that enables such minds to imagine what the solution might be, or to see if only gropingly at first what the required theory may be. It is not a matter of random guesswork, though it can be a kind of inspired guess or a series of guesses that reveals what may prove to be a valid theory for explanation of some phenomenon and for prediction in its realm.

Advancement of a subject may require first and foremost the intrusion into it of minds that intuitively perceive the phenomena in an original way and approach the problem from an untraditional point of view. With hindsight, it can be seen quite clearly just how impossible it was to solve the problem of the excess motion of the perihelion of Mercury by any hypothesis within the scope of then-existing astronomy. The imaginative conceiving of new ideas and the developing of them into a rational workable theory for new predictions ranks as mental activity at least the equal of those brought to bear in any of the realms of so-called arts, but where science is concerned it may go much further and lead on to yield a power over nature, not to mention intellectual joy and satisfaction, that no other art gives or possibly can give. It is this that makes science the noblest work of all and the greatest of the Arts. It is essentially this capability of knowledge of the future that is the keynote of real science, and the marvellous quality that gives science its uniquely powerful importance.

Before attempting construction of any theory (verbal or otherwise) to account for some phenomenon, it is of primary importance to establish that the phenomenon under discussion is actually occurring and not some wishful process swamped by the associated noise of measurement, for example. For what is to become of the theory if the phenomenon turns out later to have been purely imaginary and non-existent? Science journals today abound with elaborate theories of alleged phenomena quite inadequately established other than by often intemperate asseveration, and before the prime requirement of any theory has been found, namely an engine or cause of the phenomenon. The main evidence usually disseminated to support such "theory" often consists of no more than a sedulously conducted campaign of repetitious empty verbal propaganda assertively leading to the shallow conclusion that the theory is now "generally accepted", a claim in itself always a clear warning to regard the validity of the theory with reserve and to examine its basis (if any) calmly and independently paying no heed to the alleged numerical strength of its adherents. It is an essential part of scientific investigation to bring every detail of assumption, approximation, method, and all else to the surface, and have every component on the table, as it were, for examination and discussion: nothing should remain buried or left aside that any consideration suggests relevant until its importance or otherwise can be assessed. Equally so, in presenting scientific research publicly, it should be in such manner that a reader can recover for himself all the steps by which the results have been reached so that he can if necessary verify the conclusions for himself. In other words, none of the cards should remain face-down with assurances (or excuses) that this or that step is "all right" and can be taken for granted, because it has been put on a machine for instance, or otherwise remains inaccessible to verification. Yet many papers are deficient in this respect.

It is imperative to realise that the test of any new hypothesis or theory cannot be made by any *prior* supposedly aesthetic considerations or by moral judgement of their seeming merits or demerits in terms of existing theories. This can only be done posterior, *after* the consequences have been correctly worked out to a stage at which comparison with properly interpreted observations can be made. A theory is to be judged acceptable solely to the extent both that its results accord satisfactorily with the existing data and that future observations predicted on the basis of the theory duly come to pass with pleasing accuracy. When continuous variables are concerned, which is the most frequent case, perfect accuracy of prediction is never attainable any more than perfect accuracy of observation is possible; only if pure counting in whole numbers is entailed could this be possible, and even then of course it may not be achieved if the theory is imperfect. How little this scientific attitude to new ideas is adopted reveals itself by the sort of absurd comments they have been known to provoke, criticisms springing from beads well off the wire, such as, "The probability of the initial assumptions being true . . .", or the following unequalled gem: "The author seems unaware that the problem may be conditioned by some effect as yet unknown to science." These are just two samples from a great many that have come the way of this writer and sent out in all seriousness by chosen representatives of so-called learned societies, and if protest is made one may be informed, truthfully enough alas, that the opinions are from the most eminent referees in the field, not the most competent: the two are not always the same. It is axiomatic scientifically that no meaning attaches to such usage of the word "true": only when the initially assumed ideas have been followed through to comparison with data or to verified predictions can the stage arrive for their truth to be assessed. And just how an author can be expected to make allowance for "effects as yet unknown to science" requires an alchemy yet unknown to anyone.

Such absurdities would be avoided if the importance of keeping one's beads on the wire were sufficiently appreciated, and if it were also remembered that even though an idea should eventually turn out not to be true, in the sense that its predictions, properly evaluated of course, do not accord with observation, this alone does not constitute rigorous proof of the moral obliquity of the proposer. So reluctant are some scientists to bear such personal criticism that they shy away from the slightest risk of controversy, which new ideas frequently lead to, and devote themselves slavishly to the cloistered shelter of making routine measurements and observations, sometimes with no discernible objective. Some even make a virtue of the unimaginative collection of "facts", which they regard as the real work of science, and deride the searching for valid hypotheses as mere airy-fairy speculation. To dash a pail of water on the floor, and then set about the tedious measurement of the size and shape of all the splashes, with extensive tabulation of these, and published at great expense, would not represent in smallest degree a contribution to hydrodynamics, and obviously could never suggest the Stokes–Navier equations. Indeed, if anything, such misguided effort would far more likely prove an obstacle to any such important theoretical advance. Yet much modern work is in this sort of vein, and woe betide the theorist whose work takes no account of the resulting well-established "facts".

Any dedicated scientist will continually strive to imagine new ideas even though he recognises that most of them will probably not prove fruitful, but for every nine failures, one real success will be more than adequate reward for his efforts. But he will take care to test them out for himself by thinking them through in private, or by informal discussion with colleagues, or by observation or experiment, or when possible by formal analysis. Should these steps so reduce confidence in an idea to a point making further work upon it seem no longer worth while, then he must start over and try to imagine some new or modified idea in the hope that it may bring the data into order and suggest an appropriate theory. However, so great is the incentive to publish nowadays, that many regard it as good "scientific" practice to conduct this part of their education in public, mistakes and all. As a result, scientific literature becomes cluttered with inadequately ripe or even entirely erroneous material, which if subjected to open criticism is often, through faulty positioning of the relevant beads, quite invalidly defended, so great is the hurt to the pride of the authors and so unwilling are they to admit to the slightest error.

By way of conclusion and emphasis of the theme of this essay may be quoted what Poincaré had to say more than fifty years ago concerning an attempt to explain away the so-called ultra-violet catastrophe by means of a "theoretical" structure of matter more suggestive of poor-quality plumbing than the high-quality physical theory that eventually prevailed and revolutionised science: "It is obvious that by giving suitable dimensions to the communicating tubes connecting the reservoirs and giving suitable values to the leaks, this "theory" could account for any experimental results whatever. But this (type of descriptive story) is not the role of physical theories. They should not introduce as many (or more) arbitrary constants as there are phenomena to be explained. They should establish connections between different experimental facts, and *above all they should enable predictions to be made*." It is to precepts of this stamp, distilled from the wisdom of great thinkers, that I would wish to see the attention of research workers perennially and forcefully directed, for only by adherence to them and in no other way will science be advanced.

J. A. Wheeler

J. A. Wheeler

Professor of Physics at the University of Texas, Austin, from September 1976 and Joseph Henry Professor of Physics Emeritus at Princeton University.

A past President of the American Physical Society, member of the U.S. National Academy of Sciences and recipient of the Enrico Fermi award and the National Medal of Science.

In nuclear physics his principal work has been concerned with the alpha-particle model, nuclear rotation, resonating group structure, introduction of the scattering matrix, the mechanism of fission, the design of plutonium production reactors and thermonuclear devices, and the collective model of the nucleus. More recently he has been occupied with gravitation physics, gravitational collapse, geons, neutron stars, black holes, gravitational waves, quantum fluctuations in geometry, and the super-space description of Einstein's general relativity.

C. M. Patton

Received his A.B. in Mathematics at Princeton University in 1972 and his Ph.D. at the State University of New York at Stony Brook in 1976; American Mathematical Society Postdoctoral Fellow, Institute for Advanced Study, Princeton, New Jersey, 1977-.

IS PHYSICS LEGISLATED BY COSMOGONY?

. . . time and space are modes by which we think and not conditions in which we live

A. Einstein

A brief account of the Rutherford Laboratory Conference is followed by a review of the three levels of gravitational collapse and of the crisis confronted by all of physics in the classically predicted singularity at the end of time. One proposed way out considers that the deterministic collapse predicted by classical theory is replaced in the real world of quantum physics by a "probabilistic scattering in superspace" in which the Universe, momentarily extremely small, is heavily "reprocessed" in all its features, before starting a new cycle of big bang, expansion, constraction and collapse. This "reprocessing model" of cosmology asks no questions about cosmogony, about how the Universe came into being. The contrasting view is also considered, that the Universe comes into being at the big bang and fades away at collapse, and that all laws are forced on physics by the requirement that the Universe must have a way to come into being. The $-+++$ signature of spacetime, and the quantum principle, are two examples of how physics might be conceived to be thus legislated by cosmogony. Any hope of finding a basis in the quantum dynamics of geometry for the reprocessing model would appear to be mistaken. Five lines of evidence argue that geometry is as far from giving an understanding of space as elasticity is from giving an understanding of a solid. They also suggest that the basic structure is something deeper than geometry, that underlies both geometry and particles ("pregeometry"). For ultimately revealing this structure no perspective seems more promising than the view that it must provide the Universe with a way to come into being.

1. QUANTUM RADIANCE

Some conferences are forgotten; but how can this conference be regarded as anything but historic? Here for the first time Stephen Hawking showed us how to calculate from first principles the quantum radiance of a black hole. For the first time we had before our eyes a macroscopic and in principle observable effect born from the union of quantum mechanics and general relativity. For the first time we had a proving-ground where we could hope to see in action all the many different ways of doing the coupled wave mechanics of fields and geometry, to learn what each has to give, as we learned in earlier days about quantum electrodynamics in the proving ground of the Lamb–Retherford level shift of hydrogen. That level shift is small, as the Bekenstein–Hawking temperature ($\sim 2 \times 10^{-7}$ deg K for a black hole of solar mass) is small. Yet one decade sufficed to measure the one quantum effect with fantastic precision. Who shall say that a century will not be enough for ingenious observers from the worlds of physics and astrophysics to make at least an order-of-magnitude determination of this marvellous new effect, the quantum radiance of a black hole? And if, as Hawking suggests is at least conceivable, there are small black holes around, formed in the only way that anyone has been able to imagine, in the big bang itself, then the violent quantum burnout of one of these objects near the end of its life gives, as he points out, a second way to check on the predicted radiance.

TABLE 1. Black hole collapse, and the big bang and collapse of the Universe, as predicted by classical geometrodynamics, compared and contrasted with classically predicted collapse of the atom

System	Atom (1911)	Universe (1970s)
Dynamic entity	System of electrons	Geometry of space
Nature of classically predicted collapse	Electron headed toward point-center of attraction is driven in a finite time to infinite energy	Not only matter but space itself arrives in a finite proper time at a condition of infinite compaction
One rejected "way out"	Give up Coulomb law of force	Give up Einstein's field equation
Another proposal for a "cheap way out" that has to be rejected	"Accelerated charge need not radiate"	"Matter cannot be compressed beyond a certain density by any pressure, however high"
How this proposal violates principle of causality	Coulomb field of point-charge cannot readjust itself with infinite speed out to indefinitely great distances to sudden changes in velocity of charge	Speed of sound cannot exceed speed of light: pressure cannot exceed density of mass-energy
A major new consideration introduced by recognizing quantum principle as overarching organizing principle of physics	Uncertainty principle: binding too close to centre of attraction makes zero-point kinetic energy outbalance potential energy: consequent existence of a lowest quantum state; cannot radiate because no lower state available to drop to	"Participator" replaces the "observer" of classical physics. It is impossible in principle to separate what happens to any system, even the Universe, from what this participator does. This principle of Bohr's of the "wholeness" of nature (d'Espagnat: "non-separability") may be expected to come to the fore in a new and far deeper form

2. THE ORIGIN OF THE UNIVERSE AND THE CRISIS OF COLLAPSE

Hawking's work puts us on the road to seeing more clearly how quantum effects come into the dynamics of geometry. It brings us into closer confrontation than ever with the greatest question on the books of physics: How did the Universe come into being? And of what is it made?

Many great discoveries have been made in the past 500 years since the birth of Copernicus. None ranks higher for the light it shed on existence than the discovery by Darwin and his successors of how present life forms came into being. No discovery penetrated mysteries more widely agreed to be forever beyond the power of the mind of man to fathom. No achievement gives more hope that the next 500 years hold in store for us a still greater discovery, how the Universe itself came into being.

If there is no hope of progress towards a discovery without a paradox, we can rejoice in the paradox of gravitational collapse and the associated paradox of the big bang. Let a computing machine calculate onward instant by instant towards the critical moment, and let it make use of Einstein's standard 1915 geometrodynamics. Then a point comes where it cannot go on. Smoke, figuratively speaking, rises from the machine. Physics stops. Yet physics has always meant that which goes on its eternal way despite all the surface changes in appearances. Physics stops; but physics goes on: here is the paradox.

Physics grappled once before with a comparable paradox. In the 1910s Ernest Rutherford had shown that matter is made of highly localized positive and negative charges. Then matter must collapse in a time of the order of 10^{-17} sec. But matter does not collapse. This paradox of collapse, not the orbit of the electron in the hydrogen atom, was the overriding concern of Niels Bohr month after month. Many proposed finding a way out by giving up the Coulomb law of force at small distances, or giving up the laws of electromagnetic radiation, or both. Bohr, in contrast, held fast to both. At the same time he recognized the importance of a third law, Planck's radiation law, that seemed at first sight to have to do exclusively with quite another domain of physics. Only in that way did he find the solution of the apparent paradox. We are equally prepared today to believe that a deeper understanding of the quantum principle will someday help us resolve the paradox of gravitational collapse (Table 1).

3. GEOMETRY: PRIMORDIAL OR DERIVATIVE?

Tied to the paradox of big bang and collapse is the question, what is the *substance* out of which the Universe is made? Great attraction long attached to W. K. Clifford's 1870 proposal, and Einstein's perennial vision, that space geometry is the magic building material out of which particles and everything else are made. That vision led one to recognize perhaps earlier than might otherwise have been the case how rich are the consequences of Einstein's standard geometrodynamics, including not least constructive properties sometimes epitomized in the shorthand phrases "mass without mass" (geons), "charge without charge" (charge as electric lines of force trapped in the topology of a multiply connected space), and "spin without spin" (distinction between, and separate probability amplitudes for, the 2^n distinct geometrical structures that arise out of one and the same 3-geometry, endowed with n "handles" or "wormholes", according as one or another topologically distinct continuous field of triads is laid down on that 3-geometry). However, in the end the explorations of the Clifford-Einstein space theory of matter have taught us the deficiencies of "geometry as a building material", and Andrei Sakharov has reinforced the lesson. No one sees any longer how to defend the view that "geometry was created on 'Day One' of creation, and quantized on Day Two". More reasonable today would appear the contrary view, that "the advent of the quantum principle marked Day One, and out of the quantum principle geometry and particles were both somehow built on Day Two".

Glass comes out of the rolling mill looking like a beautifully transparent and homogeneous elastic substance. Yet we know that elasticity is not the correct description of reality at the microscopic level.

Riemannian geometry likewise provides a beautiful vision of reality; but it will be as useful as anything we can do to see in what ways geometry is inadequate to serve as primordial building material.

Then, at the end of this account, it will be appropriate to turn to the quantum principle as primordial, or as a clue to what is primordial. There our objective must of necessity be, not the right answers, but a start at the more difficult task of asking the right questions.

4. "QUANTUM GRAVITY"

What is sometimes called "quantum gravity" has a different and more immediate objective. Take as given Einstein's Riemannian space and his standard classical geometrodynamic law; and investigate the quantum mechanics of this continuous field, as one does for any other continuous field.

However workable this procedure of "quantization" is in practice for some fields and most discrete systems, we know that in principle it is an inversion of reality. The world at bottom is a quantum world; and any system is ineradicably a quantum system. From that quantum system the so-called "classical system" is only obtained in the limit of large quantum numbers. It is an accident peculiar to a sufficiently simple system that for it "the circle closes exactly"; for example, (1) quantum harmonic oscillator to (2) its limiting behaviour for large quantum numbers to (3) the "classical harmonic oscillator" to (4) the so-called "process of quantization" back to (5) the original quantum system. In contrast to this mathematics, nature does not "quantize"; it is already quantum. Quantization is a pencil-and-paper activity of theoretical physicists. This circumstance gives us always some comfort when a quantum field theory turns out to be "unrenormalizable", because we know that for nature the word "renormalization" does not even exist; nature manages to operate without divergences!

However hopeful these general considerations may be for the future they are of scant help in getting on with the immediate tasks, and tracing out the immediate consequences, of quantum gravity. No one today knows how to get quantum theory as quantum theory without having at the start the mathematical guidance of what we call a "classical" theory. No one has found a way, and nobody sees a need to find a way, to talk about quantum gravity that does not constantly make use of one or another of the concepts of canonical quantization or covariant quantization, from canonical field momentum to radiative degrees of freedom and initial value data, and from "background plus fluctuations", and "closed loops plus tree diagrams", to superspace. Abdus Salam reminded us of the important similarities and also the important differences between quantum geometrodynamics, and quantum field theory as developed in the context of Minkowski spacetime. His interesting and far-reaching survey mentioned among other points eight important "inventions" that have been made in elementary particle theory in recent years. It makes him and us happy that in quantum gravity one has as "Lehrbeispiel" a field theory that carries a far lighter burden of arbitrary elements.

For the mathematical framework in which a field theory expresses itself, one has been familiar for a long time with such alternatives as a direct description in spacetime, a description in terms of Fourier amplitudes, and a description in terms of the scattering matrix; but the latter two alternatives give a full payoff only in the context of flat spacetime. Roger Penrose and George Sparling showed in twistors quite another formalism for doing field theory, still in Minkowski geometry, but one that conceivably will let itself be extended or generalized to deal with curved space and its dynamics – if so to our great enlightenment.

The mathematical methods of quantum gravity have many beauties. Hawkings' arguments for and analysis of black hole radiance are history-making. Not only on these two counts does special interest attach to our subject today, but also because new X-ray and other astrophysical observations make it seem quite possible that the compact X-ray source Cygnus X-1 is the first identified black hole.

5. QUANTUM FLUCTUATIONS AND THE THIRD LEVEL OF GRAVITATIONAL COLLAPSE

The black hole, as "experimental model" for gravitational collapse, brings us back full-circle to the paradox that continually confronts us, and all science, the paradox of big bang and gravitational collapse of the Universe itself. The existence of these two levels of collapse reminds us, however, that theory gives us also what is in effect a third level of collapse, small-scale quantum fluctuations in the geometry of space taking place and being undone, all the time and everywhere.

Among all the great developments in physics since World War II, there has been no more impressive advance in theory than the analysis of the fluctuations that take place all the time and everywhere in the electromagnetic field. There has been no more brilliant triumph of experimental physics than the precision measurement of the effect of these fluctuations on the energy levels of the hydrogen atom. There has been no more instructive accord between observation and theory.

These developments tell us unmistakably that the electron in its travels in a hydrogenic atom is subject not only to the field Ze/r^2 of the nucleus, but also to a fluctuation field that has nothing directly to do with the atom, being a property of all space. In a region of observation of dimension L the calculated fluctuation field is of the order,

$$\Delta\epsilon \sim (\hbar c)^{1/2}/L^2 . \tag{1}$$

This field causes a displacement Δx of the electron from the orbit that it would normally follow. This displacement puts the electron at a slightly different place in the known potential $V(x,y,z)$ in which it moves. In consequence the electronic energy level undergoes a shift,

$$\Delta E \simeq [(\Delta x)^2/2] < \nabla^2 V > \text{average.} \tag{2}$$

Measuring the main part of the Lamb–Retherford shift, one determines the left-hand side of this equation. One knows the second term on the right-hand side. Thus one finds the mean squared displacement of the electron caused by the fluctuation field and confirms the predicted magnitude of the fluctuation field itself.

The considerations of principle that give one in electrodynamics the fluctuation formula (1) tell one that in geometrodynamics, in a probe region of extension L, the quantum fluctuations in the normal metric coefficients $-1, 1, 1, 1$ are of the order,

$$\Delta g \sim L^*/L. \tag{3}$$

Here

$$L^* = (\hbar G/c^3)^{1/2} = 1.6 \times 10^{-33} \text{ cm} \tag{4}$$

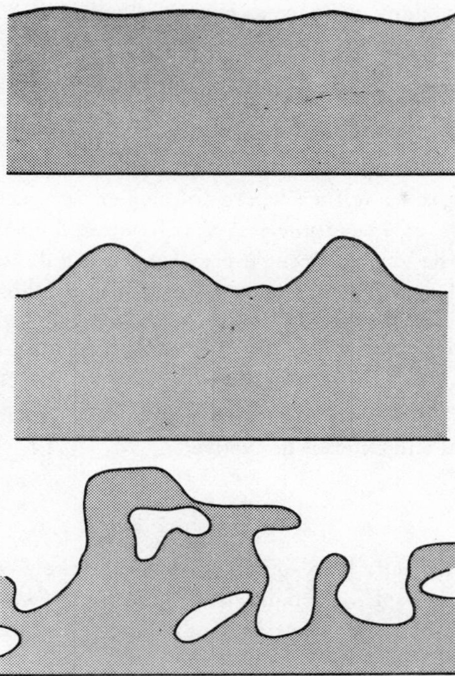

Fig. 1. Schematic representation of typical fluctuations in the geometry of spaces as picked up at three levels of observation, ranging from a large scale of observation, L, at the top, to a scale of observation comparable to the Planck length at the bottom.

is the Planck length. These fluctuations are negligible at the scale of length, L, of atoms, nuclei, and elementary particles, as the wave-induced fluctuations in the level of the ocean appear negligible to an aviator flying 10 km above it. As he comes closer, or as L diminishes, the fluctuations become more impressive (Fig. 1). Finally, when the region of analysis is of the order of the Planck length itself, the predicted fluctuations are of the order $\delta g \sim 1$. When the changes in geometry are so great, no one sees how to rule out the possibility that there will also be changes in connectivity (bottom frame in Fig. 1), with "handles" or "wormholes" in the geometry all the time and everywhere forming and disappearing, forming and disappearing ("foam-like structure of space").

One who had never heard of electricity, looking for evidence of this multiple connectivity of space, would *predict* electricity as consequence of it. Thereupon *finding* electricity in nature, he would take this discovery as evidence that space really is multiply connected in the small. Nothing prevents our rising above the accidents of history to take the same position.

The concept that electricity is lines of force trapped in the topology of a multiply connected space was put forward by Hermann Weyl in 1924, before one even knew about wave mechanics, let alone quantum fluctuations in geometry. Today quantum theory tells us that electric lines of force of fluctuation origin will thread through the typical wormhole of dimension L, carrying a flux of order

$$\int E \cdot dS \sim [(\hbar c)^{1/2}/L^2] \cdot L^2 \sim (\hbar c)^{1/2}. \qquad (5)$$

Accordingly we are led to think of space as having a kind of fluctuating foam-like structure, with everywhere positive and negative charges of order

$$q \sim (\hbar c)^{1/2} \sim 10e \tag{6}$$

continually being created and annihilated.

These fluctuation charges are not a property of elementary particles. The relevant scale of distances is twenty orders of magnitude less than nuclear dimensions. The charges are not quantized in magnitude. The charges occur everywhere, not only where there is a particle. The charges are exclusively electric in character. In evidence of this conclusion, note that the primary quantity in the quantum-electrodynamics analysis is the potential A. The equation

$$B = \nabla \times A,$$

or, in the language of forms and exterior derivative,

$$B = \mathrm{d}A, \tag{7}$$

excludes the trapping of magnetic lines of force in any wormhole. Thus, drape a closed surface S with the topology of a 2-sphere around the mouth of a wormhole, and note that the one-dimensional boundary, ∂S, of S is automatically zero. Then the magnetic pole strength, p, associated with this wormhole mouth is given by the flux integral

$$4\pi p = \int_S B = \int_S \mathrm{d}A = \int_{\partial S} A \equiv 0. \tag{8}$$

Thus the same wormhole concept that gives electricity rules out magnetic poles.

Why fluctuation charges should assemble into elementary particles, and why the net charge of each assembly should be quantized, are, like the question why should there be any elementary particles at all, beyond our power to answer today.

In times past the assumptions were tacitly made (1) that the concept of "space" makes sense at small distances and (2) that the topology, or connectivity, of space in the small is Euclidean. These assumptions left no escape from the picture of charge as a mystic magic electric jelly, or the equally unsatisfactory picture of charge as associated with a place where Maxwell's equations break down.

No one has ever been able to explain why there should be such an entity as electricity, and why there should not be such an entity as a magnetic pole, except by giving up the assumption that connectivity in the small is Euclidean, and accepting the concept of lines of force trapped in the topology of space. Nothing gives one more reason to take seriously the prediction of something like gravitational collapse going on all the time and everywhere, at small distances, and continually being done and undone, than the existence of electricity.

The value of the scale dimension L of the typical wormhole cancels out in expression (5) for the fluctuation charge itself. Moreover, the magnitude of the fluctuation charge has nothing whatsoever directly to do with the magnitude e of what on these views is an assembly of such charges. Therefore nothing can be learned from the *magnitude* of the quantum of charge about the distance down to which the concept of geometry makes sense. However, from the *existence* of charge it would seem to follow on this picture, either that the fluctuations in geometry are great enough to bring about changes in topology ($L \sim L^*$ in eqn. (3)) or that the very concept of geometry fails in such a way at small distances (distances L^* or larger) as to bring about the equivalent of a change in connectivity.

The view that large fluctuations go on at small distances puts physics in a new perspective. The density of mass-energy associated with a particle (one or a few orders of magnitude larger than the density of nuclear matter, 2×10^{14} g/cm^3) is as unimportant compared to the calculated effective density of mass-energy of vacuum fluctuations down to the Planck scale of lengths,

$$\rho \sim \frac{[(\hbar c/L^*)/c^2]}{L^{*3}} \sim \frac{M^*}{L^{*3}} \equiv \frac{2.2 \times 10^{-5}\ \text{g}}{(1.6 \times 10^{-33}\ \text{cm})^3} \sim 10^{94}\ \text{g/cm}^3, \tag{9}$$

as the density of a cloud, $\sim 10^{-6}$ g/cm^3, is unimportant compared to the density of the sky, $\sim 10^{-3}$ g/cm^3. The track of the particle looks impressive on its passage through a Wilson chamber. The white cloud, too, looks impressive in the transparent sky. However, the proper starting point in dealing with physics in the one case is the sky, not the cloud; and we are free to believe that the proper starting point in the other case is the physics of the vacuum, not the physics of the particle. To adopt this perspective does not yield any sudden illumination about either particles or the vacuum, but does at least suggest that no theory of particles that deals only with particles will ever explain particles.

6. RELATION BETWEEN THE THREE LEVELS OF COLLAPSE

Relegate to Appendix A† a few notes about the mathematical machinery for describing quantum fluctuations in the geometry and the quantum state of the geometry, and turn directly to the connections between the three levels of gravitational collapse that we have just reviewed. For definiteness adopt Einstein's view that the Universe is closed. Then we expect a spacelike singularity in the geometry at collapse similar to the spacelike singularity predicted at the big bang. Thus a black hole, once formed, cannot endure for ever. The singularity associated with it ultimately has to amalgamate with the cosmological singularity, as the icicle hanging from the roof of an ice cave, traced upward from its tip, has to join on to that roof. It does not change this conclusion to have several black holes join together into fewer and larger black holes, either early on, or as part of the final collapse itself.

The irregularities in the overall 3-geometry occasioned by black holes and other compact objects in no way prevent the systematic following out of the steady dynamics of contraction as depicted in terms of the "Kuchař–York extrinsic time",‡ τ. For any finite value of τ, however large, the spacelike hypersurface on which the 3-geometry is being studied, however close it may be to the final spacelike cosmological singularity, has still so perfectly accommodated itself to "the icicles hanging from the ceiling" that it does not quite touch the tip of any one of them. Moreover, the proper volume of this hypersurface of constant τ, as

† For reasons of brevity Appendix A has not been included in this work.

‡ This "extrinsic" time is defined in Appendix A, equation (A25). For a simple Friedmann universe of radius $a = a(t)$, at late times, one has

$$\tau = -(4/a)\,(da/dt) = [(8/3) \text{ or } 2]/(t_{\text{final}} - t), \tag{10}$$

according as the model is matter-dominated or radiation-dominated.

calculated classically, decreases indefinitely as τ increases indefinitely in the final stages of collapse. This circumstance links collapse of the Universe, and collapse to a black hole, not only with each other, but also with the third level of collapse, the collapse predicted to be all the time taking place and being undone at small distances. The classical prediction of a zero volume at infinite τ therefore surely loses force for real physics when dimensions as calculated classically have fallen to the order of magnitude of the Planck length, if not before.

7. PROBABILISTIC SCATTERING IN SUPERSPACE AND THE "REPROCESSING MODEL" OF COSMOLOGY

Why not take a model universe, for simplicity even one that derives its entire content of effective mass-energy from gravitational waves and source-free electromagnetic fields, and let the quantum dynamics of this system systematically crank ahead through the phase of collapse to whatever happens afterwards? The electron travelling towards a point centre of positive charge arrives in a finite time at a condition of infinite kinetic energy, according to classical theory, just as the Universe arrives in a finite proper time at a condition of infinite compaction. But, for the electron, quantum theory replaces deterministic catastrophe by probabilistic scattering in (x,y,z)-space. Why then for the Universe should not quantum theory replace deterministic catastrophe by probabilistic scattering in superspace? Even without the actual quantum geometrodynamic calculation, which is too difficult – and too difficult to define – for today's power of analysis, can one not conclude that any given cycle of expansion and contraction is followed, not by a unique new cycle, but a probability distribution of cycles? According to this expectation, in one such cycle the Universe attains one maximum volume and lives for one length of time; in another cycle, another volume and another time; and so on. In a few such cycles life and consciousness are possible; in most others, not; no matter; on this view the machinery of the Universe has nothing to do with man. In brief, this picture considers the laws of physics to be valid far beyond the scale of time of a single cycle of the Universe, and envisages the Universe to be "reprocessed" each time it passes from one cycle to the next (model of "reprocessing within the framework of forever frozen physical laws").

8. CONTRASTING MODEL OF "PHYSICS AS LEGISLATED BY COSMOGONY"

A universe built of geometry (and fields), with this geometry ruled by Einstein's field equation (plus quantum mechanics), is central to this "reprocessing model" of cosmology. However, an examination in §10, from five points of view, will argue that (1) "geometry" is as far from giving an understanding of space as "elasticity" is from giving an understanding of a solid; and (2) a geometrodynamic calculation extended through the regime of collapse, if extend it one ever can, will give results as misleading as those from a calculation via elasticity theory on the fracture of iron. It is difficult to escape the conclusion that "geometry" must be replaced by a more fundamental concept, as "elasticity" is replaced by a collection of electrons and nuclei and Schrödinger's equation. Whatever the deeper structure is that lies beneath particles and geometry, call it "pregeometry" for ease of reference, it must be decisive for what goes on in the extreme phases of big bang and collapse. But is it really imaginable that this deeper structure of physics should govern how the Universe came into being? Is it not more reasonable to believe the converse, that the requirement that the universe should come into being governs the structure of physics? If so, the problem of "pregeometry" is not separate from the problem of cosmogony; the two are aspects of one and the same problem.

As an illustration of what it might mean to speak of "physics as legislated by cosmogony", look at the difference between a spacetime with metric in the local tangent space of the form

$$- dt^2 + dx^2 + dy^2 + dz^2 \tag{11a}$$

and one of the form

$$+ dt^2 + dx^2 + dy^2 + dz^2. \tag{11b}$$

Even in the simplest case of an ideal spherical model universe the resulting alteration is drastic in the history of the radius, a, as a function of time: from

$$(da/dt)^2 - (a_{max}/a) = -1, \tag{12a}$$

with solution

$$a = (a_{max}/2)(1 - \cos\eta),$$

$$t = (a_{max}/2)(\eta - \sin\eta), \tag{13a}$$

to

$$(da/dt)^2 + (a_{min}/a) = +1, \tag{12b}$$

with solution

$$a = (a_{min}/2)(1 + \cosh\eta),$$

$$t = (a_{min}/2)(\eta + \sinh\eta), \tag{13b}$$

where in both cases, for simplicity, pressure is treated as negligible, as is appropriate for a mixture of stars and dust. The scale of the Universe, far from being defined by a maximum radius, is defined by a minimum radius (Fig. 2). No longer does the history show big bang or collapse. With no big bang there is no singularity, no "umbilicus", and no evident way for such a universe ever to come into being.

Shall we conclude that the only cosmology worse than a universe with a singularity is a universe without a singularity, because then it lacks the power to come alive? Is this why nature rules out a positive definite metric for spacetime? If so, this is an example of "the structure of physics as forced by the requirement for the coming into being of the universe".

The quantum principle may conceivably some day provide a second example of what it might mean to think of "physics as legislated by cosmogony". Quantum mechanics does not

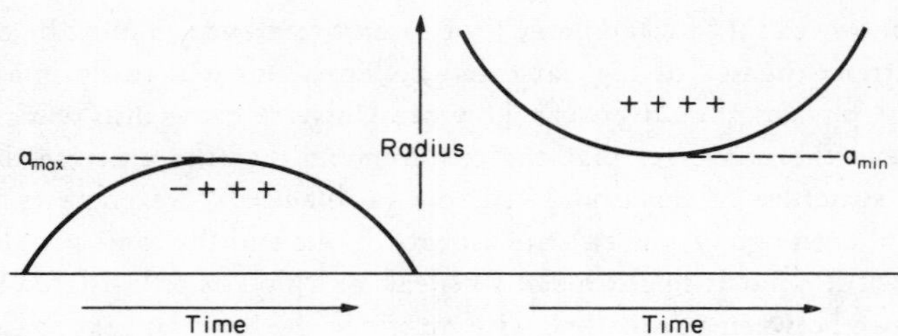

Fig. 2. Radius of an ideal spherical model universe as a function of time: left, for the $-+++$ signature of the real world; right, for $++++$ signature.

supply the Hamiltonian; it asks for the Hamiltonian; but beyond the rules of quantum mechanics for calculating answers from a Hamiltonian stands the quantum principle. It tells what question it makes sense for the observer to ask. It promotes observer to participator (Fig. 3). It joins participator with system in a "wholeness" (Niels Bohr) or "non-separability" (Bernard d'Espagnat) quite foreign to classical physics. It demolishes the view we once had that the Universe sits safely "out there", that we can observe what goes on in it from behind a foot-thick slab of plate glass without ourselves being involved in what goes on. We have learned that to observe even so miniscule an object as an electron we have to shatter that slab of glass. We have to reach out and insert a measuring device (Fig. 3). We can put in a device to measure position or we can insert a device to measure momentum. But the installation of the one prevents the insertion of the other. We ourselves have to decide which it is that we will do. Whichever it is, it has an unpredictable effect on the future of that electron. To that degree the future of the Universe is changed. We changed it. We have to cross out that old word "observer" and replace it by the new word "participator". In some strange sense the quantum principle tells us that we are dealing with a participatory universe.

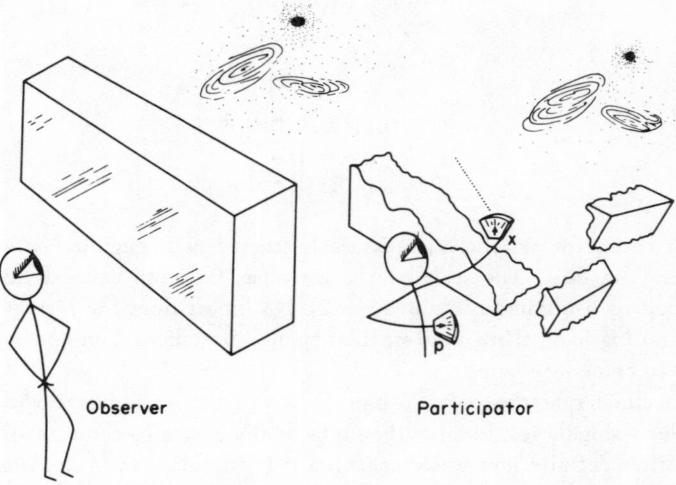

Fig. 3. The quantum principle throws out the old concept of "observer" and replaces it with the new concept of "participator". It demolishes the concept that the Universe sits "out there".

9. SELF-REFERENCE COSMOGONY

Is the necessity of this principle in the construction of the world beyond hope of explanation? Or does it, along with $-+++$ signature, originate in the requirement for the coming into being of the Universe? If so, the strange and inescapable role of observer-participator in physics cannot easily be imagined to come from anything but a strange and inescapable role for the observer-participator in cosmogony itself. The idea is very old (Parmenides of Elia, ∿ 500 B.C.; George Berkeley, ∿ 1710) that the "observer" gives the world the power to come into being, through the very act of giving meaning to that world; in brief, "No consciousness; no communicating community to establish meaning? Then no world!" On this view the Universe is to be compared to a circuit self-excited in this sense, that the Universe gives birth to consciousness, and consciousness gives meaning to the Universe (Fig. 4).

"In giving meaning to the Universe, the observer gives meaning to himself, as part of that Universe." With such a concept goes the endless series of receding reflections that one sees in a pair of facing mirrors. In this sense one is dealing with "self-reference cosmogony".

No test of such thinking would be more decisive than a derivation of the quantum principle from this "self-reference cosmogony", as one long ago derived the formula for the energy of a moving electron from "relativity". For developing a mathematical description of this new kind of "meaning circuit" one has as guides: (1) what one already knows from quantum mechanics; (2) what one has learned from the central role of self-reference in the revolutionary discoveries of recent decades in mathematical logic – the only branch of mathematics that has the power "to think about itself"; and (3) experience in analysing physical situations that transcend time and bring past, present and future together under one roof.

As contrasted to the cycle-after-cycle "reprocessing model" of cosmology – which is indifferent to how the world came into being – self-reference cosmogony has these features: (1) one cycle only; (2) the laws and constants and initial conditions of physics frozen in at the big bang that brings the cycle and dissolved away in the final extremity of collapse;† and (3) a guiding principle of "wiring together", past, present, and future that does not even let the Universe come into being unless and until the blind accidents of evolution‡ are guaranteed to produce, for some non-zero stretch of time in its history-to-be, the consciousness, and consciousness of consciousness, and communicating community, that will give *meaning* to that universe from start to finish.

Of all evidence that might distinguish between the two models, one the reprocessing of a forever existing universe, the other the coming-into-being and fade-out of a one-cycle-only universe, nothing attracts inquiry more immediately than geometry. Is space really a 3-manifold? And is its quantum dynamics really ruled by Einstein's field equation? "Yes" is the working conception of the reprocessing model. "No" has to be the answer of self-reference cosmogony. True mathematical 3-geometry "is simply there"; it has no way to "come into

† On this view of "big bang as coming into being, collapse as dissolution" there is nothing "before" the moment of commencement of the Universe, and nothing "after" the moment of collapse, not least because already in the 3-geometry-superspace formulation of quantum geometrodynamics all meaning is denied to the terms "before" and "after", and even to "time", in the analysis of (1) what goes on at sufficiently small distances and therefore (2) what happens to the Universe itself when *it* is sufficiently small.

‡ Chance mutation, yes; Darwinian evolution, yes; yes, the general is free to move his troops by throwing dice if he chooses; but he is shot if he loses the battle. Deprived of all meaning, stripped of any possibility to exist, is any would-be universe where Darwinian evolution brings forth no community of evidence-sharing participators, according to the view of "self-reference cosmogony" under examination here.

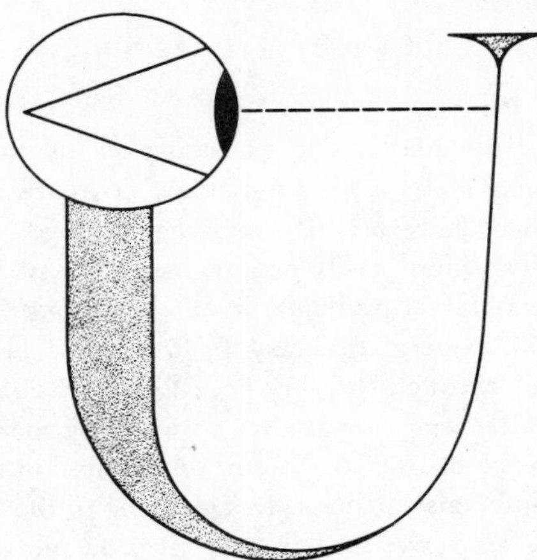

Fig. 4. Symbolic representation of the Universe as a self-excited system brought into being by "self-reference". The Universe gives birth to communicating participators. Communicating participators give meaning to the Universe.

being". If the world contains any such ingredient, the idea is contradicted from the start that all physics, all structure, and all law arise from the requirement that the Universe must have a way to come into being. Thus the relevance is clear of a closer look at the role of geometry in physics.

To suppose that the attention of all investigators is now limited, or ever will be limited, to the two cited concepts of the Universe, would be to underestimate the exploratory character of the community of science, that makes it so effective in uncovering new evidence, winning unexpected insights, and achieving final consensus. More ideas and more cosmologies will surely receive attention, not least because our understanding of the tie between cosmogony and the basic structure of physics ("structural cosmogony") is today so clearly in its medieval infancy. The road ahead can hardly help being strewn with many a mistake. The main point is to get those mistakes made and recognized as fast as possible! It may help in this work to reassess the concept of geometry, regardless of its bearing on reprocessing cosmology or self-reference cosmogony.

10. REASSESSMENT OF "GEOMETRY"

Five considerations suggest that 3-geometry does not give a correct account of physics and ought to be replaced by a more basic structural concept ("pregeometry"): (1) Mathematical space doesn't tear, but physical space must tear. (2) Mathematical space has only geometrical properties ("surface geology") but physical space is full of virtual pairs of all kinds of particles ("underground strata"). (3) It is difficult to avoid the impression that every law of physics is "mutable" under conditions sufficiently extreme, and therefore that geometry itself – a part of physics – must also be mutable. (4) Einstein's law for the dynamics of geometry follows from nothing deeper than the "group" of deformations of a spacelike hypersurface (Hojman, Kuchař and Teitelboim) and therefore tells nothing more about the underlying structure than crystallography tells about the structure of an atom. (5) Gravitation, as the "metric elasticity of space" (Andrei Sakharov), is as far removed from the deeper physics of space as elasticity itself is from the deeper physics of a solid.

"Doesn't Tear; Must Tear"

Nothing speaks more strongly than the existence of electricity, and the non-existence of magnetic poles, for small-scale fluctuations in the connectivity as well as in the geometry of space. What are the consequences of such changes in connectivity? When a handle thins and breaks, two points part company that were once immediate neighbours. But in quantum physics no change is sudden. There must be some residual connection between these two points. Moreover, there is nothing special about these two points. If there is a physical tie between them, there must also be a physical tie between every point and every other point that is very foreign to differential topology. The concept of nearest neighbor would seem no longer to make any sense; and with it gone, even the foundation for the idea of dimensionality disappears. Farewell geometry!

"Surface versus Stratigraphic Geology"

When sufficient electromagnetic energy is imploded into a sufficiently small volume of space, pairs of particles emerge. The "stratigraphic structure" built into the vacuum, and thus revealed, cannot be disregarded merely because it does not seem to lend itself to description in geometrical language, however different it is in kind from the Riemannian curvature ("surface geology") that describes gravitation. Moreover, if gravitation-as-curvature is universe-wide, so is this stratigraphy, to our astonishment. How did a quasar at red-shift $z = 3$, so separated from us in place and time, "know enough" to show by its spectrum the same atomic properties as an atom in the here and now? The elementary particles in the source at the time of emission had been around only $\sim 1 \times 10^9$ yr since the big bang, whereas they would have had to reach an age $\sim 2 \times 10^9$ yr to get the first information from us about particle properties here. Zel'dovich gives a beautiful description of pair production, showing how the particles come into evidence at a spacelike separation of the order of magnitude of the Compton wavelength. No way is evident of reconciling with the principle of causality the identity of stratigraphy between points with spacelike separations of 10^9 ℓyr and 10^{10} ℓyr except to say that they also once had separations of the order of a Compton wavelength. Whatever this consideration does to explain how mass-spectral information imprinted on space at different points can be identical, it does nothing to explain how the imprinting itself is possible. That is a property of empty space quite irreconcilable, so far as we can see, with any known concept of geometry whatsoever. Farewell geometry.

Mutability

Physics can be viewed as a staircase. Each tread marks a new law. Each riser symbolizes the achievement of conditions sufficiently extreme to overpower that law. Density seemed a conserved quantity until one discovered that sufficiently great pressures alter density. Valence won ever wider fields of application until one found that temperatures can be raised so high that the concept of valence loses its usefulness. The fixity of atomic properties is transcended in thermonuclear reactions. The principle of the conservation of lepton numbers and baryon numbers, discovered to be central to elementary particle physics, is transcended in the *classical* physics of black holes. The black hole obeys in its transformations the laws of conservation of mass-energy, electric charge, and angular momentum. However, none of these would-be conserved quantities has any meaning when one turns from a finite system to a closed universe. Finally, with the classically predicted collapse of the Universe, space and time themselves are transcended as categories, and the framework would seem to collapse for everything one has ever called a law, not least the concept of geometry itself. Farewell geometry.

Groups Conceal Structure

Pile up a 50-m-high pyramid of wrecked automobiles and steel bed springs, attach a spring to the top, drive it with slowly changing period, and measure the characteristic frequencies. Each mode of vibration is orthogonal to all the others, but not one clue does this fact, or all the details of the frequency spectrum, provide as to the constitution of the vibrating system. Nor does the cubic symmetry of salt reveal anything about the constitution of the sodium atom. Turning from the sodium atom to other selected atoms, and analysing the pattern of the lowest score or so of energy levels – H. P. Dürr has shown – one would conclude that the system in question has SU(3) or higher symmetry; but nothing could be more ridiculous than to conclude from this circumstance that the outer structure of the atom is made up from quarks or from hadrons.

More generally, much that appears to be "structure" turns out on more detailed analysis to be mere consequence of the principle that "a symmetry is spontaneously broken when for a physical problem invariant under a group G there exist solutions (which can be grouped into orbits of G) which are only invariant under a strict subgroup of G"; hence features of physics like the Jahn–Teller effect, loss by crystals of certain symmetries, spontaneous magnetization, and the Cabibbo angle for the weak current. Groups hide structure! Similarly in general relativity.

All the laws of gravitation follow from the Einstein–Hamilton–Jacobi equation, and this equation follows from a beautiful argument of group theory (invariance with respect to the "group" of deformations of a spacelike hypersurface; in evolving dynamically on σ from σ' to σ'', the system must end up in the same state on σ'' whether σ was pushed ahead faster, first "on the left", then "on the right", or faster first on the right, then on the left). The very fact that group theory is the heart of this derivation of the dynamics of geometry, more than concealing from view a structure deeper than "ideal mathematical geometry", would seem a prime sign that there must be such an underlying structure. Farewell geometry.

Gravitation as the "Metric Elasticity of Space"

Is geometry the magic building material out of which everything is made? To try to build particles out of geometry is as mistaken, Sakharov in effect argues, as to try to build atoms out of elasticity; the plan of construction goes the other way around. The two elastic constants of a homogeneous isotropic material not only summarize but also hide from view the second derivatives of the hundred atomic and molecular bonds that are the source of that elasticity. Sakharov proposes to look at gravitation in similar terms. Space, in his view, if we may coin an analogy, is like an empty sausage skin. It puts up no resistance to being bent until it is pumped full of sausage meat. This sausage meat is the zeropoint energy of particles and fields. When space is curved, correction terms arise in the renormalized invariant Lagrangian density for each field, proportional to the four-dimensional Riemann scalar curvature invariant,

$$\delta L_{\text{one field}} \sim \hbar c\, {}^{(4)}R \int_0^{k_{\text{cut off}}} k\, dk; \tag{14}$$

and with the cut-off wave number taken to be of the order of the reciprocal of the Planck length (to give the right answer!) this agrees in form and general magnitude with the Lagrangian for the gravitational field itself. The constant of gravitation, on this view, measures the "metric elasticity" that space acquires from its content of particles and fields.

Though a step forward, it was not a simplification to replace two elastic constants of a solid by the potential energy curves of a hundred molecular bonds. The real simplification came when one recognized the solid as nothing but a system of charged masses moving in accordance with the laws of quantum mechanics. The thousand remarkable features of chemistry and chemical physics found their explanation, not in a thousand specially tailored laws, but one simple picture. Neither can we believe that Sakharov has gone the whole road when he explains the resistance of geometry to bending as a consequence of the complicated physics of dozens of varieties of particles and fields. Neither particles explained in terms of geometry, nor geometrodynamics explained in terms of particles, gives the truly simple concept, we have to believe, but both entities expressed in terms of a structure deeper than either, call it pregeometry or call it what one will. Farewell geometry.

In conclusion, it would appear from these five considerations of tearing, stratigraphy, mutability, group concealment, and metric elasticity that the concept of "ideal mathematical geometry" as applied in physics is too finalistic to be final and must give way to a deeper concept of structure. Towards the finding of this "pregeometry" no guiding principle would seem more powerful than the requirement that it should provide the Universe with a way to come into being. It is difficult to believe that we can uncover this pregeometry except as we come to understand at the same time the necessity of the quantum principle, with its "observer–participator", in the construction of the world. Not "machinery", but a guiding principle, is what we seek. It would seem a great mistake to expect a "cheap solution" of the crisis of collapse. Does this mean that we must decipher almost everything before we will be able to understand anything? Certainly of problems of less than cosmic magnitude in gravitation physics and quantum gravity there are more, and more of greater interest, outstanding today, than at any time we can name. They range from renormalization to the mathematics of topology change, and from the hydrodynamics and gravitational radiation associated with the collapse of a rotating star, to the Hawking radiation and the polarization of the vacuum around a black hole. As for the problem of "structural cosmogony", the "great" problem, we are reminded that every great crisis in science has created its own hard conditions – and opportunities – for making progress. In looking for a satisfactory model to bring into harmony the rich evidence that nature lays before us, perhaps we can derive some comfort from the words of the engine inventor, John Kris, "Start her up and see why she don't go".

Acknowledgement

This article originally appeared in *Quantum Gravity*, edited by C. J. Isham, R. Penrose and D. W. Sciama, Clarendon Press, Oxford, 1975. It included two appendices, bibliography, references and acknowledgements, all of which are omitted here in the interest of brevity.–Ed.

I. W. Roxburgh

Professor of Applied Mathematics at Queen Mary College, University of London, and Chairman of the London University Astronomy Committee.

Member of the Royal Astronomical Society and the International Astronomical Union, Member of the Council of the Royal Astronomical Society 1968-71.

Research interests include the History and Philosophy of Science, Cosmology, Evolution of Stars, Solar Physics and Gravitational Theory.

THE COSMICAL MYSTERY— THE RELATIONSHIP BETWEEN MICROPHYSICS AND COSMOLOGY

Is the Universe of which we are a part unique, or is it just one out of a whole family of possible Universes? Is the world as it is because there is no other way for it to be, or because a Creator arbitrarily chose to create it this way rather than another? Can physics help to answer these questions?

The laws of physics so far uncovered by man contain arbitrary dimensionless numbers whose values determine the properties of the Universe; change these values and the properties of the Universe change, so at present physics suggests an ensemble of possible Universes. But are these numbers arbitrary or is it that we have yet to discover the interrelatedness of nature and how this uniquely determines their values? It would be arrogant indeed to think that our generation of scientists has uncovered the "real" laws of nature, so the position is still unclear, but there are clues that an interrelatedness exists, waiting to be uncovered: these arbitrary numbers are either about one, or the huge number 10^{39}, surely this is not chance, there must be some relation between such huge numbers, but what is it?

THE PROBLEM

Our present description of the physical world is governed by empirically determined constants like the mass and charge of an electron, or the rate of expansion of the Universe, a partial list of these is

$c = 3.0 \times 10^{10}$ cm sec^{-1} : velocity of light,
$\hbar = 1.5 \times 10^{-27}$ erg sec : Dirac–Planck constant,
$e = 4.8 \times 10^{-10}$ erg$^{1/2}$ cm$^{1/2}$: charge on a proton,
$m = 1.6 \times 10^{-24}$ g : mass of a proton,
$m_e = 9.1 \times 10^{-28}$ g : mass of an electron,
$G = 6.7 \times 10^{-8}$ erg cm g^{-2} : Newtonian constant of gravity,
$H = 1.6 \times 10^{18}$ sec^{-1} : Hubble constant for the expansion of the Universe,
$\rho = 4 \times 10^{-31}$ g cm^{-3} : mean density of the Universe.

These constants have dimensions, that is they are expressed in terms of some arbitrary chosen standard units, and so their numerical values have no significance, by changing the reference unit we change the numerical value of the constants of Nature. But from these constants we can form pure numbers, independent of any reference standard, these pure numbers contain the real empirical content of the laws of physics. From the constants listed above we can form the numbers

$$\frac{e^2}{\hbar c} = 7.3 \times 10^{-3} \quad \text{: fine structure constant,} \tag{1}$$

$$\frac{m_e}{m_p} = 0.54 \times 10^{-3} \quad \text{: mass ratio of electron and proton,} \tag{2}$$

$$\frac{e^2}{Gm_p m_e} = 2.3 \times 10^{39} = C_1 \quad \text{: ratio of electrical to gravitational force in a hydrogen atom,} \tag{3}$$

$$\frac{m_e c^3}{e^2 H} = 10.6 \times 10^{39} = C_2 : \quad \text{age of the Universe in atomic units,} \tag{4}$$

$$\frac{8\pi\rho c^3}{3m_p H^3} = 1.2 \times 10^{78} = C_3 : \quad \text{no. of particles in the observable Universe.} \tag{5}$$

From the last three we can derive the result

$$\frac{8\pi}{3}\frac{G\rho}{H^2} = 0.08 = C_4 . \tag{6}$$

The first three results are accurately determined, the others are less well known due to the uncertainty in the measurement of both the Hubble constant and the mean density of the Universe.

These numbers characterise the world we live in, if they had different values the Universe would be different, what are we to make of these results? The quantum physicist working in microphysics does not understand why the first two ratios have their measured values, but he believes that one day he will, these are not "God-given" arbitrary parameters but the only possible values they could have, it is just that as yet we do not know why. But what are we to make of the other constants, why are they so large, and why are (3) and (4) nearly equal to each other, and approximately the square root of (5)? Did God choose to set these constants equal to an arbitrary very large number like 10^{39}?

The Standard Explanation

The standard theory for "explaining" gravitation and cosmology is Einstein's General Theory of Relativity. In this theory the constant of gravity is God given, and as the Universe evolves so do the constants C_2 to C_4 change in time; in fact

$$C_1 = \text{constant},\ C_2 \propto f_1(t),\ C_3 \propto f_2(t),\ C_4 \propto f_3(t) \tag{7}$$

where f_1, f_2 and f_3 are known functions of time that contain an arbitrary constant, God given. The observed coincidence

$$C_1 = C_2 = (C_3)^{1/2} \tag{8}$$

is then a transitory phenomena, we just happen to be living at the time when this is approximately true. The argument can be made more sophisticated, by arguing that in order for life to evolve and ask why these coincidences occur, the coincidences must occur! That is for $C_2 \ll C_1$ the Universe would not have evolved to the state where galaxies, stars, planets, and life would develop; for $C_2 \gg C_1$ the Universe would have evolved so far that there would no longer be stars providing the energy for life, only if $C_2 \simeq C_1$, would we expect life to exist, it is therefore not surprising that we observe $C_2 \simeq C_1$, if it were not so we would not be here![1]

This argument has several weaknesses: firstly, it requires the arbitrary constant that occurs in the functions $f_i(t)$ to be zero, or sufficiently small; secondly, it does not explain why C_1 has the particular value of 2.3×10^{39}. God just created the Universe with this value, it could have been different. But this is far from satisfactory, a complete theory of physics must explain why there exists the number 10^{39} in the description of the physical world, and why life evolved at all; the existence of life is a physical phenomenon that needs an explanation just like the existence of stars and galaxies. If conditions could have been different with a different choice of constants, why does the actual Universe have those values that give rise to life?

The alternative point of view is the same as that of the quantum physicist's view of the microscopic numbers (1) and (2). We do not yet know why there exists a number 10^{39} but it is knowable, it is not an arbitrary constant. This is the view advanced by Dirac[(2)] some years ago, and it leads naturally to the observation that as the age of the Universe in atomic units is 10^{39}, then the reason that $C_1 = C_2 = (C_3)^{1/2}$ now, is that they are interrelated; we may not as yet understand why, but as the only number of this magnitude is C_2 what else can be the explanation? But if this is correct now, it must also be correct in the future, and have been correct in the past, so that for all time

$$\frac{e^2}{Gm_p m_e} \simeq \frac{m_e c^3}{e^2 H} \simeq \left(\frac{8\pi\rho c^3}{3m_e H^3}\right)^{1/2}. \tag{9}$$

But this cannot be true with the standard theory since C_1 is constant whereas C_2 and C_3 vary in time, therefore the standard theory is wrong. But what is to replace it?

THE DIRAC COSMOLOGY

One way to proceed is to retain the standard representation of the expanding Universe according to which there is a standard Reimanian metric governing the behaviour of particles and photons

$$ds^2 = dt^2 - S^2(t)\left(\frac{dr^2}{(1-Ar^2)} + r^2 d\Omega^2\right). \tag{10}$$

This introduces another dimensionless number, the curvature of space at a given time, measured in atomic units

$$C_5 = \frac{Ae^4}{m^2 c^4} \tag{11}$$

which is necessarily constant in units where e, m and c are constant. We have no reliable estimate of C_5, it could be positive, negative or zero. For such a model we readily deduce that

$$H = \frac{\dot{S}}{S}, \quad \rho = \rho_0 \frac{S_0^3}{\dot{S}} \tag{12}$$

and so

$$\frac{e^2}{Gm_e m_p} = \frac{mc^3 S}{e^2 \dot{S}} = \left(\frac{8\pi\rho_0 c^3}{3m_p}\dot{S}^3\right)^{1/2} \tag{13}$$

and so

$$S(t) = S_0 t^{1/3}, \quad G \simeq \frac{e^4}{3m_p m_e{}^2 t}, \quad \rho = \frac{m_e{}^2 m_p c^3}{8\pi e^4 t}. \tag{14}$$

The constant of gravity G decreases with time as the Universe expands.

An Alternative Representation

General relativity is a theory of gravitation and inertia, the theory is complete without reference to the electrical properties of matter, this suggests that if we retain general relativity (or a modification of it) then it may be valid except with reference to electrical properties. Accordingly we solve the large number coincidences with G, m and c constant and find

$$G \approx \frac{1}{6\pi\rho t^2}, \; S(t) = S_0 t^{2/3}, \; e^2 \approx \frac{m_e c^3 t}{\left(\frac{4\pi}{3} n t^3 c^3\right)^{1/2}} \tag{15}$$

and since $\rho S^3(t)$ is a constant we have

$$\hbar c \; \alpha \; e^2 \; \alpha \; t^{1/2}. \tag{16}$$

As the Universe expands, the Dirac–Planck constant increases.

If these results can be successfully predicted by a theory then the number 10^{39} is nothing but the present age of the Universe, as the Universe evolves, all the large numbers vary in unison because they are just the same thing.

What about the constant C_5? It could be anything, it is another "God-given" constant. But if we accept that physics is interrelated and there are no arbitrary constants it must either be related to some existing number or be zero. It cannot be related to 10^{39} since C_5 is constant, the only possibility is that it is a small number of order unity. But this would give the radius of curvature of space of the order of 10^{-13} cm, the size of an elementary particle. We therefore expect $C_5 = 0$, but our theory should force this conclusion.

MACH'S PRINCIPLE AND GRAVITATION

In the general theory of relativity the constant of gravity G is arbitrary – this is highly unsatisfactory, if we know the position and motion of all the particles in the Universe we can predict the future evolution without the introduction of an arbitrary constant. This concept (Mach's principle)† was the driving force behind Einstein's search for a relativistic theory of gravitation, but it failed, the theory does not make unique predictions. This is readily understood, Einstein's theory is a differential theory and differential equations have many

† This idea, though nowadays attributed to Ernst Mach, has its origins in the writings of Gottifried Leibniz and George Berkely criticising Newton's concept of absolute space.

solutions, if we wish to develop a unique theory we have to go to integral equations not differential ones. This has been recognised in recent years and in particular Hoyle and Narlikar[3] have developed theories along these lines. Can their theory be modified to provide the uniqueness we require? Technically the modified theory is given by

$$\delta \left(\int (\phi^2 R + 6g^{ij}\phi_{ji}\phi_{ij}) \sqrt{-g}\, d^4x + G \sum_i \int m_i \phi \, ds \right) = 0, \tag{17}$$

$$\phi = \sum_i \int \widetilde{G}\_ (\bar{x}, x)\, dx \tag{18}$$

where $G\_$ is the retarded Scalar greens function given by

$$\Box_{\underline{z}}\, \widetilde{G}\, (\bar{x}, x) + \frac{1}{6} R\, \widetilde{G}(z,x) = [-g(x)]^{-\frac{1}{2}} \delta^4 (z,x). \tag{19}$$

This theory is formally equivalent to Einstein's theory in differential form, but as it contains an integral formulation of the mass field ϕ, not all the solutions of Einstein's equations are necessarily solutions of this theory, the integral acts as a filter. If it turns out that there is only one cosmological solution of the field equations — the Einstein de Sitter cosmology in which

$$6\pi\, G\rho t^2 = 1 \tag{20}$$

which for our purposes we interpret as

$$G = \frac{1}{6\pi\rho t^2} \tag{21}$$

the constant gravity is determined by the cosmological distribution of matter. The other solutions of Einstein's equations, hyperbolic or elliptic universes, may not satisfy the integral formulation. C_5 is therefore zero, and the coincidences C_4 is satisfactorily explained.

MICROPHYSICS AND COSMOLOGY

It was Arthur Eddington[4] that drew attention to the uncertainty in physics caused by the Universe and its connection with microphysics. In making measurements in science we have to measure relative quantities, relative coordinates of the position of a particle and a reference frame, but what determines the reference frame? The best we can do is to take the centroid of all the particles in the Universe, or rather those within the horizon of an expanding Universe, this has the distinct advantage that the uncertainty in locating this centroid is known from the statistics of large numbers, without knowing the probability distribution of the individual particles. It is the Gaussian

$$f(x_0) = \frac{1}{\sqrt{2\pi\sigma^2}}\, e^{-x^2/2\sigma^2}. \tag{22}$$

The uncertainty in position has an associated uncertainty in momentum, for a wave packet with a standard deviation σ, the distribution of momentum from standard wave mechanics is

$$\varpi(p) = \frac{1}{(2\pi\varpi^2)^{1/2}} e^{-p^2/2\varpi^2}, \quad \varpi = \frac{\hbar}{2\sigma}. \tag{23}$$

What is σ? If we have N particles each of which can be anywhere in a sphere of radius R then

$$\sigma = \frac{R}{3\sqrt{N}}. \tag{24}$$

For a Universe full of photons moving with speed c, then at time t the distance to the horizon is $R \approx 2t$, and photons can have travelled a substantial fraction of R in the time t, so approximately

$$\sigma \approx \frac{ct}{\sqrt{N}}.$$

In the world there is an equivalence between mass and energy, can we understand mass as a concealed form of energy? If we took a model of the Universe with pure radiation and compared it with one of pure matter would we reveal this relationship between them? This was essentially Eddington's line of attack, we shall borrow his ideas but apply them to an expanding model universe (or Uranoid) comparing radiation- and matter-only models. This is quite straightforward using Einstein's equations in the mass field integral form we find in general

$$\begin{aligned} \text{matter: } & 6\pi G\rho_m t_m^{\ 2} = 1, \\ \text{radiation: } & \frac{32\pi G}{3}\rho_R t_R^{\ 2} = 1, \end{aligned} \tag{25}$$

where ρ_m, t_m, and ρ_R, t_R are the density and age of the matter and radiation models. For the two models to have the same number of particles in the horizon of the same dimensions we have $t_R = 3t_m/2$, hence

$$\frac{\rho_m}{\rho_R} = 4. \tag{26}$$

Eddington in fact compared two static Einstein Universe and obtained 4/3 for this ratio.

What can we say about the mass density of radiation? The radiation-only uranoid is equivalent to the mass-only uranoid but with the particle velocities increased to c and the rest mass going to zero, i.e. an infinite temperature gas. But the momentum distribution and mean momentum of the particles is well behaved, since due to the fluctuations in the physical frame any momentum distribution relative to a geometrical frame is modified by the weight factor $\bar{\omega}(p)$ which is the momentum distribution corresponding to the Gaussian distribution of position given in (22), thus the mean momentum of the photons is given by

$$\bar{p^2} = \frac{3\hbar^2}{4\sigma^2}. \tag{27}$$

But for a gas of photons the energy density is just the number of photons per unit volume n, multiplied by the mean energy per photon of $(p^2 c^2)^{1/2}$, hence

$$\rho_{\text{rad}} = \frac{\sqrt{3}}{c} \frac{n\hbar}{2\sigma}. \tag{28}$$

But for a radiation uranoid σ is $ct/\sqrt{N}$, hence

$$\rho_{\text{rad}} = \frac{\sqrt{3}}{2} \frac{n\,\hbar}{c^2 t_R} \sqrt{N}. \tag{29}$$

On the other hand, for the matter-only uranoid $\rho_m = nm_0$ where m_0 is the mass of the particle, where n is the same for the matter and radiation uranoids since N and R are the same, hence

$$\frac{\rho_m}{\rho_R} = 4 = \frac{2m_0 c^2 t_{\text{rad}}}{\sqrt{3N}\;\hbar}. \tag{30}$$

Now the number of particles in the horizon is $4\pi nc^3 t^3/3$ and so using the constancy of the fine structure constant we find

$$e^2 = \frac{1}{137}\hbar c \simeq \left(\frac{m_0}{137m_e}\right) \frac{m_e c^3 t}{[4\pi nt^3 c^3/3]^{1/2}} \tag{31}$$

which is just the relation (15) provided m_0 is of the order of the pion mass.

We cannot hope to do any better than this at this stage, we would hope like Eddington that m_e and m_p are determined in terms of this mass constant m_0, but how is still a mystery.

CONCLUSIONS

The preliminary results obtained by this simple analysis are encouraging, they give the correct dependence of G and e on cosmic time t, they give the correct order of magnitude for the cosmic numbers. 10^{39} is the current age of the Universe and the other numbers of this magnitude are causally related to the age. Of course we have not solved all problems, we need to show why in this theory the Universe has the properties it has, galaxies, stars, planets and people at the present time of 10^{39}. This is the same problem as in the standard theory, and a similar reasoning, with similar uncertainty would lead to the conclusion that life emerges when $C_2 \approx 10^{39}$.

But there are also empirical checks on the theory, if as measured in atomic units e, m, c, the constant of gravity changes $G \sim 1/t$, then the properties of stars and the orbit of the Moon round the Earth or the Earth round the Sun should show a secular change. This change would be of the order of 1 part in 10^{10} years and some indication of this has been suggested by the work of van Flandern.[5] It can also be measured in the laboratory and an experiment is currently underway at the University of Virginia to look for this. If it is found it will strongly support the contention that the Universe is unique, we may not have a full theory for the interrelatedness of nature, the connection between cosmology and microphysics, but perhaps we are on the right lines.

REFERENCES

1. R. Dicke, *Nature*, **192**, 440 (1961); B. Carter, *Proc. I.A.U. Symposium* no. 63, p. 291 (1974).
2. P. A. M. Dirac, *Nature*, **165**, 199 (1937).
3. F. Hoyle and J. Narlikar, *Proc. Roy. Soc.* A, **282**, 191 (1964).
4. A. S. Eddington, *Fundamental Theory*, C.U.P. (1966).
5. T. C. van Flandern, *Science*, p. 44 (1976).

William H. McCrea, F.R.S.

Educated at Trinity College, Cambridge, and Göttingen University; former Bye-Fellow of Gonville and Caius College, Cambridge; Professor of Mathematics in Queen's University, Belfast, 1936-44 and in London University (Royal Holloway College) 1944-66; Professor of Astronomy, Sussex University 1966-72, Emeritus since 1972. Past President of the Mathematical Association and the Royal Astronomical Society. Awarded the Royal Astronomical Society Gold Medal 1976. Research in mathematics, mathematical physics, astrophysics and cosmology including problems of the origin and evolution of the solar system and particularly the significance for such problems of the large-scale structure of the Galaxy.

ORIGIN OF EARTH, MOON AND PLANETS

The problem of the origin of the solar system is perhaps the most notable of all unsolved problems in astronomy. It has unique relevance to ourselves and our place in the cosmos; it ought to be solvable in the present state of general astronomical knowledge; the correct solution, if found, should be recognizable as correct; numerous attempts to solve it have, in fact, been made; no problem of the formation of *any* astronomical body has ever yet received an accepted solution. All other astronomical problems lack some of these features. There are, for example, problems of origins of much greater magnitude but they are of less immediate concern to ourselves; also the greater the scope the more difficult it is to formulate a problem, and the less certain that a solution may be achievable or even recognizable. Elsewhere (McCrea, 1977) I have tried briefly to describe some wider aspects of the astronomical setting of the problem. In the present article I seek to discuss the problem on its merits.

SOLAR SYSTEM

The composition of the solar system is summarized in Table 1. There is the Sun, a rather

TABLE 1. Solar System

	Mass or mass-range Earth-masses	Mean density or density-range, g/cm³
Sun	3.3×10^5	1.4
Terrestrial planets: Mercury Venus, Earth, Mars, Pluto	0.06 to 1.0	4 to 5.5
Major planets: Jupiter, Saturn	95 to 320	0.7 to 1.3
Outer planets: Uranus, Neptune	15 to 17	1.6 to 2.3
Asteroids (minor planets)	5×10^{-4} total mass	
Interplanetary gas, dust, meteoroids, comets		

$$\frac{\text{Mass of planetary system}}{\text{Mass of Sun}} \sim \frac{1}{700} \qquad \frac{\text{Angular momentum of planetary system}}{\text{Angular momentum of Sun}} \sim 200.$$

Planet	Known natural satellites: Regular satellites: Main	Known natural satellites: Regular satellites: Minor	Miscellaneous satellites	Angular momentum of main satellites / Angular momentum of planet
Mercury	–	–	–	–
Venus	–	–	–	–
Earth	Moon	–	–	5
Mars	–	–	2	–
Jupiter	Io, Europa, Ganymede, Callisto	1	8	1/100
Saturn	Titan	6 + rings	3	1/80
Uranus	–	5 + rings	–	–
Neptune	Triton	–	1	~1/20
Pluto	–	–	–	–

Total mass of main satellites ~ 0.16 Earth mass.

average normal star, which is 1/3 million times as massive as the Earth; there are nine planets, the heaviest, Jupiter, being nearly 6000 times as massive as the lightest, Mercury, and the densest, the Earth, has nearly 8 times the mean density of the least dense, Saturn. There are thirty-three known natural satellites shared between six of the planets. There are much smaller bodies in considerable variety and numbers. Proceeding in this way, the range of properties is bewildering. It becomes necessary therefore to ask what, if any, basic regularities may be discerned. Fortunately, it is possible to recognize quite a number of these, as follows:

Mass and composition. The six principal planets Venus, Earth, Jupiter, Saturn, Uranus, Neptune *all have roughly the same heavy-element content*, which is therefore a standard feature. One way of looking at this is that Venus and Earth are very similar in mass and mean density; bodies similar to the other four of these planets could be got simply by taking Earth or Venus and surrounding it by suitable quantities of the lightest elements hydrogen and helium. A probably more significant description is this: the composition by mass of the Sun is about 70 per cent hydrogen, 28 per cent helium, 2 per cent heavier elements; of all the planets Jupiter has most nearly the same composition; were perhaps 50 per cent of the light elements hydrogen and helium removed from Jupiter, the result would be quite like Saturn; were some 90 per cent removed the result would be like Uranus and Neptune; were 99 per cent removed, the result would be like Earth and Venus; in fact, the principal planets are all derivable from six proto-planets all having the same chemical composition and not far from the same mass. The "lesser planets" Mercury, Mars, Pluto are bodies of chemical composition somewhat like Earth and Venus, but of the order of only one-tenth the mass.

Rotation. The spin-axes of the planets are inclined to the orbital axes at mostly quite considerable angles. In two or three cases the present rotation-rate is obviously subject to the tidal influence of other bodies in the system. However, in six cases the rotation-period is between about 10 and 25 hours; that is to say, *we may recognize a roughly standard natural rotation-period* with a median of, say, 15 hours.

Revolution. All the planets revolve about the Sun in the same sense in orbital planes that are within a few degrees of each other, with fairly small orbital eccentricities. Roughly speaking, each orbit has getting on for double the size of the next smaller orbit; were a theory to insist upon a totally different arrangement of sizes it would be unacceptable; it cannot be claimed, however, that there is a regularity of sizes ("Titius-Bode formula") to which a cosmogonic theory as such must unquestionably conform. Incidentally, the Earth's orbital plane is not the equatorial plane of the Sun, but is inclined to it at about 7 degrees.

Satellites. The six main satellites of Earth, Jupiter and Saturn, even as they exist, *are the most standard bodies in the solar system* – the satellite Triton of Neptune is probably in this category, but values of its parameters are still somewhat uncertain and it is best to leave it out of the discussion; three of them have about the same density as the Moon, which is believed to be composed mainly of silicates; it is inferred that the other three have mantles (or admixtures) of ices, and that the silicate portions of all six are even more standard bodies. These main satellites move in orbits about the planets that have small, but not all negligible, eccentricities, and whose axes in two cases have large inclinations to the rotation axes of the planets.

The twelve very much smaller satellites classed as "minor" are all within 10^5 and 6×10^5 km from the three relevant planets, round which they revolve in orbits of very small eccentricity and very small inclination. In the three cases in which mean densities have been well estimated the values lie in the range 1.1 to 1.3 g/cm^3; the material is probably the same as the mantle-material (when present) of the main satellites. Saturn's rings appear to be closely related to its inner (minor) satellites. *The evidence is that the minor satellites are products of a standard process.*

The fourteen miscellaneous known satellites are all very small, and they are clearly distinct from regular satellites – although maybe some were formed along with the minor satellites and suffered subsequent disturbance. The fact that *three planets have no known satellites* must be significant.

Finally, in the solar system as a whole *nearly all the angular momentum is associated with a very small part of the mass*; in the satellite systems the reverse is generally the case.

The main regularities are those in the italicized statements, they strongly indicate that the solar system was formed as the one system and that we should look for a theory of its origin as such. The principal requirement of a theoretical model is to reproduce these features. At the same time, it has got to reproduce them as they actually occur as *approximate* regularities – a theory that required, say, all the planets to have orbits precisely in one plane with precisely zero eccentricity might be more objectionable than one making no particular prediction about these characteristics.

We know only the one planetary system, and this is obviously a chief source of difficulty in discovering its origin. Within that system, however, we know six satellite systems – or more significantly nine, of which three are almost certainly void. Therefore one has a better chance of discovering the origin of a satellite system than that of a planetary system. But the formation of satellite systems is obviously closely bound up with that of the associated planets, so anything learned about satellite formation should throw light upon planet formation. This is why it is so important to take account of satellites, even though they be lightweight members of the system as a whole. At the same time various properties, particularly in regard to angular momentum, show that a satellite system is *not* like a planetary system in miniature. Indeed, if a model that has been proposed for planet-formation should turn out to be good for satellite-formation, then that probably shows it to be bad for planet-formation!

As regards the possibility of observing another planetary system, planets like ours associated with any other star like our Sun would be utterly undetectable by any available means. Van de Kamp (1975) has collected a number of cases where a less massive star shows variability of motion possibly attributable to the presence of a specially massive planet, but nothing is yet known that helps the present study.

RAW MATERIAL

Current models in general postulate raw material having about "solar" composition. At the relevant time it has to be cold ($\lesssim 100°$ K); at such a temperature and at the relevant density, much of the heavier-element content would have condensed to form minute grains of silicates, metals, graphite, ices, to the extent of about 1 per cent of the material by mass. The remainder would still be in the gaseous state and would consist mostly of hydrogen and helium, with about another 1 per cent of heavier elements not incorporated in the grains.

Were all the material of even the entire solar system (say 2×10^{33} g) spread evenly through the volume of a sphere with radius about the distance of Neptune (say 5×10^{14} cm) the density would be only about 4×10^{-12} g/cm^3. The processes to be considered must take place through a region of some such size, and they cannot involve much more than this amount of matter; therefore on any model the initial mean density of the raw material is extremely low – less than a good terrestrial vacuum.

The most significant way of classifying models is into those that treat the Sun and the requisite raw material in isolation, and those that consider them necessarily as part of some larger assembly.

Solar nebula. A model in the first of these classes always reaches a stage when the Sun, either while in course of formation or sometime later, is surrounded by a well-formed *solar nebula.* This is a body of the raw material having mass-distribution and momentum-distribution such that when some subsequently adopted process is inferred to form the planets from it they will turn out to have the correct masses and motions.

One proposed mechanism for producing the nebular material is that the gravitational attraction of the Sun would entrap it after drawing it out by tidal action from a passing diffuse star. In recent years, Woolfson (1969) and his co-workers have shown by computation that this is dynamically feasible; the basic suggestion is much older.

Another mechanism is for the material to be a residue from the formation of the Sun from an original a solar nebula – as long ago proposed by Laplace – after there has been outward transfer of spin-momentum, say by magnetic coupling, or after considerable loss of mass without too great loss of spin, as proposed by A. G. W. Cameron (1976) and his group.

Finally, as the Sun travels through interstellar clouds it may capture cloud-material in its gravitational field and such material may settle down into orderly rotation around the Sun. In this context, the process was first considered many years ago by Schmidt (1959).

Star formation and protoplanets. An important part of the Milky Way Galaxy consists of stars and clouds of gas and dust that form a disc, and that a remote observer would see as a spiral galaxy, like the Andromeda Galaxy. The stars and clouds are all in orbit round the galactic centre, producing the phenomenon of "galactic rotation". The motion as a whole is such that there exist regions where there is a traffic-jam; these regions *are* the spiral arms. Clouds running into a jam sometimes suffer a shock which causes them to be considerably compressed, as readily seen in the features known as "dust-lanes" in photographs of other galaxies (Lynds, 1974). Clouds normally recover as they emerge from the jam; occasionally, however, a cloud gets so compressed that gravitational effects within it overcome its tendency to re-expand, and instead it condenses to form a cluster of new stars. This also is seen in our own and other galaxies in that what are recognized as young clusters or young stars are in fact observed in the appropriate locations. All this is quite direct inference from much observation and it involves no speculation.

Astronomers are convinced that they are witnessing the formation of new stars in certain parts of the sky; the phenomena are compatible with the foregoing general picture. Unfortunately there is still no agreed model for the way in which a cloud of interstellar material does become resolved into stars. It seems simple to say that a cloud breaks into fragments and each fragment contracts under its self-gravitation; the trouble is that by the time it had contracted to the density of a star it would be spinning with equatorial speed exceeding the speed of light! The escape from this difficulty must have to do with the fact that a lot of stars are formed all at once. It may be that something very drastic happens to the cloud as a whole before the stars are formed from it. Or it may be that at a critical stage the cloud as a whole "clots" or "coagulates" into stars; the picture then would be that blobs of cloud-material coalesce here and there into what become centres of gravitational condensation leading primarily to stars; after these main condensations have been formed, there would be some residue of blobs that have too much angular momentum about every one of the main condensations for them to fall into main condensations – some will disperse and some may just manage to hold together and survive as what we shall call *protoplanets*. These will be shared amongst the gravitational fields of the newly formed stars, or at any rate such of those stars as possess no stellar companion. In the process of assembling any protoplanetary system, it will tend to flatten out, in the fashion that, for instance, the particles of Saturn's rings have settled into a plane, but on a coarser scale in the case of protoplanets (McCrea, 1960). Here the point is that when we recognize that stars are formed in clusters, it is more natural to infer that some of them acquire a family of protoplanets, rather than a solar nebula.

PLANET FORMATION

Once the raw material has been deposited around the Sun in the shape of either a solar nebula or protoplanets there are two possibilities for the formation of normal planets from it:

Accumulation. Most models that invoke a solar nebula envisage planets are being produced from it by a process of accretion or *accumulation.* Some such models give most attention to mechanical, thermodynamical and thermochemical processes taking place before the appearance of the planets, with the object of specifying a nebula that is self-consistent in regard to its density distribution and gas dynamics, its heating and cooling at any point, the condensation of grains, the evaporation of light gases across its boundary, and so forth. If it be then accepted as self-evident that planets will indeed form from belts of this nebula, there seems to be no doubt that the dependence of chemical composition of actual planets upon distance from the Sun can be reproduced in a convincing manner. But if the formation of planets be thus taken for granted, then the model is not asked to predict the number, sizes, spins, satellite-features, etc., of these planets.

Of course nobody really takes it to be self-evident that planets must be formed at all under the inferred conditions, and so some models give major attention to this aspect. For example, Dole (1970) has carried out extensive computer simulations of the formation of planets, according to a stated prescription, in a solar nebula specified by several adjustable parameters. With particular values of these parameters, he succeeded in reproducing the actual planetary system with astonishing fidelity. Rather reluctantly, one has to state three main criticisms:

(i) The solar nebula described by Dole's parameters is not shown to be self-consistent in the above-mentioned sense – so far as one knows, nobody has married a fully worked out process of forming planets to a fully worked out structure of the solar nebula.

(ii) Dole's computations expressly give no information about the duration of the processes; it is quite unknown whether the postulated nebula would endure long enough for the processes to attain consummation.

(iii) Dole's processes would not get off the ground – to venture an incongruous use of the current idiom – without his postulated nuclei for planet-accumulation which are an arbitrary element in his prescription.

In order for the accumulation ever to get started, the view developed in recent years

(Goldreich and Ward 1973) is that after the solar nebula has reached the state already mentioned, yet another process supervenes – under the gravitational field of the nebular material itself, the grains fall through the gas and settle into a layer in its equatorial plane. Accidental irregularities in this layer, if sufficiently extensive, will contract gravitationally upon themselves; condensations initiated in this way are then considered to set in motion a train of processes such as that treated by Dole. Those who follow a theory like this usually claim that satellites arise by a further application of the theory.

Protoplanet evolution. As we have seen, protoplanets seem to arise naturally in the context of the formation of a stellar cluster, although they *might* also arise by fragmentation of a solar nebula. Discussion shows that, by whatever means it does arise, a protoplanet would be initially a rather standard body of the raw material, of about the mass of Jupiter, in a cold diffuse state but sufficiently dense to hold itself together by its self-gravitation; it is in orbit round the Sun, but it will evolve largely as an independent body. Like any naturally occurring body it has non-zero spin-momentum about its mass-centre, and this is conserved according to ordinary mechanical principles.

The diffuse protoplanet will proceed to contract under its own gravitation, the gravitational energy released being radiated away so long as the material is sufficiently transparent. But as the size diminishes the body inevitably spins faster; ignoring for the moment rotational distortion, at some radius R it will start spinning off material round its equator. The body will go on contracting and leaving material behind in the equatorial plane until its remaining material becomes relatively incompressible. For a mass like Jupiter's this must occur at mean density about that of Jupiter, i.e. about 1 g/cm^3. It is still rotationally unstable, but for material of small compressibility this will be relieved once and for all by fissional break-up in which the main body casts off a portion about one-tenth its mass. Thus there would be, as calculation shows, about half of the original material near the equatorial plane out to distance R and most of the rest of the mass forming a stable planet in the middle. *This is the proposed origin of Jupiter and Saturn.*

These ideas go back more than 50 years to the work of Sir James Jeans which has in part been redeveloped more recently by Lyttleton (1960). More modern work has been done on the whole subject but most of it is applicable rather to the rotational flattening and possible break-up of a star than to a planet or protoplanet. One can either accept that the treatment outlined above is valid for the present application or one can regard it as a rough approximation to what would be expected from a theory along modern lines adapted to this application.

Part of the story is that *the material left near the equatorial plane is that from which (regular) minor satellites are formed* in this model. The process would be the accumulation process envisaged in the preceding section as a process for forming planets from a solar nebula; here we appeal to it on a much smaller scale and we require only very low efficiency; it would explain why these satellites have (closely) circular orbits in the equatorial plane of the planet concerned. Also on this interpretation R is about the outer radius of the set of these satellites in each case and this serves to set a value on the original spin, and this in turn determines how much of the original mass is left to form the main resulting planet. On the model the final spin is significantly less than the critical spin that caused rotational break up, but still of that order, *which implies a rotation period of order 10 hours as actually found.*

There is, however, another process that to a greater or lesser extent must operate from the outset: the grain material must tend to sediment towards the centre of the protoplanet. In the solar nebula model the same process causes this material to settle towards the central plane, but it would be more effectual in a protoplanet. The evolution of a protoplanet as already described applies if the sedimentation rate is slow compared with the contraction rate. However, we may consider the other extreme case in which the light gases continue to occupy about the original volume of the protoplanet while only the heavier material contracts towards the centre. Much the same treatment as before applies now to *this* material. But there are two adjustments: as the heavier material contracts through the excess of lighter material it leaves behind most of its spin-momentum; and the density at which the contracting material becomes effectively incompressible, and break-up occurs, is greater than before – presumably about the mean density of the Earth which is composed of just such material. *On the model the surviving main body is indeed the Earth or Venus.*

We have now rather literally to tidy up this model. In the case just considered the diffuse gas and some discarded grain-material are left behind. If the protoplanet is near enough to the Sun such material would be removed by tidal action and by the solar wind. On the model there would therefore be no possibility of the occurrence of (regular) minor satellites of a terrestrial planet. On the other hand, if the protoplanet is far enough from the Sun some diffuse gas and any other material left beind in the sedimentation process may proceed to contract as in the process first considered. *This would be the present explanation of Uranus and Neptune* and it would allow them to have minor satellites – Uranus actually has them, Neptune has no regular minor satellite.

If a body in orbit about the Sun undergoes rotational fission the main portion remains in a not much different orbit; Lyttleton has shown that the smaller portion goes into a separate orbit if the original orbit is at about the Earth's distance from the Sun or nearer, while it would escape altogether if the original orbit is at about Jupiter's distance or further. The implication is that *Mars is the smaller portion from the fission of proto-Earth, and Mercury that from proto-Venus.* The inferred fission of proto-Jupiter and proto-Saturn would have produced no surviving smaller planets. Lyttleton's conclusions apply, however, only if no other matter intervenes; if the lighter material of proto-Neptune is retained as suggested, this will retard the escape of the smaller fission portion which then need not be lost to the solar system. *The implication is that such was the origin of Pluto.*

When a fluid body is on the point of fissional break-up into two portions, these are connected by a "neck" of this fluid. Fission is the demolition of this connection; directly it happens some of the connecting material may fall back into the two main bodies, but some will form "droplets" strung out between them. The description is essentially Lyttleton's; he has shown how some droplets would be expected to go into orbit about the larger main portion, some would escape, none could go into orbit about the smaller main portion because any setting out to do so would collide with that body (collision fragments might survive as tiny satellites). A suggestion apparently due originally to Lyttleton is that *the Moon may be such a droplet.* If it is, then *all seven main satellites would have the same sort of origin* and it is satisfactory that the model requires fission in all relevant cases. Venus might have had a moon in this way, but owing to solar tidal friction its day has become longer than its month could be and so "lunar" tidal friction would cause a moon to fall into the planet. Finally, various escaping bodies have been mentioned; had any of these suffered collisions amongst themselves then there would have been plenty of fragments to account for asteroids and meteorites, and may be also miscellaneous satellites.

The protoplanet model appears therefore to account for the main regularities in the solar system and to provide for the variety of its properties both positive and negative – it explains, for example, not only why certain planets have satellites of certain sorts but also why others do not have them.

The model qualifies nevertheless, along with the rest, for inclusion in an encyclopaedia of ignorance. One cannot claim that things had to be like this.

DISCUSSION

There are compensations in writing about ignorance: there can be no completeness, and no author is likely to complain if one fails to quote his particular contribution to the sum total of our ignorance. However, there are embarrassments: if ignorance is complete, there is nothing to be said about it anyhow, and if knowledge is fairly complete it does not qualify for this book. So any discussion that does qualify necessarily concerns the in-between region – that which appears to be just ahead of the edge of knowledge. And here it is that one's personal preferences and expectations have most influence. So one can only try to make clear why one has adopted a particular presentation, while freely allowing it to be biased.

My first concern is to show why there is such profusion of ignorance about the origin of the solar system. It is primarily because the subject is so complex, and I thought I could demonstrate this only by trying to show the reader how people are actually trying to tackle it – if possible to give a *feeling* for the subject. I hope this brings out the difference from ignorance in other fields; in pure mathematics, for instance, much ignorance may depend upon failure to prove or disprove one single theorem; in biology, progress in some branch may be held up because some quite specific question in chemistry is still unanswered; there are problems in the foundations of physics where no one knows even how to begin, and so forth. Nevertheless it might be more true to say, not that ignorance in cosmogony is different, but that it includes all possible kinds of ignorance encountered elsewhere. And for that very reason, as should be apparent from the account given in this paper, the problem of the origin of the solar system gives rise to a series of problems each of which is of a more usual scientific sort. Happily considerable progress is being made in these individual problems. There is even considerable agreement that the results are indeed those needed for solving the overall cosmogonic problem. Where there is most uncertainty and disagreement is in the way in which those results can be combined into a coherent whole.

A minor but still significant question is the role of tradition; this happens to be well illustrated here. Interest in the origin of the solar system goes back to long before modern ideas on the structure of the Galaxy, the behaviour of interstellar clouds and the formation of stellar clusters. If raw material from interstellar space was out of the reckoning, then it could be got only out of a star or as a residue in the making of the Sun; this led to the two traditional approaches. As indicated in this paper, they still provide the starting places of much current work. It is pardonable to ask to what extent this is simply because even in science, tradition dies hard.

REFERENCES

Cameron, A. G. W. and Pollack, J. B. (1976) in *Jupiter*, ed. T. Gehrels (University of Arizona Press).
Dole, S. H. (1970) *Icarus* 13, 494-508.
Goldreich, P. and Ward, W. R. (1973) *Astrophys. J.* **183**, 1051-61.

Lynds, B. T. (1974) An atlas of dust and H II regions in galaxies. *Astrophys. J.* Supplement number 267.
Lyttleton, R. A. (1960) *Monthly Not. Roy. Astron. Soc.* **121**, 551-69.
McCrea, W. H. (1960) *Proc. Royal Soc.* **A256**, 245-66.
McCrea, W. H. (1977) in *Origin of the Solar System*, ed. S. F. Dermott (John Wiley).
Schmidt, O. (1959) *A Theory of the Origin of the Earth; Four Lectures*, London, Lawrence and Wishart.
van de Kamp, P. (1975) *Ann. Rev. Astron. and Astrophys.* **13**, 295-333.
Woolfson, M. M. (1969) *Progress Phys.* **32** 135-185.

General Reference

Dermott, S. F. (ed.) (1977) *Origin of Solar System*, John Wiley.
Reeves, H. (ed.) (1972) *Symposium on the Origin of the Solar System*, Paris, CNRS.

Michael Rowan-Robinson

Lecturer in Mathematics at Queen Mary College, University of London.

Main interest has been the application of the "new" astronomies, particularly radio and X-ray astronomy, to cosmology. A special interest has been quasars, their distances and evolutionary properties. Increasingly involved since 1974 with far infra-red astronomy, working with the Queen Mary College far infra-red group at Kitt Peak, Arizona.

GALAXIES, QUASARS AND THE UNIVERSE

In extragalactic astrophysics we suffer from three types of ignorance: ignorance in principle, ignorance due to observational limitations, and ignorance due to inadequacy of theory and observation.

An example of ignorance in principle is associated with the finite time that light takes to reach us from distant galaxies. While in our locality we have evidence about our remote past, for distant galaxies from which the light set off long ago we have no evidence about how they look *now*. We can thus never test one of the most basic philosophical ideas underlying modern cosmological theory, that the universe is *homogeneous*, each observer seeing the same picture, on average, as we do. There is a considerable logical gulf between the earth not being the centre of the universe and perfect homogeneity. Virtually no theories populate this gulf, although the universe itself clearly does. The sceptic (a rare bird among cosmologists) would like to see a model of the universe less dependent on metaphysical abstractions like the principle of homogeneity. It is true that perfect *isotropy*, together with the assumption that we are not in a special place, implies homogeneity. Unfortunately, the distribution of matter in the universe is certainly not isotropic. Just look up at the night sky! Can the lumpiness of the matter distribution really be ignored?

We cannot count observational limitations due to the earth's environment as sources of ignorance in the long term, for surely most of these can be overcome, if we so desire. The earth's ionosphere reflects all radio waves of frequency $\lesssim$ 30 MHz, but lower frequency radiation can be detected from satellites. Virtually no radiation with wavelengths between 20 μm and 1 mm reaches sea-level, but dry high-altitude sites, aircraft, balloon (and soon satellite) observations are opening up the far infra-red and sub-millimetre bands for us. No cosmic X-rays reach the surface of the earth, but rocket and satellite experiments have already mapped the X-ray sky. The angular resolution of ground-based optical telescopes is limited by atmospheric fluctuations to 1$''$ arc, but this can be overcome by balloon or satellite observations. Indeed at any time there will be a finite achievable sensitivity, angular resolution, spectral and time resolution at each wavelength, the transcending of which will lead to new detail in our knowledge and occasionally to entirely new types of phenomena.

However, there are some genuine observational limitations which define areas of ignorance for the forseeable future. The very lowest energy radio waves ($\lesssim 10^4$ Hz) and the highest energy γ-rays ($\gtrsim 10^{24}$ Hz) cannot reach us from outside our own Galaxy. The interstellar medium restricts our horizon in far ultraviolet light to only about 100 light years, compared with the 10^{10} lt.yr or more accessible in the visible. The absorbing and scattering power of minute grains of dust in the interstellar medium hides from our view most of the plane of the Milky Way, and the cores of dense clouds where stars are forming, except to radio and infra-red waves. But when we consider how the electromagnetic spectrum has been opened up for astronomy in the past 30 years, these areas of ignorance seem to be continually dwindling.

It is when we turn to the third type of ignorance, that due to the inadequacy of theory and observation, that we have greatest cause for modesty in our claims of achievement in astronomy. True we have some sort of overall scenario for the evolution of matter in the universe, and this represents a great advance on the situation as recently as a decade ago. It is, however, a scenario riddled with untested assumptions.

Firstly we assume, with no evidence, that Einstein's General Theory of Relativity applies to matter on the large scale. An associated assumption is that the gravitational constant and other physical constants do not change with time. Then we assume that matter is distributed homogeneously and isotropically through the universe. We assume, again without evidence, that the redshifting of the spectral lines of galaxies, which is found to be proportional to distance (the Hubble law), is due to the expansion of the universe.

The picture is that about 2×10^{10} years ago the universe exploded out from a state of infinite density, temperature and pressure (the Big Bang). For the first 10^6 years the dominant form of energy was radiation, with the less significant matter in thermal equilibrium with it. The temperature fell as the ratio by which the universe expanded and at around 10^9 K nuclear reactions ceased, leaving protons, electrons, neutrinos, about 27 per cent helium by mass, about 10^{-5} deuterium (^2H), and traces of other elements.

When the temperature dropped to about 3000 K, after 10^6 years, protons and electrons recombined to form neutral hydrogen and the universe became suddenly transparent, allowing matter and radiation to decouple. The radiation travelled freely through the universe until, redshifted by a further factor of 1000, it reaches the observer today as a microwave background radiation with the spectrum of a 2.7 K blackbody. The matter, on the other hand, was now free to condense into galaxies, presuming the basically uniform gas to be dimpled with protogalactic irregularities in density. After about 10^9 years the collapse of the fragments was complete. Some formed almost entirely into stars as they collapsed, yielding elliptical galaxies. Others formed only partly into stars, the residual gas forming a rotating disc: spiral density waves rotate through the disc triggering star formation in their wake to yield spiral galaxies like our own.

Both types of galaxy form a dense core, or nucleus, which may be continually fuelled by gas ejected from evolving stars or supernovae (the final explosive stage of massive stars). In this nucleus a dense star cluster, supermassive star or giant black hole, forms from time to time, acting as a powerful accelerator of relativistic particles. These give rise to a variety of types of "active" galaxy: radio-galaxies with strong sources of non-thermal radiation centred on the galaxy or located in double sources far from the galaxy, galaxies with non-thermal optical cores (Seyferts, N-galaxies or, when the core outshines the whole optical galaxy, quasars).

This then is the framework within which we can understand most of what we know about galaxies, quasars and the universe. Yet there are so many gaps in the picture, so many areas of doubt or controversy, that the sceptic asks if it is really a picture with any predictive power at all.

The first question, and in many ways the most crucial for the whole Big Bang picture, is whether the 2.7 K blackbody background radiation is indeed the relic of an opaque, radiation-dominated phase of the universe. Recent balloon-borne experiments have shown that the spectrum does indeed start to turn down at wavelengths short of 1 mm, as expected for a blackbody spectrum. The spectrum and the great smoothness of the radiation (isotropy on the small scale) make it extremely difficult to suppose that it is due to a superposition of faint discrete sources. The high degree of isotropy on the large scale ($\pm$ 0.1 per cent) is the only solid evidence for isotropic (and homogeneous) models of the universe and soon the local motion of the solar system should be clearly measurable, an experiment as fundamental as that of Michelson and Morley. One disquieting fact about this background is that its energy-density is rather similar to that expected from the integrated starlight from galaxies, but no satisfactory method of thermalizing this radiation to give a 2.7 K blackbody has been devised. If nature has not found a way, then we simply happen to be alive at the epoch of this coincidence. The neo-Aristotelians might counter with the "anthropic principle": the universe is as it is because if it were not, we would not be here to observe it. Still more inexplicable is the coincidence between the energy-density of the microwave background and those due to cosmic rays, and also to magnetic field, in our Galaxy.

Another important plank in the Big Bang platform is the "primordial" abundance of helium, found in the oldest stars in our Galaxy. While this could have been formed in an early, long-vanished generation of stars (and such an early burst of star-formation can also be argued for from the absence of stars with very low abundances of heavier elements which *have* to be formed in stars), it is generally assumed to be another relic of the "fireball" phase. *Ad hoc* explanations are needed for one or two stars in our Galaxy in which *no* helium is seen in their surface layers. Another accepted candidate for cosmological honours is deuterium since though it is easily destroyed in stellar interiors, it is hard to make. The observed abundance requires the average density of the universe now to be about 10^{-31} g cm^{-3}, implying that the retarding effect of gravity on the expansion of the universe is almost negligible. This is also about the observed average density of matter in galaxies.

Before we leave those questions directly connected with the fireball phase of the Big Bang, we have to ask why we live in a universe in which matter dominates over antimatter. For in the conventional picture, matter and antimatter must have coexisted in thermal equilibrium in almost identical amounts when $T > 10^{12}$ K. Why was there that slight excess of protons and electrons over antiprotons and positrons to survive annihilation? Do I hear the anthropic principle being wheeled out again?

The formation of galaxies provides one of the thorniest problems in cosmology today. Despite intensive work, no solution has been produced which does not amount to saying: a galaxy forms because the initial conditions of the universe preordained that it would. Significant density fluctuations are an essential feature of the creation. The same applies to the formation of clusters of galaxies. This seems rather a devastating blow to the philosophy that the universe started off simple and structureless, and then subsequently developed the structure that we see. An interesting development of recent years has been the idea that the universe started off extremely *in*homogeneous and *an*isotropic, and then evolved towards isotropy and homogeneity, retaining only those irregularities needed to form galaxies and clusters. A totally different approach is to suggest that the nuclei of galaxies and quasars are sources of continuous, spontaneous creation of matter, and that galaxies grow rather than condense. Such ideas were more popular when the steady-state cosmology, which brought the unmentionable act of creation into the astrophysical arena, was in vogue.

There is a real paradox associated with rich clusters of galaxies, that they do not seem to be gravitationally bound by the matter we see in them. The velocities of the galaxies ought to be fast enough for them to escape from the cluster, causing its dissolution. Could there be enough invisible or undetected matter, e.g. planets, isolated stars or star-clusters, dwarf galaxies, intergalactic gas, or for that matter rocks and burnt-out space-ships, to bind the cluster? The most popular candidate (on the basis that galaxy formation cannot be 100 per cent efficient) is residual intergalactic gas. There is evidence for such gas, both from its X-ray emission and from its effect on radio-sources, but not in the required quantities. This leads naturally to the question of a general intercluster gas. Such gas would have to be ionized and at a temperature of about 10^6 K to avoid detection to date. If deuterium is made in the fireball, the density of intercluster gas cannot be much above 10^{-31} g cm^{-3}.

Stars – clusters of stars – galaxies – clusters of galaxies: does the hierarchy end there? While complexes consisting of several clusters of galaxies are known, there is no evidence that this is a general feature of the cluster distribution. Yet although we can see galaxies at distances of up to 10^{10} lt.yr, so that we are looking backwards in time at least half-way to the Big Bang, the *spatial* fraction of the universe explored so far by observation is almost certainly negligible. If what we have seen so far is only a chunk of some Metagalaxy, then all we can say is that other metagalaxies must be very distant, to avoid too great an anisotropy in the radio source distribution.

Surely when we turn from the large scale to our own Galaxy, we must be on more solid ground. But even here we are faced with at least one immense lacuna in our knowledge, the problem of star formation. By mapping out the location of the youngest stars we see that they are found preferentially in the spiral arms. These arms appear to constitute a spiral wave of higher than average density rotating through the disc of gas at an angular velocity slightly higher than the disc itself. This spiral instability in the gravitational field of the Galaxy seems to be driven by the motion of the stars, but a variety of completely different explanations have been suggested. That they represent outflow of gas from the galactic nucleus, for example, or that they represent tubes of enhanced magnetic field intensity. Recent work in the infrared and in molecular line astronomy has begun to reveal the dense, cool clouds of molecules (mainly hydrogen) and dust which are the potential sites of star formation. The passage of the spiral density wave appears to trigger the collapse of such a cloud and its fragmentation into a loose association or cluster of stars. But the details remain a mystery for the moment. We still do not know whether the heavy elements (carbon upwards) in these clouds are mainly in the form of molecules or are condensed onto dust grains, and the nature and role of the dust grains is still a matter for speculation. This ignorance of the details of star formation embraces also the formation of planetary systems – how common are these? – and the origin of that most significant grain of dust (for us at least), the earth.

Once formed, a star like the sun evolves by transforming hydrogen to helium through thermonuclear reactions and, in the later stages of its life, helium to carbon, nitrogen, oxygen and so on. However, even the respectable edifice of stellar evolution theory has begun to look rather shaky, with the failure to detect the expected flux of neutrinos emitted in certain nuclear reactions in the core of the sun. The final stages of a star's life are still obscure. Theory tells us that only stars of mass less than about 1.4 times that of the sun can find a stable, cold, final state as a white dwarf or neutron star (degenerate, compact objects which simply cool off indefinitely). The identification of pulsars (pulsating radio sources with periods ranging from milliseconds to several seconds) as rotating neutron stars was a triumph for theoreticians who had predicted the existence of neutron stars more than 30 years previously. Do more massive stars manage to shed their "excess" mass in more or less violent outbursts (novae, supernovae), or must they inevitably collapse inside their "horizon" to give a black hole? The X-ray source Cyg X-1 has aroused tremendous excitement due to its possible explanation as a black hole orbiting another, visible star. Certainly there appears to be a dark, massive, compact companion onto which gas from the visible star is falling and getting heated up to produce the X-rays. Pulsars and X-ray sources in binary star systems are examples of totally unexpected discoveries from the new astronomies of radio- and X-rays, which turn out to clarify our ideas about how stars evolve. The newest astronomies of infrared and of γ-rays will doubtless have further surprises in store.

The first discovery of the oldest of the new astronomies, radio-astronomy, was that some galaxies are the sources of powerful non-thermal radio emission. Despite decades of intensive observational and theoretical work, there is no agreed model for the origin of the energy in the nucleus of the galaxy or for the generation of the double sources often symmetrically placed at distances of up to 10^6 lt.yr from the galaxy.

This problem of violent events in galaxies was exacerbated by the discovery of quasars, radio sources identified with quasi-stellar objects with large redshifts, often variable both at radio and optical frequencies. Their redshifts are up to 10 times those found in the faintest galaxies, so that their optical luminosities have to be 100 times greater, if the Hubble law (distance proportional to redshift) is satisfied. This led to doubts about the cosmological nature of the redshifts, still not resolved. Associations with galaxies of much lower redshift suggested that the main contribution to quasar redshifts might be some intrinsic effect in the quasar. However, the consensus (for what that is worth) is now that the redshifts *are* cosmological, and that quasars are simply a more violent form of the activity seen in Seyfert galaxies. The latter also have an active, quasi-stellar core, but one that does not outshine the whole galaxy, as quasar optical cores must do. Naturally we must keep looking for decisive evidence of the link between quasars and galaxies, and that quasar distances are as huge as their redshifts suggest.

We must be glad that we have some sort of overall scenario for the evolution of matter in the universe, but we must remain continually critical of it. To be a true scientist is only to have an inkling of the full extent of man's ignorance.

SUGGESTED FURTHER READING

D. W. Sciama, *Modern Cosmology*, Cambridge University Press, 1971.

M. Rowan-Robinson, *Cosmology*, Oxford University Press, 1977.

Frontiers in Astronomy, readings from *Scientific American*, W. H. Freeman, 1970.

G. B. Field, H. Arp and J. N. Bahcall, *The Redshift Controversy*, W. A. Benjamin, 1973.

G. R. Burbidge and E. M. Burbidge, *Quasi-stellar Objects*, W. H. Freeman, 1967.

Confrontation of Cosmological Theories with Observation, I.A.U. Symposium No. 63, ed. M. S. Longair, D. Reidel, 1975.

Fig. 1. The spiral galaxy, M51. The spiral arms are delineated by bright, newly formed stars. (Photograph from the Hale Observatories.)

Fig. 2. The rich cluster of galaxies in Hercules, containing both spirals and ellipticals. (Photograph from the Hale Observatories.)

Fig. 3. The nearby radio-galaxy, Cent A (NGC 5128). Note the dust lanes, unusual for an elliptical galaxy. The overall extent of the double radio source is more than a hundred times that of the optical galaxy. The nucleus of the galaxy is a source of X-rays. (Photograph from the Hale Observatories.)

Douglas Gough

Lecturer in Astronomy and Applied Mathematics at the University of Cambridge and Fellow of Churchill College, Cambridge.

Research interests include solar oscillations, the theory of pulsating stars, convection theory and thermal properties of the Moon.

THE SOLAR INTERIOR

A decade ago the Sun was thought to be well understood. Astrophysicists had been lulled into complacency by the knowledge that the theory of the structure and evolution of stars could be made to agree with the gross properties of the Sun, and on the whole could be made to reproduce the broad features of the observations of other stars. But the relevant solar data were few: just the luminosity and radius. Detailed features of surface heliography were generally thought to be minor perturbations to a simple theoretical model. This may indeed be true, but we are now aware that the theoretical model is not quite correct, and that understanding surface features might provide us with the tools necessary for probing the solar interior.

CONTRACTION TO THE MAIN SEQUENCE

Like all other stars, the Sun is believed to have condensed from a large diffuse gas cloud. Its composition was therefore that of the cloud; principally hydrogen and helium, with all other elements totalling just 1 or 2 per cent by mass. The early stages of stellar collapse are not well understood, but there is strong evidence that a star eventually settles down to almost hydrostatic balance, with pressure supporting it against its own gravitational attraction. Under such circumstances it can easily be shown that there is a balance between the total gravitational energy Ω, which is negative, and the kinetic energy T of the stellar material, such that

$$2T + \Omega = 0.$$

This is the virial theorem of Poincaré and Eddington and is valid provided no other forces, such as electromagnetic forces, play a significant role in the hydrostatic balance. If the gas of which the star is composed is not degenerate, in the quantum mechanical sense, T is basically the energy in the random thermal motion of the gas particles and so provides a measure of the mean temperature.

The total energy of the star is

$$E = T + \Omega$$

which, by the virial theorem, is simply $-T$. As energy is radiated from the surface of a star, E decreases and consequently T rises. This is perhaps the most important property of a self-gravitating body, and leads to a natural accentuation of the temperature difference between the body and its surroundings. The star slowly contracts; compression raises the temperature of the gas, but only half the energy gained by the gas from the gravitational field is lost to outer space. The central regions of the star are compressed the most, under the weight of its outer envelope, and become hotter than the material near the surface. The evolution is controlled by the rate at which heat can be transported down the temperature gradient and radiated away at the surface.

This slow, so-called Kelvin–Helmholtz contraction might be halted by the gas becoming degenerate or by thermonuclear reactions replacing gravity as a source of energy to maintain the pressure in the stellar interior. In the former case pressure becomes almost independent of temperature. The star cools and shines less and less brightly without significant change in shape or size; it eventually goes out, and becomes like a planet. More massive stars liberate more gravitational energy per unit mass for the same relative contraction, and consequently become hotter. Nuclear reactions set in before the gas becomes sufficiently compressed to be degenerate, and the star adjusts its structure so that the thermonuclear energy generated in the core just balances the energy radiated at its surface. Such a star is said to be on the main sequence, and is in a phase of its evolution that is generally believed to be the simplest to understand. The Sun is in its main sequence phase now.

SOME THINGS THAT ARE KNOWN ABOUT THE SUN

The mass M (2.0×10^{33} g) and radius R (7.0×10^{10} cm) of the Sun can be inferred from the orbits of planets and the apparent size of the solar disc in the sky. From a knowledge of the solar constant and the distance between the Earth and the Sun, the radiant energy output or luminosity L (3.8×10^{33} erg sec^{-1}) can be deduced. Similarly the neutrino output L_ν from the nuclear reactions has been inferred, assuming all but a negligible fraction of the neutrinos incident on the earth originate in the Sun. The Sun's surface radiates almost as a black body, so its effective temperature of 5570° K, which is determined from its area and L according to Stefan's radiation law, is typical of matter temperatures in the photospheric layers from which most of the radiation is emitted. The relative abundances of most of the elements in the solar atmosphere have been obtained by analysing spectral lines in the emitted radiation, but the atmosphere is too cool for the absolute abundances X and Y of the most common elements, hydrogen and helium, to be determined accurately.

The solar surface rotates with a period of about a month. Other characteristic times that could be relevant to understanding the main sequence evolution can be deduced from the data listed above. First, a dynamical timescale, which may be characterized by $2\pi\sqrt{(R^3/GM)}$, where G is the constant of gravitation; it is simply the orbital period of a body near the solar surface. Its value is 2 hr 47 min and is of the same order, though somewhat longer than the free-fall time from the surface to the centre of the Sun. From the balance implied by the virial theorem it can be deduced that it is also of the same order as the sound travel time. Any deviation from hydrostatic balance would lead to a change in the dimensions of the Sun on this timescale. Another important time scale is that for heat to be transported from centre to surface. At the present time this is not very different from the characteristic Kelvin-Helmholtz contraction time E/L, because the Sun has not changed its structure significantly on the main sequence. Estimating E to be $0.3\,GM^2/R$, the value for a homogeneous sphere, yields a time of about 10^7 yr. The Sun is not actually homogeneous, but this estimate is of about the right magnitude. The time taken to convert the entire hydrogen content of the Sun into helium at the present luminosity can likewise be estimated from the above data and the binding energy of helium. It is about 10^{11} yr and estimates an upper bound to the total time the Sun can remain on the main sequence. The age of the Sun is thought to be about 4.7×10^9 yr. Since this is much greater than the Kelvin-Helmholtz time it is clear that the Sun has spent almost its entire life on the main sequence.

MAIN SEQUENCE EVOLUTION OF THE SUN

Conditions in the solar interior are inferred by constructing theoretical models. Simplifying approximations must of course be made. Because the Sun is seen not to change its dimensions appreciably on the dynamical time scale, and because the same is true too of other stars similar to the Sun but apparently younger, it is assumed that hydrostatic balance is always maintained. Rotation and magnetic fields are normally ignored, which is a good approximation if their surface values typify the values in the interior. This implies that the Sun is spherically symmetrical, which is a state of minimum energy, and is verified by observation to be true at the surface to about one part in 10^5. It is also normally assumed that the Sun is not contaminated by infalling interstellar matter.

From an assumed structure at the beginning of the main sequence, a model representing a star with the Sun's mass is considered to evolve for a 4.7×10^9 yr. Uncertainties in the theory occur in the opacity, which determines the rate at which radiation diffuses, in the nuclear reaction rates and in the degree to which the equation of state deviates from the perfect gas law. Just as significant are the uncertainties in the initial conditions. Because the thermal diffusion time is only a small proportion of the solar age, thermal balance is achieved early in the main sequence lifetime, so it is not necessary to know accurately the initial thermal structure. But it is important to know the initial composition. There is evidence that whilst contracting onto the main sequence the Sun experienced a phase of turbulent convection throughout its entire volume. Although this has been questioned it is normal to adopt an initial main sequence model that is chemically homogeneous, with a composition characterized by the constant abundances X_0, Y_0 of hydrogen and helium, the remaining elements having relative abundances equal to their current surface values.

During evolution hydrogen is converted into helium in the central regions. The models are found to be stable to convection in all but the outer 20 per cent by radius, so it is assumed that the products of the nuclear reactions remain *in situ*. Thus a helium-rich core develops. The number of particles per unit mass decreases, with a consequent reduction in pressure, which allows further slight gravitation contraction, an increase in density and temperature and thus an increase in the nuclear reaction rates, which are sensitive especially to temperature. This takes place on the nuclear time scale, much more slowly than thermal diffusion, and so the increased thermonuclear energy generation is reflected by an equal increase in the luminosity. Typical models predict that the luminosity has increased on the main sequence from about 70 per cent of its present value.

The convection zone in the outer envelope of the Sun presents a severe difficulty, because understanding of convection even in laboratory conditions is inadequate. A simple phenomenological theory is used with adjustable parameters that can tune the outer layers of the solar model.

The object of the calculation is to reproduce the relevant observations of the Sun. Until the end of the 1960s, these were just the present luminosity L and the radius R. With X_0, Y_0 and the parameters of the convection theory uncertain, it was possible to find many acceptable models. Indeed for any plausible abundance $Z = 1 - X_0 - Y_0$ of heavy elements, a value of X_0, can always be found that reproduces L and R once the convection theory is appropriately adjusted. Furthermore, the value of X_0 so obtained is in the vicinity of 0.7, which is similar to the hydrogen abundance of the solar wind and to the abundance deduced from analysing H and He spectral lines formed in the atmospheres of hot stars.

In the last decade two new observations have cast serious doubt on the validity of the simple models described above. The first measured the neutrino luminosity, and the second the oblateness of the solar image.

THE SOLAR NEUTRINO PROBLEM

In 1968 R. Davis and his collaborators dealt a severe blow to the theory. They attempted to detect neutrinos produced by certain reactions in the proton-proton chain, the chain of nuclear reactions believed to operate in the solar core. They were unable to detect any, but from a knowledge of the sensitivity of their apparatus were able to put an upper bound on the neutrino luminosity L_ν of the Sun which was a factor 10 below contemporary theoretical predictions.

The experiment depended on neutrino capture by ^{37}Cl stored in a 400,000-litre tank of perchloroethylene; the product is ^{37}Ar which was extracted chemically and subsequently detected as it suffered beta decay. It has become customary to measure L_ν in units depending on this experiment: a solar neutrino unit, or snu, is 10^{-36} captures per ^{37}Cl atom per second. Later refinements of experimental technique appear now to have led to a definite value for L_ν. Taking some four years of measurements into account yields 1.2 ± 0.5 snu, though measurements of order 4 snu have been obtained using the latest most sensitive detectors.

These results led theorists to check their calculations. The microphysics was thoroughly revised, and uncertain parameters were pushed to the limits of plausibility in order to produce as low a value of L_ν as possible. Now values just slightly in excess of 4 snu can be obtained with the so-called standard theory, which might be considered tolerable if all but the very latest measurements are ignored. If, however, the average of 1.2 snu is believed, the problem still remains.

In attempting to account for the discrepancy astrophysicists have questioned in turn all the assumptions of the theory. Of course, the most obvious starting-point is the nuclear physics. Since the results of the neutrino experiment were first announced all but one of the nuclear reactions in the proton-proton chain have been recalibrated. The remaining one is the first reaction of the chain. It proceeds via a weak interaction and is the slowest reaction of all, thus controlling the operation of the entire chain. The rate is too small to measure, and must therefore be calculated theoretically. Although many nuclear physicists seem to have confidence in the calculation, there is some experimental evidence from a different though similar reaction that the calculation contains a flaw. If that is the case, the neutrino problem disappears, but other investigations attempting to solve the problem have raised new questions about the Sun that now remain to be answered.

One such inquiry concerns the initial chemical composition of the Sun. It has been proposed that some fractionation process separated the elements as the Sun contracted from its parent gas cloud, and that subsequently no homogenizing fully convective phase took place to destroy the birthmark. Solar models with an initial inhomogeneity tailored to reduce the neutrino flux all have the property that in some region the density increases upwards. They are therefore subject to the Rayleigh–Taylor instability that occurs when one tries to create an interface between two fluids with the denser fluid on top. This deficiency might have been avoided had the composition gradients been smeared out somewhat. Because the material nearer the centre of the Sun is more highly compressed, it is possible to imagine a distribution in composition such that the mean molecular weight increases upwards without the density doing so. It would then be Rayleigh–Taylor stable, but it would be subject to the diffusively controlled fingering instability that occurs, for example, in the terrestrial oceans where hot salty water rests on colder, denser fresh water. Because chemical diffusion is slower than heat transfer, fingers of fluid thin enough to destroy their temperature differences but not so thin as to lose their chemical identity can penetrate vertically, liberating gravitational potential energy enough to drive the motion further. The process has been studied in the laboratory, but unfortunately it is not yet well enough understood to make reliable extrapolation to stellar conditions possible. Therefore it is not certain whether fingers would have had time to homogenize the Sun.

Another hypothesis is that substantial accretion of matter has taken place during the main sequence history of the Sun. Accretion of gas is unlikely, because it would be prevented by the solar wind, but interstellar gas clouds are known to contain dust which could fall through the wind. An obvious consequence of this hypothesis is that the resultant contamination of the solar surface, rich in heavy elements, would obscure the interior composition. Furthermore, solar models with interior heavy element abundances lower than the observational limits on the surface composition yield neutrino fluxes as low as the bounds set by Davis. Of course, this situation too would be subject to the fingering instability. It has been suggested also that the gravitational energy released during the intervals of accretion would temporarily increase the solar luminosity and upset the Earth's climatic balance, inducing glaciation. Dust clouds exist in the spiral arms of the galaxy, and the interval between successive passages of the Sun through the spiral arms is of order 10^8 yr, which is similar to the period between the major terrestrial ice ages. Moreover, the fact that the Sun is currently at the edge of a spiral arm is consistent with the Pleistocene glaciation of the last 10^6 yr. These coincidences are intriguing, but it is difficult to reconcile the large amount of accretion required by the theory with the estimates of the amount of dust in the spiral arms.

A third investigation concerns the stability of solar core. Although the core is stable to convection, regions in which thermonuclear energy is being generated are potentially unstable to oscillatory gravity waves or acoustic waves. Each element of fluid can be thought of as a thermodynamic engine, generating heat preferentially at high temperatures, and thus doing work that drives the oscillations to greater and greater amplitude. Calculations suggest that only after equilibrium concentrations of the intermediate products of the proton-proton chain have been built up over a sufficiently large volume of the core can the instability occur. This takes about 10^8 yr. It is then presumed that the oscillations break down into turbulence and mix material rich in fuel from the edge of the core into the centre. This temporarily enhances the energy generation rates and upsets the balance of the reaction chain in such a way as to quench the instability. The Sun is then quiescent again until nuclear equilibrium is once more achieved, and the whole process repeats.

An immediate consequence of the enhanced energy generation in the core is an expansion of the star. It is the reverse of the gravitational contraction that occurs when nuclear energy sources are absent or inadequate. The expanding sun cools and the luminosity decreases by a few per cent. The surface luminosity differs from the nuclear energy generation rate until thermal balance is restored on the Kelvin–Helmholtz time scale of about 10^7 yr. During this time the solar core, temporarily cooled by its expansion, liberates fewer neutrinos than usual. Once again, if the luminosity fluctuations are presumed to induce ice ages on the Earth, the beginning of the Pleistocene epoch marks the most recent mixing of the Sun's core. Because that occurred less than 10^7 yr ago, it is inferred that the Sun is presently in one of its comparatively rare states of thermal imbalance, and the neutrino flux is anomalously low.

Further work is necessary before these ideas can be reliably tested. The ^{37}Cl neutrino experiment detects mainly high-energy neutrinos produced in a side reaction of the proton-proton chain, and does not provide a direct measure of the heat generated. An experiment using a gallium detector has been proposed that will measure the lower energy neutrinos from the thermal energy-producing chain. This could provide a new check on the nuclear physics, and might enable one to decide whether or not the solar luminosity is currently balanced precisely by thermonuclear sources.

THE SOLAR OBLATENESS

The surface of the Sun is observed to rotate. This is inferred from the motion of sunspots and from line-of-sight surface velocities measured from the Doppler shifts of spectral lines. The rotation is not rigid: the equatorial regions rotate with a period of about 25 days whereas near the poles the period is of order 35 days or more. Moreover, sunspots and some other magnetically related phenomena that are presumably anchored by magnetic fields to material beneath the surface rotate somewhat more rapidly than Doppler measurements indicate, at least in the equatorial regions. This suggests an increase in rotation rate with depth. Nevertheless, it is generally believed that the rotation in the deep interior is not a great deal faster than it is at the surface.

An important property of rotating bodies is that centrifugal forces cause them to bulge at the equator. The effect on the Sun is clearly not very great, because the Sun appears circular to the eye. Indeed, if rotation were uniform throughout the Sun with a period of 25 days the polar and equatorial radii would differ by only about 1 part in 10^5. Consequently for most practical purposes the Sun can safely be assumed to be spherically symmetrical. This implies, in particular, that the solar gravitational field is also spherically symmetrical, and independent of how mass is distributed within the Sun. This important result has enabled astronomers to calculate with great precision the orbits of the planets without requiring detailed knowledge of the solar interior. However, it was realised as long ago as the middle of the last century that the orbit of Mercury was not what would be expected from Newton's inverse square law of gravitation. The discrepancy, after taking due account of perturbations from other planets, can be expressed as a precession of the perihelion of the orbit at a rate of 43 seconds of arc per century. Though small, this precession was a cause of anxiety to astronomers, and led them to predict the existence of another planet, Vulcan. The planet was never discovered. In 1915 Einstein predicted a precession of just the right value from the correction to the inverse square force law implied by his theory of general relativity, and the matter appeared to be settled.

In 1961 Einstein's explanation was challenged. C. Brans and R. H. Dicke proposed an alternative theory of gravitation which predicted a smaller precession rate for planetary orbits. It was suggested that only by chance had Einstein predicted the correct value, for the centre of the Sun is in a state of rapid rotation which distorts the gravitational field and so adds a contribution to Mercury's precession. To explain the observed precession by the Brans–Dicke theory the rotation period of the solar core would have to be about a day, and general relativity would then yield a precession rate too great by about 4 seconds per century. Of course, such rapid rotation would also distort the shape of the Sun's surface. To check this Dicke, together with H. M. Goldenberg, measured the shape of the solar image and, arguing that the edge of the sun coincides with a gravitational equipotential surface, found it to be oblate by just the amount required to explain Mercury's orbit with the new theory.

The experiment initiated much discussion. If the interpretation were correct, why should the solar core rotate so rapidly? A possible answer came from observations of other stars, younger though otherwise similar to the Sun. Although it is not possible to resolve their surfaces, Doppler broadening of their integrated light can be measured and rotation rates inferred statistically if it is assumed that the only variable in an apparently homogeneous class of stars is the angle of inclination of the rotation axis to the line of sight. What emerges from such an analysis is that surface rotation decreases with age. This is not a surprising result if one bears in mind that the Sun is continually ejecting a diffuse stream of gas, or wind, as are presumably other similar stars. There is a weak magnetic coupling which transfers angular momentum from the Sun to the wind and slows the Sun down. The present torque can be measured. Plausible estimates of how it has varied rationalizes the observed dependence of stellar rotation with age and suggests that the Sun might well have arrived on the main sequence rotating with a period of about a day. The current rotation period of the solar core might still be of that order if the coupling with the rest of the Sun is extremely weak. But is that so?

Dicke estimated how fast the core ought to slow down on the assumption that viscous shear stresses provide the only vehicle for angular momentum transfer beneath the outer convection zone. The characteristic time scale exceeded the age of the Universe. But the relevance of the calculation was disputed because a similar argument applied to a stirred cup of tea predicts a deceleration time as long as an hour. It is well known that tea stops rotating in about a minute. The explanation stems from an imbalance of pressure and centrifugal forces near the base of the teacup where viscous forces prevent rotation of the tea; this drives a circulation in meridional planes, originally discovered by V. W. Ekman, that advects angular momentum through the body of the fluid faster than is possible by viscous shear alone. It is also responsible for the convergence of the tea leaves in the middle of the cup. So perhaps the solar core is similarly decelerated and general relativity is correct. Interestingly, the dynamics of the Ekman circulation has been discussed by Einstein, in a paper on the meandering of rivers. It is not easy to extrapolate the analysis of cups of tea to the Sun, however, particularly as the Sun is not in a rigid container and is not of uniform density. None the less, it does seem likely that a similar circulation is set up, though how it redistributes angular momentum is less clear. Thus, even granted that the inside of the Sun feels the deceleration of the outside, theory cannot yet predict what the rotation of the inside ought to be.

One of the problems encountered in measuring the shape of the Sun's gravitational equipotentials by observing the shape of the solar image is in deciding just what one means by the edge of the Sun. The difficulties have been stressed recently by H. A. Hill and his collaborators, who observed that the spatial variation of the intensity of radiation near the edge of the Sun depends on latitude: near the equator it declines more sharply than it does at the poles. Hill interpreted this as resulting from an excess brightness just above the photosphere, rather than from variations in matter density, which could arise from the preferential dissipation of atmospheric waves above the photosphere in the equatorial regions. Taking this into account Hill inferred a value of the oblateness consistent with almost uniform rotation of the solar interior. But that is not all. In the course of his observations Hill discovered that the radius of the Sun was changing with time: the Sun was oscillating on a dynamical time scale. Higher-frequency oscillations had been well studied before, with periods corresponding to those of waves confined to the atmosphere alone, but these new oscillations have periods characteristic of motion whose amplitude penetrates deep into the Sun. Analysis of the data has yielded a spectrum of frequencies that agrees well with theoretical frequencies of normal modes of oscillation of a solar model computed with the standard assumptions listed earlier in this article.

The existence of these oscillations has an interesting implication regarding the solar rotation. It may be they that transmit energy above the photosphere and generate the equatorial excess brightness. Their latitude dependence has not yet been measured. However, if they are responsible for much of the apparent oblateness the oscillations must have greatest amplitudes near the equator. Such modes also have the property of being asymmetrical about the rotation axis, and of propagating around the Sun in longitude. In particular there are classes of such modes, whose members each have frequencies that are almost identical, but with slight differences induced by the Sun's rotation. Two such modes oscillating together would beat, and the dissipation would give the Sun the appearance of having equatorial bulges that rotate in longitude at a rate depending on the modes that are beating. Such a phenomenon has recently been noticed by Dicke, who observed that the apparent oblateness varies with a period of 12.2 days. At present the measurements are not detailed enough to enable the modes that are beating to be identified, or even to confirm or refute the hypothesis that it is indeed a beat phenomenon that is being observed, but if the hypothesis is correct it is clear that the rotation rate of at least some region of the solar interior exceeds the rotation of the surface, though not by a large amount. Hopefully future observations of this kind will make more precise deductions possible.

CONCLUDING REMARKS

The neutrino flux and oblateness measurements have shocked astrophysicists out of their comfortable belief that the Sun is a simple, easily understood body. None of the theoretical models of the time was able to explain the neutrino observations, which provided a stark reminder that the ability of a theory to rationalize data does not prove that theory. Even if it turns out that the neutrino luminosity is not as low as has been feared, and, as now seems likely, that the oblateness of the gravitational equipotentials is no greater than would be expected from assuming that the rotation of the solar surface typifies the rotation of the Sun as a whole, these observations have been of great importance because they have stimulated much deeper thought into the problems posed by trying to infer the structure of the solar interior. Furthermore, they have led to the discovery of a spectrum of coherent oscillations which may provide a powerful diagnostic tool for probing the Sun, just as seismic oscillations have given important information about the interior of the Earth. In particular, it may be possible to determine whether the Sun has been contaminated by accreted interstellar material.

Why the oscillations should even be present is not understood, though it is likely that they are triggered by turbulence in the convection zone. The rough agreement between the theoretical and observed frequencies suggests that current thinking is not altogether incorrect, but the differences do appear to be significant. Perhaps the discrepancies will be removed by minor changes in the theoretical models, but alternatively they may be linked with other phenomena, such as magnetic fields, hitherto ignored by most theorists. L. Mestel has pointed out that quite strong magnetic fields may reside deep in the interiors of some stars, even though the surface fields may be quite small. If sunspots could be understood, they might provide a clue to the interior field of the Sun. The cyclic behaviour of the Sun's surface magnetism, with a period of 22 years, is a solar phenomenon without a convincing explanation, and might eventually provide additional diagnostics.

The discussion of the solar oblateness has brought to the forefront an interesting problem: the history of the Sun's rotation. Even though the question of angular momentum loss by the Sun had previously been considered, it was not until after Dicke's oblateness measurement that mechanisms coupling the core to the surface were seriously investigated. Magnetic fields and the coherent solar oscillations are capable of angular momentum transport. So is rotationally induced material circulation. The possibility of the latter raises new questions concerning mixing of the products of nucleosynthesis into the solar envelope.

Perhaps the new solar data have generated more questions than answers. They have certainly reinforced our awareness of our ignorance of the solar interior.

FURTHER READING

Bahcall, J. N. and Davis, R. Jr. (1976). Solar neutrinos: a scientific puzzle. *Science*, **191**, 264.

Bahcall, J. N. and Sears, R. L. (1972). Solar neutrinos. *Ann. Rev. A. & Ap.* **10**, 25.

Bretherton, F. P. and Spiegel, E. A. (1968). The effect of the convection zone on solar spin-down. *Ap. J.* **153**, 277.

Christensen-Dalsgaard, J. and Gough, D. O. (1976). Towards a heliological inverse problem. *Nature*, **259**, 89.

Dicke, R. H. (1970). Internal rotation of the Sun. *Ann. Rev. A. & Ap.* 8, 297.

Hill, H. A. and Stebbins, R. T. (1975). The intrinsic visual oblateness of the Sun. *Ap. J.* **200**, 471.

Tayler, R. J. (1970). *The Stars: their Structure and Evolution*, Wykeham, London.

Paul Charles William Davies

Lecturer in Applied Mathematics, King's College, University of London.

Research activities have covered several areas of fundamental physics including quantum theory, relativity and cosmology.

Recent research has been on quantum field theory in curved space, and particularly concerned with quantum particle production effects near black holes and during the initial moments after the cosmological big bang.

CURVED SPACE

For those who think of space as emptiness, the assignment of an adjective, especially one as enigmatic as "curved", might be regarded as cryptic. To the mathematician, and more so to the physicist, empty space may be devoid of matter, but it is by no means devoid of properties. I shall, at the outset, sidestep the dangerous mire dealing with the centuries-old controversy over whether space (and time) is really a substance existing in its own right, or simply a linguistic convention describing the relations between material objects. The question of the curvature of space may only be settled observationally by the use of light rays and material structures, and it matters little here whether the results obtained are to be regarded as properties of the latter or not.

What does "curved space" mean? Curvature is a familiar property of everyday life. But there is a subtlety; *lines* can be curved, and *surfaces* can be curved. Two quite distinct entities may share a common property. We describe with the same adjective (and scarcely a thought) things as fundamentally dissimilar as a railway line or the surface of the Earth. The distinction here is one of dimensionality. A point on the one-dimensional railway line needs but one number (e.g. the distance from Euston station) to locate it uniquely, while a point on the two-dimensional surface of the Earth needs two numbers (e.g. latitude and longitude) for its location. Here I shall explore the question: can curvature apply to *three-dimensional* volumes, as well as two-dimensional surfaces and one-dimensional lines, and if so, what can be learnt from observation and mathematical theory about it?

Although the examples of railway line and Earth describe the curvature of material entities, this is not an essential feature; for example, the path of an arrow is a set of points in space which possesses curvature. The problem of curved volumes is just the problem of curved space, empty or otherwise.

In mathematics, curvature belongs to the subject of geometry. School geometry is called Euclidean, after the Greek geometer Euclid. Everyone remembers a few of the theorems – the angles of a triangle add up to two right angles being an elementary one. Euclidean geometry works well on *flat* sheets of paper. It does not always work on *curved* surfaces, such as that of the sphere. A glance at Fig. 1 shows how triangles on a sphere may have three right angles!

Until the nineteenth century nobody questioned the fact that although, for example, the Earth's surface was curved, the surrounding space itself was subject to the rules of Euclidean

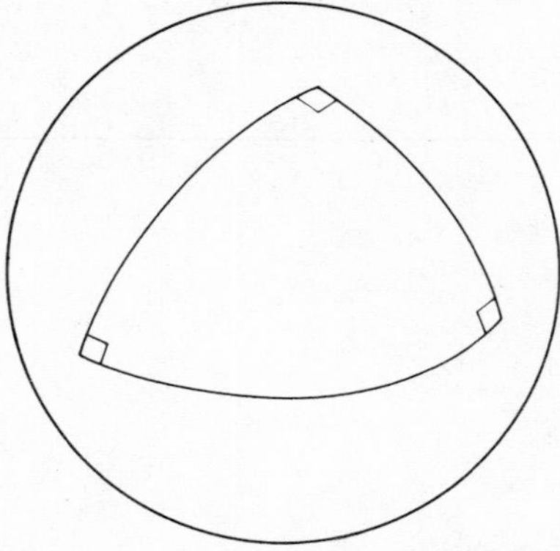

Fig. 1. Two-dimensional spherical space. The usual rules of Euclidean geometry do not work here. A triangle may have three right angles.

geometry. Indeed, ever since Kepler physicists had frequently employed the Euclidean theorems across the empty space of the solar system to describe the paths of the planets around the Sun. "Flat" space geometry seemed to work well in these regions remote from the Earth. With one exception. Mercury, the nearest planet to the Sun, seemed to have an orbit which was slightly distorted. This distortion could not be accounted for by the disturbance of the other planets, and although exceedingly small (a change of only 43 seconds of arc per century) remained a total mystery until the beginning of this century.

Already the nineteenth-century mathematicians Gauss, Lobatchevsky and Riemann had developed non-Euclidean geometry for volumes, but it took the genius of Einstein to construct a *physical* theory which exploited these mathematical developments. Einstein was interested in the motion of material bodies, and had constructed a theory of uniform relative motion, called special relativity, in 1906. In attempting to generalise relativity to accelerative motions, he realised that the unique nature of gravity singled it out from all other physical forces which accelerate material bodies. This is because (small) bodies which fall freely under gravity all travel along the same path in space, no matter what their mass or constitution. This suggested that gravity was not really a force at all, but a geometrical property of space itself (strictly speaking, space and time together). Einstein's revolutionary proposal was to identify gravity with a departure from Euclidean geometry – in popular parlance, a bending or curvature of space. According to this viewpoint the planets do not move in a curved path around the Sun, as Newton supposed, with the Sun exerting a gravitational *force* upon them to bend them away from their natural tendency to move in straight lines. Instead, the Sun's gravity is interpreted as a distortion of space (and time) in its vicinity, and the planets merely follow the "easiest" route – the path which minimises their mechanical action through the curved space. This easiest route turns out to be very nearly the same as the "forced" route taken by the planets according to Newton's theory of a gravitational force. But not quite. Mercury's orbit has to be displaced by 43 seconds of arc per century. This was Einstein's great triumph.

If the space around the Sun is not precisely Euclidean, it might be expected that the images of objects seen beyond the Sun would be somewhat distorted. In 1919 Sir Arthur Eddington checked this prediction during a solar eclipse, by observing a displacement in the positions of stars in the region of the sky near the eclipsed Sun.

For many years this remarkable theory of Einstein, called general relativity, appeared to be limited to describing minute corrections to Newtonian gravity. This was because our knowledge of astronomy did not suggest that there were ever situations in the Universe where gravity was strong enough to cause a really drastic bending of space. However, during the last 10 years, this prospect has become entirely plausible. We shall see that it leads to some of the greatest outstanding mysteries of physical science.

Although the present density of gravitating matter in the Universe is very low (about one star of solar mass per billion cubic light years) the Universe is very big (present telescopes see as far as a billion light years) and the effect of space curvature can be cumulative, with each star causing its own tiny distortion. As long ago as 1915, Einstein realised that on a cosmological scale the cumulative space curvature might become so great that it would alter the *topology* of space. To understand this remark, return to the two-dimensional analogy. If the curvature of a surface (or sheet) is always in the same direction, and everywhere roughly the same amount, the surface will eventually join up with itself. The sphere is a good example of this. Although in a small enough region the geometrical properties of a spherical surface are not greatly different from those of a flat sheet, the *global* structure is distinctly different. For one thing, the spherical surface is *finite* in size, though nowhere does it possess a boundary or barrier. One

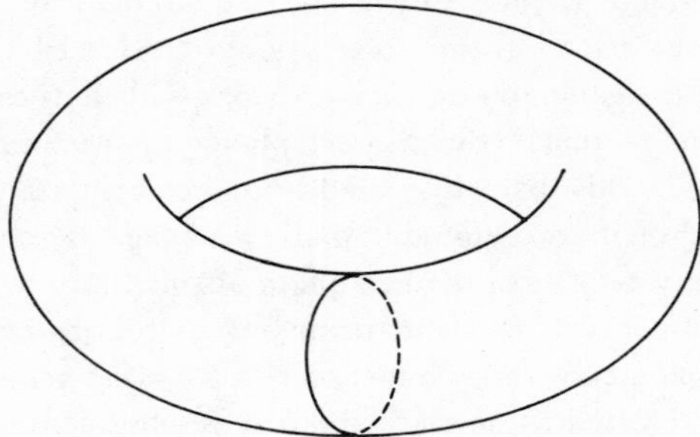

Fig. 2. Two-dimensional toroidal (doughnut) space. Unlike spherical space there are closed curves which cannot be continuously shrunk to a point.

consequence of this is that if we were to travel along the shortest route between any two points on such a surface, and then continue "straight" on, we should eventually return to our starting-point from the opposite direction.

The sphere is by no means the only finite and unbounded curved surface. The torus (see Fig. 2) is another example. This is distinguished from the sphere by the property that there exist closed lines on the torus which cannot be shrunk continuously to a point. Still more complex properties could be achieved by putting twists into the surface before joining them up, so that a journey round certain closed paths would change right-handed orientations to left-handed orientations.

Einstein's novel proposal was that space curvature closes up the Universe into the three-dimensional analogue of the spherical surface. Such a universe would then be finite in volume, and could be circumnavigated by a sufficiently tenacious traveller. It might even be possible to see the back of one's head by looking out into space through a large enough telescope! Still more bizarre would be a "hypertoroidal" space (like a three-dimensional doughnut) or even a "twisted" space, in which our adventurous traveller would return with his left and right hands transposed!

In spite of the Alice-in-Wonderland quality of these considerations, they are by no means idle curiosities. A determination of the global nature of space is one of the foremost tasks of modern cosmology, and a variety of observational techniques are being employed on this task.

The portion of the Universe accessible to even our largest telescopes is too small to reveal the global structure directly, so a combination of theory and observation has to be used. Since Einstein's original proposal, it has been discovered that the Universe is expanding – growing

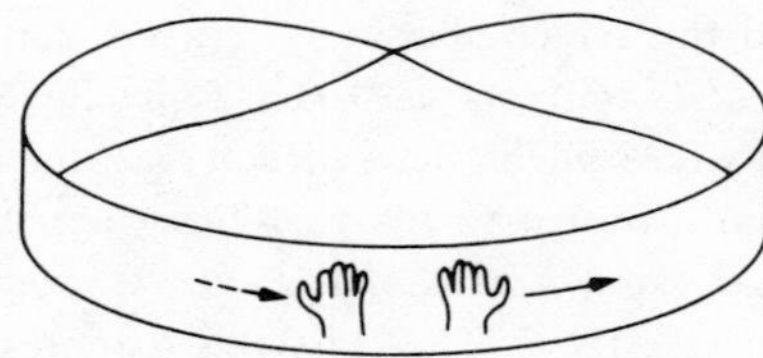

Fig. 3. Two-dimensional twisted space (called a Möbius strip). If both surfaces of the strip are identified as the same space, then a journey once round the loop changes right handedness into left handedness.

with time. If it is also spatially closed, space would then be analogous to the surface of a balloon being slowly inflated. It turns out that the *motion* of the Universe is related to the geometry of space through Einstein's theory of general relativity. Thus, it may be possible to determine if space is closed and finite by observing how the Universe moves.

This is no mean task. Distances between the galaxies only grow at about one-millionth of 1 per cent each century. However, because of the enormous size of the observable Universe, the light received from the farmost galaxies started out many millions of years ago, and it is possible to infer how the expansion rate then compares with now. Until recently, all indications were that the rate of expansion was gradually slowing down, and would eventually stop. This not only implied that space was indeed closed up on itself as Einstein had suggested, but also that the Universe would eventually collapse as recontraction followed the cessation of expansion. Recently, these observations have been questioned on a variety of grounds, with the result that the global geometry of space and the eventual fate of the Universe are still in the balance. What is needed is either a very good estimate of the amount of gravitating matter in the Universe, and this includes much which is inconspicuous (dark gas and dust for example), or else a better knowledge of the way galaxies change their brightness with age. The latter enables a better estimate of galactic distances, and therefore motions, to be made. The next decade or so should see this question answered.

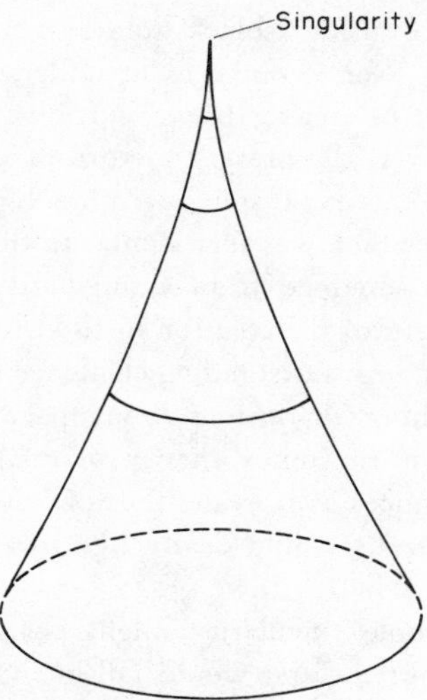

Fig. 4. A singularity in space. The circles on the cone-like structure measure the curvature of this two-dimensional space. As the apex is approached, these circles become progressively more strongly curved, until they reach a point of infinite curvature at the tip, where the space ceases to exist.

In 1967 J. Bell and A. Hewish of Cambridge University discovered a new type of star, called a pulsar on account of the regular radio pulses which they emit. It is widely believed that pulsars are incredibly compacted stars, so dense that they are only a few kilometres across and can spin many times a second. The compaction of matter in these stars is great enough for their atoms to be entirely crushed into neutrons. From what is known about nuclear matter, these neutron stars appear to be teetering on the edge of disaster. So intense is the gravity at their surface, that any incipient neutron stars more massive than our sun would not reach a stable structure at all, but would instead implode catastrophically, and in a fraction of a second disappear entirely from the Universe!

The explanation for this astonishing phenomenon lies in the violent curvature of space which accompanies the escalating gravity near a collapsing star. As the radius of the star shrinks, the space curvature at the surface soon becomes strong enough to bend back even light rays – the swiftest of all entities. After the last light ray has escaped the star becomes a prison – nothing whatever can escape to the outside, for nothing can travel faster than light. From a distance, the star appears to have been replaced by a hole in space – a black hole. Inside the black hole, the inexorable shrinkage continues; in fact, no force in the Universe can prevent the collapsing star from crushing itself ever more forcefully. With each microsecond, the space becomes progressively more curved and the density of the star rises.

The burning question is: what happens in the end? Does the implosion stop, or does the star literally shrink to nothing?

According to the general theory of relativity, the star must continue to shrink until space becomes so curved that a *singularity* forms. This situation may be visualised using the two-dimensional analogy of a surface which has been curved to form a cone-like structure. The curvature of the surface rises without limit as we pass towards the point of the "cone". At the point itself something catastrophic occurs – the curvature is, loosely speaking, infinite, and the "space" comes to an end.

Can it really be true that space inside a black hole is so curved that it too comes to an end? This is not just an idle question. Even if our present understanding of stars is badly wrong, a black hole could in principle be made by a sufficiently resourceful and determined technological community. Moreover, the present picture of an expanding universe leads almost inevitably to the conclusion that this expansion began in a big bang some 10 to 20 billion years ago. The condition of the Universe then was very similar to the inside of a black hole, though in reverse. The Universe apparently emerged from a singularity before which neither space nor time existed. This scientific picture of the creation of the Universe cuts right across philosophy and religion, and is surely today's most outstanding challenge in physical science.

Whether or not these far-reaching conclusions about space and time terminating are believed depends on the extent to which Einstein's theory of relativity can be trusted to apply as successfully over very small distances as it evidently does over cosmic dimensions. This is the question which theoretical physicists would dearly like to answer. Let us see how well it has already been answered.

One way in which a black hole singularity might be avoided is if the implosion is not symmetric. Then different parts of the star would fall in slightly different directions, thereby "missing" each other at the centre. Alternatively, some drastic modification might occur in the properties of the contracting matter (which is, after all, compacted way beyond all terrestrial experience). Remarkably enough, neither of these two possibilities proves particularly fruitful. Some amazing mathematical theorems, largely due to the British mathematicians S. Hawking and R. Penrose, demonstrate that, under a very wide range of circumstances, a singularity will inevitably form. The sort of modification to the properties of matter necessary for singularity avoidance are most unacceptable to physicists, involving negative energy matter and so forth.

It might then be asked: will Einstein's theory break down in the extreme conditions which precede the singularity? We cannot give the definitive answer "no" without falling into one and being crushed out of existence. But the answer "yes" could be near at hand. There is a sense in which the general theory of relativity is known to be wrong as it stands. This is because it is incompatible with the laws of microscopic physics, known as *quantum mechanics.* The quantum theory has been responsible for averting another catastrophic collapse – the implosion of atoms as a result of electric attraction. According to pre-quantum ideas of mechanics, all atoms should be unstable and collapse, but they obviously don't. The resolution of the paradox is provided by the quantum theory, which explains the stability of atoms by supposing (heuristically) that microscopic particles are subject to sudden fluctuations in their energy and position. These fluctuations have the effect of preventing the particles in an atom from approaching each other too closely. So, could this effect also operate in gravitational collapse, when the star has shrunk down to atomic dimensions?

The answer is, not directly. The quantum properties of matter alone are probably not strong enough to overcome gravity. But there is still the possibility that the quantum properties of *gravity itself* might be important. Fluctuations in gravity imply fluctuations in the curvature of space. If such microscopic fluctuations really occur, they could profoundly modify the structure of space on a very small scale, for the smaller the distances of interest, the larger the fluctuations become. Eventually, they would be so violent that space would literally start to tear itself apart. The sort of dimensions on which such a traumatic effect would be operative is minute indeed – a hundred billion billion times smaller than the nucleus of an atom! At this scale, space would no longer have uniform properties but would assume a sponge-like structure – with bridges and holes interconnecting in a fantastic and complex web. Just what happens to a catastrophically collapsing body which is squeezed into this spongy domain is anybody's guess. Whether space and all of physics ends at a quantum singularity, or whether space gives way to some new entity, and if so what, are questions eagerly under attack by theoretical physicists.

During the last 20 years, mammoth efforts have been made to formulate a rigorous theory of quantum gravity to give a proper theoretical basis to these conjectures. Unfortunately, not only does success seem very far away, but some physicists even doubt whether the quantum theory is compatible at all with the general theory of relativity. Such an incompatibility would presage a new theory more fundamental than either of the old, involving unfamiliar concepts out of which curved space must be built as a "large"-scale approximation. To find out just what these concepts might be, and what strange new perspectives they will provide on the nature of the Universe, we must await tomorrow's discoveries.

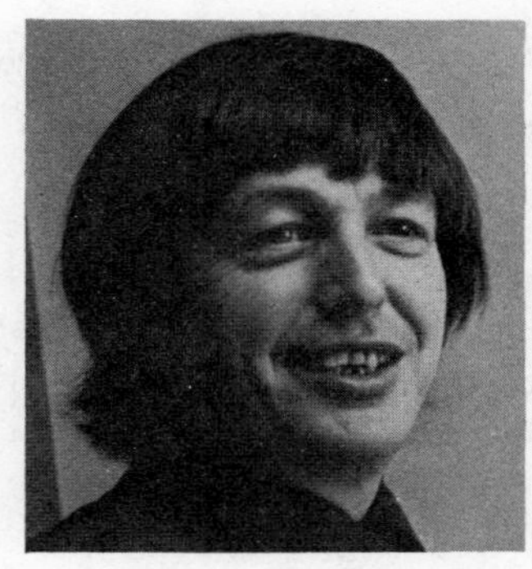

I. W. Roxburgh

Professor of Applied Mathematics at Queen Mary College, University of London, and Chairman of the London University Astronomy Committee.

Member of the Royal Astronomical Society and the International Astronomical Union, Member of the Council of the Royal Astronomical Society 1968-71.

Research interests include the History and Philosophy of Science, Cosmology, Evolution of Stars, Solar Physics and Gravitational Theory.

IS SPACE CURVED?

Ever since the development of general relativity gravitation has been attributed to the "curvature" of space, in the words of John Wheeler: "There is nothing in the world except empty curved space. Matter, charge, electromagnetism are all manifestations of the bending of space. Physics is Geometry."[(1)] While this is a rather extreme example of the geometrodynamic philosophy it is nevertheless shared by the scientific community, and the belief that "space is curved" has entered and remained in our conceptual framework of physics. But is space curved? Indeed what if any meaning can be given to such a statement? I shall argue that such a statement is meaningless, empty of any empirical content, space is curved or flat depending on the choice of the scientist.

Until the discovery of non-Euclidean geometry in the eighteenth and nineteenth centuries the geometry of space was assumed to be the flat Euclidean geometry discovered by the Babylonians, Egyptians and Greeks, finally formalised by Euclid in the *Elements*. Drawing on practical experience of measurement and surveying they discovered that the sum of the angles of a triangle was always 180°, that the square of the hypotenuse of a right-angled triangle was always equal to the sum of the squares of the other two sides, that the circumference of a circle was always the same multiple of the diameter. These results were acquired through measurement, using a ruler or angle measuring device – it was an empirical geometry. From Euclid onwards the situation was changed, by showing that all the empirical results of geometry could be deduced logically from a small number of definitions and axioms, it set the example for axiomatic mathematics and demonstrated that the behaviour of the physical world was governed by a few simple statements, or laws.

Then came the discovery of non-Euclidean geometry – not as the result of measurement but as the consequence of logico-mathematical inquiry, by changing one of the axioms in the Euclidean scheme an alternative astral, or hyperbolic geometry was discovered, just as logically sound as the Euclidean example. We see geometry now as simply a mathematical structure, given certain definitions and certain axioms – then the application of standard logical reasoning produces a geometry. There is no end to this procedure – any number of different mathematical geometries can be (and have been) derived by mathematicians.

But is there one of them that corresponds to the real world – a real geometry of space? Shortly after his discovery of hyperbolic geometry, Karl Fredrick Gauss set out to try and answer this by direct surveying techniques, from points on the mountains of the Brocken, the Höher-Hagen and the Insleberg, he measured the sum of the angles of a triangle to see if they were equal to or less than 180° and space was flat. Gauss like others since have been convinced that the determination of the curvature of space is an empirical issue, there is this substance called space that has intrinsic characteristics – one of which is curvature – and the experimental scientist can in principle measure that curvature.

This belief that space has substance of its own has a very long history, but was finally accepted into physics with the success of Newtonian physics and the Newtonian philosophy of nature. But it was challenged in Newton's own time by Gottifried Leibniz who advocated a relational theory of space and time;[(2)] only the relationships between objects or time ordering has any empirical meaning – to introduce a concept of absolute space with no properties other than that of being absolute he considered a redundancy, indeed a challenge to the absolute power of God. But in spite of the cogency of Leibniz's arguments, Newton's natural philosophy prevailed – because it worked – no alternative to Newton's physics was available and scientists accepted the philosophy as well as the physics of the Newtonian scheme.†

† This was a problem Huygens wrestled with for many years but he was unable to account for Newton's absolute centrifugal forces.

Criticism of the Newtonian philosophy arose again in the writings of Ernst Mach that so influenced Einstein in his search for general relativity, absolute space was meaningless, inertia in matter must mean inertia relative to other matter not relative to absolute space.[3] But absolute space remained in physics through the revolution of Special relativity in which absolute space remained as the backcloth defining the inertial frames in which the new Lorentz invariant physics was valid. While special relativity has a major impact on physics, conceptually it is little different from the old Galilean Newtonian physics, replacing Galilean invariance with respect to inertial frames by Lorentz invariance but still retaining the absoluteness of space and inertial frames.

With the advent of general relativity Einstein hoped to give expression to the views of Leibniz and Mach in a quantitative way, to show how the inertial properties of matter were indeed caused by the rest of the matter in the Universe, absolute space was to be abolished in favour of relative space, there was to be no absolute space – no privileged reference frame – anyone's view was as good as anyone else's and only relative behaviour entered physics. Unfortunately this programme did not succeed, absolute space remains in general relativity, but even more, general relativity described the inertial gravitational field in terms of the curved space-time geometry developed by Riemann following on the work of Bolyai, Lobatewski, and Gauss.[4] This was an ideal tool for the expression of these views, the inertial and gravitational properties could be described in terms of a curved space, the curvature produced by matter, and bodies moving along "straight lines" in this geometry. The theory, like Newton's before it, works, that is its predictions agree with experiments, and just as in the Newtonian period scientists have accepted not only the theory, but the philosophical backcloth of curved space-time, just as they accepted Newton's absolute space. We live in a curved space – the scientist's task is to determine the curvature.

But what is this stuff called space whose curvature is to be determined – how do we measure it? We can, like Gauss, set up a triangulation experiment and measure the angles of a triangle – the answer will not be 180° – but this does not mean space is curved. The experiment is done with light rays and theodolites – the empirical result is a statement about the behaviour of light rays – not about space. It is, as it must be, an experiment about the relationship between objects in space not about space itself. The same is necessarily true about any experiment; from it we learn of the relationship between objects not of the background we call space. Indeed it is part of the philosophy of relativity that it is the relative positions and motions of matter that determine the behaviour of other matter; space, or space-time, is an intermediary that we bring into the formalism for ease of representation, but in any empirical statement about the world the representation is eliminated, in the end scientific theory is reduced to numerical predictions of the kind $A = 2B$, where A is the predicted quantity and B a reference quantity. But the method of representation has been given an existence in its own right – we talk of curved space, or curved space-time, as though this is something intrinsic to the world instead of the mathematical representation it actually is.

The difference between the mathematical representation and empirical physics was carefully elucidated by Henri Poincaré[5] at the beginning of this century, and led him to his conventionalist philosophical stance. Clearly distinguishing between experimental data, and the mathematical framework used for expressing them, Poincaré announced his dictum "No geometry is more correct than any other – only more convenient".[5] To illustrate this point consider an accurate version of Gauss' surveying experiment, the sum of the angles of the triangle is greater than 180°, as predicted by general relativity and in agreement with curved non-Euclidean geometry. How are we to interpret this result? The relativist announces that space is curved, but his actual experiment is with light rays and carefully stated is that the sum of the angles between the light rays is, say, 180.1°, it makes no mention of space. The experiment can equally be interpreted by saying that space is flat and Euclidean but that light rays do not move on straight lines. To build up a description of the world the scientist will then look for objects that influence the motion of photons – he will find them – the mass of the Earth, the Sun, the Moon, etc., he can now construct an equal representation of the physical world taking space to be flat and Euclidean. No experiment can decide between his representation and that of the relativist for they make the same predictions about empirical consequences but they choose to express it in different languages, for one space is flat, for the other it is curved – yet there is no way of choosing between them. The curvature of space is therefore in the mind of the scientist, to be chosen at will in the description of the physical world. The geometry or mathematical framework is like a language – an empirical truth about the world can be expressed in many languages, the truth is independent of the means of expressing the truth. The physical world is no more German because Einstein expressed his theory in German, than it is curved because he expressed it in a curved space-time.

We can build on this analogy by drawing up a dictionary to translate from the curved space-time description to the Euclidean description.

Curved space	*Euclidean space*
Space curvature	Fields in flat space
Straight line in curved space	Curved line in flat space
Matter causes curvature	Matter causes fields
Field equations for curvature	Field equations for fields

It follows, therefore, that not only can physics be described in flat Euclidean space but indeed in any space that the scientist chooses. It may be that the theory can be expressed more succinctly in one geometry, but that does not make that geometry "correct", only convenient. The curvature of space is at the behest of the scientist.

Nowhere has the "curvature" of space had a firmer hold than in cosmology, the large scale structure of the Universe. General relativity predicts three possibilities, a positive curved space, negatively curved, or flat, and experimental cosmologists strive to determine this curvature. But again, this curvature is an illusion. A given experiment yields a particular answer about the properties of objects in the Universe – it cannot tell us anything about the curvature of the Universe, only about the results of experiments and we can choose how to interpret these. This was realised many years ago by E. A. Milne(6) who showed that his particular model of the Universe could be described either as a static infinite negatively curved Lobatewski space, or as an expanding motion of galaxies in a flat Euclidean space, depending on the units of measurement, both representations are equally valid and make the same predictions about any experiment, but by changing the units of measurement, that is our conventions, we change the curvature, only if we accept one set out of an infinity of possible sets of conventions do we recover the curved space-time of General Relativity.

We finish this argument with a technical example. The hyperbolic curved space representations of the Universe are normally expressed by the geometrical line element of Riemannian geometry as

$$ds^2 = dt^2 - R^2(t)\left(\frac{dr^2}{1+r^2} + r^2 d\theta^2 + r^2 \sin^2\theta d\phi^2\right)$$

where

$$\begin{aligned} R(t) &= A(\cosh \psi - 1) \\ t &= A(\sinh \psi - \psi) \end{aligned}$$

By a simple change of coordinates this can readily be transformed into[7]

$$ds^2 = \left(1 - \frac{T_0}{(\tau^2 - \rho^2)^{1/2}}\right)^4 (d\tau^2 - d\rho^2 - \rho^2 d\theta^2 - \rho^2 \sin^2 \theta d\phi^2),$$

that is some multiple of a flat space. The motion of a photon given by $ds = 0$ and therefore follows a straight line in the Euclidean space. A body of mass m_0 follows a geodesic in the Riemannian space

$$\delta \int m_0 ds = 0.$$

In the Euclidean space it no longer follows a straight line but a path given by

$$\delta \int m_0 ds = \delta \int m_0 \left(1 - \frac{T_0}{(\tau^2 - \rho^2)^{1/2}}\right)^2 d\eta = \delta \int m_0 \phi d\eta.$$

The particle's path is determined by the "field" ϕ in the Euclidean space, ϕ is determined by field equations which yield

$$\phi = \left(1 - \frac{T_0}{(\tau^2 - \rho^2)^{1/2}}\right)^2 .$$

There is no empirical way of choosing between these alternatives – they predict identical results, yet in the General relativistic representation we say the Universe is curved, in the second that it is flat.

Is space curved? The answer is yes or no depending on the whim of the answerer. It is therefore a question without empirical content, and has no place in physical inquiry.

REFERENCES

1. J. A. Wheeler, *Geometrodynamics*, p. 225, Academic Press, 1962.
2. See the *Leibniz–Clarke Correspondence*, ed. H. G. Alexander, Manchester University Press, 1965.
3. Ernst Mach, *The Science of Mechanics*, 1883, English Translation, Open Court Publishing Co., 1960.
4. See H. Grunbaum, *Philosophy of Space and Time*, 2nd edition, North Holland, 1974.
5. H. Poincaré, *Science and Hypothesis*, English Translation, Dover, 1929.
6. E. A. Milne, *Kinematic Relativity*, Oxford, 1948.
7. I. W. Roxburgh and R. K. Tavakol, *Monthly Notices of the Royal Astronomical Society*, **170**, 599 (1975).

Bruno Bertotti

Professor of Quantum Mechanics at the University of Pavia, Italy.

Worked under the supervision of E. Schrödinger at Dublin Institute for Advanced Studies. Was for a long time active in the contribution of science and scientists to disarmament, and attended Pugwash Conferences on Science and World Affairs.

Scientific interests cover the theory of gravitation, plasma physics and space physics. During recent years particular interests have been in gravitational wave detection and the use of space missions to measure the gravitational field of the Sun.

THE RIDDLES OF GRAVITATION

1. THE PECULIARITY OF GRAVITATION IN THE HISTORY OF SCIENCE

The catalogue of our ignorance has *two*, not one, gates: there is the obvious exit gate, through which questions answered and settled by experimental and theoretical developments march out and disappear into the textbooks and the applications; but there is also a more important, albeit less perspicuous, *entrance gate*, through which *new* riddles come to life in the scientific world. In fact it is an important feature of the history of science (stressed, for example, in the book by T. S. Kuhn, *The Structure of Scientific Revolutions*) that very often the problems set to a scientific community at any given time had been considered beforehand solved or irrelevant; and this process of growing ignorance is usually met with great resistance, since it involves abandoning well-established certainties and efficient formal methods.

After a new area of ignorance has been accepted and defined, its gradual filling up is a fairly standard process, in which the usual interplay between theory and experiments, between hypothesis and verification engages a thriving scientific community; but the appearance of a new riddle is a most delicate and unforeseeable step, like the birth of a new baby in a family: it throws confusion and bewilderment in its plans and opens the way to entirely new developments. This phase deserves careful attention: it is important to know why some places more than others in the established pattern of scientific knowledge are likely to break and to open a gap into a new riddle; and why sometimes it takes such a long time before the riddle becomes really active.

I wish to submit the study of gravitation has served in the past, and is likely to do so in the future, as a *midwife* for the birth of new riddles in the physical sciences. We can quote at least three examples in the past of a feature of gravitation which has triggered a development of much wider scope and consequences to other disciplines. Whether this more advanced level of our field will bear similar fruits in the future, we do not know for certain; but it is important to spot those unanswered problems of gravitation which betray possible, new perspectives. We shall see that, in spite of the precise formal structure and the present excellent agreement with experiments, the theory of gravitation places such deep questions to justify the qualification of "riddle".

2. THE PAST RIDDLES OF GRAVITATION

The first attempt to a dynamical theory goes back to Aristotelis, who distinguishes two kinds of motion, those that have no intrinsic justification ("unnatural"), like the shooting of an arrow, and those which can be understood from the nature of the bodies ("natural motion"). Each body strives to go to its natural place and it will indeed go there, if unimpeded. The basic example of natural motion is fall under gravity: heavy bodies, like stones, belong to the earth and tend to reach it; air and flames belong to the sky and tend to go up. The concept of natural motion is extended to celestial bodies, although they have nothing to do, in this conception, with gravitation. The restricted and basic experience of free fall has suggested a concept which is assumed to be a universal property of things and is applied to the whole philosophy of nature: every being has a "natural" state and will tend to evolve towards it. This interpretation of free fall is the origin of deep riddles: what is "up" and "down", after all?; the strife to deal with them was exceedingly fertile. For example, Aristotelis in his *Physica* noticed that in vacuum there can be no privileged direction of motion ("up" and "down" can be said only in relation to a reference body); hence he concluded that space cannot be empty, since it would not allow natural motion.

The idea of a common law governing both the fall of bodies and celestial motion is due to Newton (1687) and is based upon a *gravitational force of attraction* between two point-like masses inversely proportional to the square of their distance. It had a tremendous success and even nowadays is the basis for the exceedingly accurate predictions of celestial mechanics and for the planning of space navigation of artificial satellites; but it opened up the embarrassing riddle of the "action at a distance" (with the more precise latin words, *actio in distans*). Newton himself was greatly puzzled by this and said: "That one body may act upon another at a distance through a vacuum, without the mediation of anything else . . . is to me so great an absurdity that I believe that no man, who has in philosophical matters a competent faculty for thinking, can ever fall into it." In spite of this riddle, in the eighteenth century the bad example set by gravitation teemed and other important instances of *actio in distans* were established in electricity and magnetism. People tried for a long time to save everything with complicated and unobserved ethereal media, carrying the action from point to point at a finite speed; but at the end modern field theory emerged, according to which a body is supposed to generate in the space around it an abstract quantity, defined mathematically at every event; this quantity – a scalar, a vector, a tensor, etc. – is not directly observable, but determines the force acting upon a test particle. The riddle has served its function: the area of ignorance it had opened up to investigation has been essentially filled up by electromagnetism and allied theories.

3. THE RIDDLE OF GEOMETRICAL SIMPLICITY

The next, fundamental advance was due to Einstein (*Theory of General Relativity*, 1915) and consisted in reducing the theory of gravitation to geometry: the world is a Riemannian, four-dimensional manifold, where the separation between two neighbouring events can be measured by precise clocks.† In this manifold privileged lines are defined, as those which connect two events with the shortest (or longest) path; these lines are the possible trajectories of test particles. The gravitational force is hidden in the geometrical structure of the manifold and is described quantitatively by its curvature. Like Newton's, Einstein's theory had a striking and unexpected success: minute deviations from Newton's law of motion have been predicted and verified with a good accuracy (a few per cent) in a large number of experiments. Its simplicity and conceptual beauty have set a standard which has never been met in other physical theories, as one can see from the following points. All physical quantities have a geometrical meaning and can be measured independently with clocks. The dynamical laws for the motion of bodies need not be postulated separately from the field equations which determine the geometrical structure of the manifold, but follow from the latter; this is a much simpler system of postulates than, for example, in electromagnetic theory, where one can, in principle, change the law of motion without changing the field. The large amount of freedom in constructing the possible equations of ordinary field theory contrasts with the almost uniqueness of the field equations for gravitation, severely constrained by geometrical requirements. Finally, the topological properties of the Universe at large are in principle

† See Note A, p. 97.

included in the fundamental physical laws: for example, whether one comes back to the same place or not if he moves straight ahead without ever deviating, follows from the theory in a given solution.

The success of Einstein's theory has produced the "riddle of simplicity", in particular, the riddle of geometrical simplicity: one wonders whether the paucity of postulates and the use of self-contained mathematical structures is indeed a universal key for interpreting and modelling physical experience. An extensive trend of physical investigation was started and shaped by this hope and consisted in studying new, abstract mathematical structures and attempting to endow them with physical interpretation, relying upon experience only as far as fundamental properties are concerned. To quote C. Misner and J. Wheeler (*Annals of Physics*, 2, 525 (1957)): "Two views of the nature of physics stand in sharp contrast: (1) The space time continuum serves only as *arena* for the struggles of fields and particles. (2) There is nothing in the world except empty curved space. Matter, charge, electromagnetism, and other fields are only manifestations of the bending of space. *Physics is geometry*." These attempts have been particularly intense in the 1940s and the 1950s; one of them, led by Einstein himself and by E. Schrödinger, aimed at formulating a new geometrical theory, to include both electromagnetism and gravitation.† Somehow, people felt, classical electromagnetism was inferior to the theory of gravitation, it attained a more superficial level of "reality". Most of these attempts have miserably failed and many physicists were led to rely more upon experimental clues than on aesthetical values in constructing new theories. The riddle of geometrical simplicity – the hope to understand on mathematical grounds why the world is as it is – is still far from fulfilment and keeps its function of stimulating the ingenuity of theoretical physicists.

4. THE RIDDLE OF PROPER TIME

Einstein's theory of gravitation is based upon a mathematical concept – proper time – which is used to define the invariant separation between two events. Every other physical concept in the theory can be derived from it: spatial intervals, the geodesic character of world lines of freely moving bodies, the gravitational field as described by the curvature of the Riemannian manifold, and so forth. In order to measure it one uses a "good clock", usually taken as the best available on the market as far as accuracy and stability; but this hides a deep inconsistency, already pointed out by E. Mach (*The Science of Mechanics*, Dover editions, p.273): "It is utterly beyond our power to measure the changes of things by time. Quite the contrary, time is an abstraction, at which we arrive by means of the change of things. . . . A motion may, with respect to another motion, be uniform; but the question whether a motion is in itself uniform, is senseless. With just as little justice, also, may we speak of an 'absolute time', of a time independent of change. This absolute time can be measured by comparison with no motion; it has therefore neither a practical, nor a scientific value and no one is justified in saying that he knows aught about it. It is an idle metaphysical conception."

This vicious circle is quite clear in Einstein's theory: a clock is needed to test free fall, hence also the dynamical laws on which its functioning itself is based. For the most advanced applications people nowadays use atomic clocks based upon the "hyperfine structure" of spectral lines: their "ticking rate" is given by the frequency of interaction between an electron spin and a nuclear spin in a cesium or a hydrogen atom. This frequency is expressed in terms of fundamental constants by

† See Note B, p. 98.

$$\frac{m_p e^4}{h^3} \quad \frac{m_e}{m_p} \left(\frac{e^2}{h\,c}\right)^2,$$

where h is Planck's constant, e the electron charge, c the velocity of light, m_e and m_p the masses of the electron and the proton, respectively. The good functioning of an atomic clock is based upon the validity of electromagnetism and quantum theory in a small domain; in particular, when the clock is used to measure gravitational fields, one must assume that the dynamical laws in a small region of space are not affected by gravitation. This assumption, called the *Strong Equivalence Principle* (SEP) has observable effects and has been tested with a good accuracy, thus giving a sound, practical basis for the definition of proper time; but violations could occur and, indeed, an important one may follow from the analysis of the lunar motion by T. Van Flandern (see *Scientific American*, February 1976). Leaving aside technical details, he claims to have found that the gravitational "constant" G diminishes with respect to atomic time, with a time scale comparable to the age of the Universe. If this result is confirmed, either we have to alter Einstein's theory (for which G is rigorously constant), or we must assume that one or more fundamental constants entering the above formula undergo a cosmological change, possibly connected with the expansion of the Universe. In this alternative the laws of microscopic physics must change with time (and probably also with space, since the cosmos is not uniform) and do not lead any more to a good proper clock.

Van Flandern's claim can be phrased also by saying simply that atomic time does not agree with "gravitational time", that is to say, the time measured by the revolution of two bodies bound by gravitational force, like the Moon and the Earth; the connection of this with Mach's argument is clear and leads one to think that the very concept of proper time – an absolute entity of metaphysical flavour – is essentially unfit to attack these very important aspects of our experience. One should look for a theory in which *all clocks are treated on the same footing* and physical meaning is given only to their relative rates: there are different and equivalent "times", each related to different interaction mechanisms. In normal conditions, of course, all (well functioning) clocks agree with a very good accuracy and would lead to proper time. The difficulty is, of course, that we know how to describe a dynamical interaction only with respect to a time variable given to us *a priori*. It is clear that such a theory must be very different from general relativity; it might be based, perhaps, on the concept of *conformal invariance*, according to which all the Riemannian manifolds, different only by an arbitrary change of the separation between two neighbouring events, are considered physically identical.

We should mention at this point the deep relationship between the last two "riddles": the reduction of gravitational physics to geometry and the strong equivalence principle. In a geometrical theory the motion of those bodies which are so small and light that do not affect the structure of space-time, is given once for all and cannot depend on their nature and their mass: it follows from the geometry. This assumption is called the Weak Equivalence Principle (WEP). We can show that a violation of SEP may imply a violation of the WEP. Consider, for example, a composite test body made up of two parts, kept together and very near by a force which depends upon the gravitational potential, in contradiction with the SEP. Its binding energy W – the energy required to separate its parts – is therefore also dependent on the gravitational energy and changes from place to place. The theory of special relativity tells us that the binding energy W contributes to the mass m of the body with a term W/c^2; hence the work needed to raise the body in a gravitational field has an additional contribution due to the change in the mass m, which is larger for bodies more strongly bound. The motion of the body then depends upon the ratio W/m and is not given by geometrical properties alone.

The WEP has been verified experimentally with an exceedingly good accuracy; this implies then a (less accurate) test of the SEP for those forces which contribute to the binding energy.

The proper time riddle and riddle of geometrical simplicity call for more and better experiments to provide the theoreticians with clues and constraints. An important, recent result related to it was obtained with the help of the special mirror placed by the American astronauts on the Moon: by sending out to it a very short pulse of laser light and measuring the two-way transit time, it is possible to get the distance between the ground station and the Moon with the surprising accuracy of less than 10 cm. This makes it possible to test various theories for the lunar motion and, in particular, to check if a large, celestial body fulfils the WEP, as Einstein's theory assumes. This test has not, so far, given any discrepance.

5. THE RIDDLE OF INERTIA

In Mach's book *The Science of Mechanics* another criticism of Newton's theory dynamics is raised, which has deep consequences for the physics of gravitation. When in the second law one equates the product of (inertial) mass by acceleration to the force, he must make specific assumptions concerning the frame of reference: the law is usually founded upon the empirical basis of the distant galaxies being, on the average, at rest. This works very well in practice: it determines the zero of the angular rotation of our coordinates with an accuracy of about 1" per century. But it raises the theoretical problems, how is the frame of reference determined if distant galaxies would rotate differently in different directions, and why do they affect in such an essential way the local dynamics?

This connection was indeed acutely present in Einstein's mind when he was constructing the theory of general relativity. There a single entity – the metric field – determines the dynamics and the measurements of space and time, in particular the frame of reference; the metric on turn depends upon the distribution of matter in the Universe. In Einstein's theory it can be expected that, for example, the rotation of a shell of matter produces in this interior a dynamical change similar to the change induced by a rotation of the frame of reference. This typical Machian effect is indeed a consequence of the theory of general relativity (Lens Thirring effect). A large amount of work has been done along this line; we should also quote the papers by D. Sciama, D. Lynden Bell and others who have reformulated Einstein's field equations in a way in which the matter distribution determines *uniquely* the metric. In the conventional theory of general relativity it is necessary to supplement the matter distribution with the prescription of the conditions which the metric must fulfil at infinity. They are equivalent to conditions arbitrarily imposed in the frame of reference and constitute a distinctly non-Machian feature of general relativity.

It has been pointed out by MacCrea (*Nature* **230**, 95 (1971)) that the adoption of special (and *a fortiori*, of general) relativity reduces drastically Mach's problem. In special relativity no material body can have a velocity larger than c; hence if the Universe has a radius R its largest rotation velocity is c/R, an exceedingly small quantity. Similarly, the largest acceleration we can admit is of order c/T where T is the age of the Universe. Therefore the assumption that distant matter obeys the rules of special relativity implies assigning with a very good accuracy the inertial frame of reference. (In general relativity the conclusion is not so strict, but rather elaborate and *ad hoc* constructions are needed to shirk it: for example, a gravitational field could twist the rays of light so as to make matter at rest far away appear rotating to a local observer.) *Accepting the theory of relativity, therefore, means excluding* a priori *a rational explanation of the most surprising and very accurate agreement between the inertial frames of reference* (in which there is no centrifugal force) *and the standard of rest of distant galaxies.*

Perhaps this was a reason why Mach felt reluctant to accept Einstein's theory of relativity; in this extreme view it should be considered only as a (good) approximation which one arrives at when the dynamical behaviour of the Universe is taken into account and the frame of reference is thereby determined. The usual sequence of theories

1. Newton's
2. Special Relativity
3. General Relativity
4. Machian General Relativity

should perhaps be replaced by

1. Newton's
2. Machian theory of gravitation
3. Special Relativity as an approximation
4. General Relativity.

Considering the immense success of and the lack of any serious challenge to Special Relativity in more than 70 years, it takes a good deal of courage from anybody who tries to follow this unorthodox course. A rudimentary, but interesting attempt along this direction was made recently by J. Barbour and B. Bertotti (*Nuovo Cimento*, 38B, 1 (1977)). They have formulated a dynamical theory of N point-like masses in which only observables appear: an arbitrary time variable and the $\frac{1}{2}N(N+1)$ distances between them; when the masses are divided in two groups, one far away and all around and a local one, one recovers Newton's equations of motion for gravitating bodies. Thus, Galilei invariance follows from an appropriate cosmology.

In an Encyclopaedia of Ignorance it is certainly appropriate to take the risk to mention such an unusual and unconventional point of view; only (long) time will show the real depth of this last riddle of gravitation, if there is any.

Notes

A. It is both possible and interesting, even for a layman, to know the technical details of this mathematical structure. A four-dimensional manifold is a continuum of "points" labelled by four real numbers x, y, z and t. Each point corresponds to an event of space-time. The qualification "Riemannian" (from the mathematician B. Riemann) is given to a manifold in which to every pair of neighbouring events (x, y, z, t) and $(x + dx, y + dy, z + dt, t + dt)$ is associated a real number, quadratic function of the small increments dx, dy, dt and dt of the coordinates:

$$\begin{aligned}\Phi = g_{xx}\,dx^2 + 2g_{xy}\,dx\,dy + 2g_{xz}\,dx\,dz + 2g_{xt}\,dx\,dt \\ + g_{yy}\,dy^2 + 2g_{yz}\,dy\,dz + 2g_{yt}\,dy\,dt \\ + g_{zz}\,dz^2 + 2g_{zt}\,dz\,dt \\ + g_{tt}\,dt^2 .\end{aligned}$$

This is a generalization of the distance

$$\sqrt{(dx^2 + dy^2 + dz^2)}$$

between two points in ordinary space with Cartesian coordinates (x, y, z) and $(x + dx, y + dy, z + dz)$ (Pythagoras' theorem). When Φ is positive, its square root gives the time interval between the two events; when Φ is negative, $\sqrt{-\Phi}$ is the spatial distance between two events. The length of line segment is obtained by summing the contributions $\sqrt{\Phi}$ (or $\sqrt{-\Phi}$) of each its infinitesimal segment.

Every information about physics is contained in the ten function of x, y, z and t, g_{xx}, g_{xy}, g_{xz}, g_{xt}, g_{yy}, g_{yz}, g_{yt}, g_{zz}, g_{zt} and g_{tt}, collectively denoted as the "metric"; in particular, the "curvature" of the manifold indicates the amount of matter present at each event and describes its gravitational field.

B. In this theory the array of the metric field is completed as follows:

$$\begin{matrix} g_{xx} & g_{xy} & g_{xz} & g_{xt} \\ g_{yx} & g_{yy} & g_{yz} & g_{yt} \\ g_{zx} & g_{zy} & g_{zz} & g_{zt} \\ g_{tx} & g_{ty} & g_{tz} & g_{tt}. \end{matrix}$$

The six added quantities correspond to the electric field (three components) and magnetic field (three components); the sixteen functions of space-time thus obtained form a single entity and describe in a unified manner gravitation and electromagnetism.

Gary Benzion, Cornell Daily Sun

Thomas Gold, F.R.S.

Director, Center for Radio-Physics and Space Research and John L. Wetherill Professor, Cornell University.

Co-author with Hermann Bondi of the *Steady-State Theory of the Origin of the Universe.*

Worked in Cambridge at the Cavendish Laboratory and at the Royal Greenwich Observatory. Formerly Professor of Applied Astronomy at Harvard University.

Published research has been in the areas of astronomy, physics and biophysics.

RELATIVITY AND TIME

We think we have discovered, by the most direct observation, that there exists a three-dimensional space and a uniform flow of time. However, I wonder whether this is a good or indeed a true representation of reality.

Relativity Theory has given us the details of the relation between space and time. It has also taught us that all future events to which we may ever have access can be seen, from a suitably moving observer's viewpoint, as simultaneous with the present (or arbitrarily closely so). The flow of time is abolished, but the relationship between events is still unchanged.

Perhaps three-dimensional space is only one of a variety of ways in which our environment can be described. What we really know in our minds is, after all, only an algebraic system, and we note that it has a close correspondence to a particular type of geometry.

The *flow* of time is clearly an inappropriate concept for the description of the physical world that has no past, present and future. It just is. Perhaps space is also an inappropriate concept, and an algebraic system can be invented, whose rules are simpler and yet describe the relations between events. Perhaps this will allow us to see Relativity Theory as an obvious structure, not as something complex and difficult. Perhaps the laws of physics will have a place in this, not be added as an afterthought to a subjective and possibly capricious definition of the universe in terms of the strange notions of space and time.

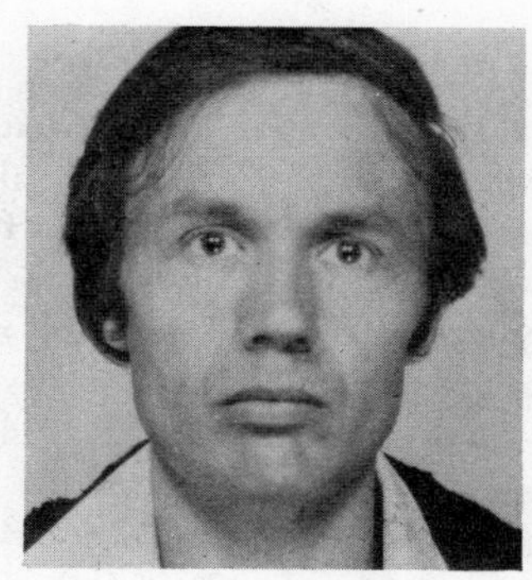

A. J. Leggett

Reader in Theoretical Physics at the University of Sussex.

Educated at Oxford and carried out research at the Universities of Illinois and Kyoto.

Main current field of research is the macroscopic manifestations of quantum-mechanical behaviour shown by superconductors, superfluids and possibly other systems. Recently has been particularly involved in research on the new ultra-low-temperature phases of liquid helium. He has also maintained a lasting interest in the fields of intersection of physics and philosophy, and regularly teaches courses in this area.

THE "ARROW OF TIME" AND QUANTUM MECHANICS

In those exciting but frustrating fields of knowledge, or perhaps one should say ignorance, where physics tangles with philosophy, the difficulties usually lie less in finding answers to well-posed questions than in formulating the fruitful questions in the first place. The attempts which follow to do this for one particular area could charitably be described as at best quarter-baked, and may well reflect the ignorance and confusion of the author rather than that of the scientific community as a whole.

Most natural scientists probably have a deeply ingrained belief that it should be possible to give a complete description of the laws of nature without explicit reference to human consciousness or human intervention. Yet at the heart of physics – for long the paradigm for natural science – lie two problems where this assumption is still subject to furious debate. One is the question of the "arrow of time" or more correctly the apparent asymmetry of physical processes with respect to time,[1, 2] the other the problem of measurement in quantum mechanics.[3] In this essay I speculate on the relationship between these at first sight disconnected problems and the vast areas of human ignorance which may possibly lurk behind them.

At the level of classical mechanics, electrodynamics and (with one minor proviso which is unimportant for present purposes) elementary particle physics, the laws of physics as currently accepted indicate no "preferred" direction of physical processes with respect to time: in technical language, they are invariant under the operation of time reversal. Crudely speaking, every process in (say) mechanics which is possible in one direction is also possible in the reverse direction: for example, if a film were taken of a system obeying the laws of Newtonian mechanics without dissipation (such as is constituted, to a very good approximation, by the planets circling the Sun), and if such a film were then projected backwards, it would be impossible to tell this from the film alone. Moreover – a related but not identical point – the laws of physics at this level give no support to the idea that the present (or past) "determines" the future rather than vice versa. Indeed, while it is certainly true that from a knowledge of the positions and velocities of the planets at some initial time we can in principle calculate the values of the same quantities at a later time, the converse is equally true: from the values at a later instant we can in principle calculate the earlier values just as accurately. Thus, within the framework of mechanics alone, the idea that the earlier events "cause" later ones rather than vice versa is a (possibly illegitimate) importation of anthropomorphic concepts into the subject: and the same is true for electrodynamics and elementary particle theory.†

Any extrapolation of these ideas to the whole of physics would, however, obviously run violently counter to common sense. It is a matter of observation (not of interpretation!) that there exist very many spontaneously occurring processes in nature whose time inverses do *not* occur spontaneously, for instance the melting of ice in a glass of warm water or the gradual loss of energy of a bouncing ball. In such cases a film of the process run backwards would be immediately recognizable as such. Moreover, most people, at least among those brought up in the modern Western intellectual tradition, would certainly say that the present (and past) can influence the future but not vice versa. Perhaps, however, it is worth noting even at this stage that this has not always been the dominant belief and may indeed even now not be the dominant one on a world scale: and while, historically speaking, these societies whose beliefs could be crudely called teleological or fatalistic have not usually embraced natural science in

† It is a misconception (though one surprisingly widespread among physicists) that experiments in elementary particle physics have "proved causality" in a sense which would determine a unique direction of the causal relationship in time. On close inspection it turns out (as usual in such cases) that in the interpretation of the experiments the "direction of causality" is already implicitly assumed.

the sense in which we understand it, it is worth asking whether this correlation is a necessary one.

Let us return for the moment to the occurrence in nature of "irreversible" processes, that is, those whose time inverses do not occur spontaneously, and which could therefore apparently be used to define a unique "direction" of time. Such processes are, of course, part of the subject matter of thermodynamics and statistical mechanics, and the conventional explanation of the apparent asymmetry despite the time-symmetry of the underlying microscopic laws goes, very crudely, as follows: if a system is left to itself, its degree of disorder (technically, its entropy) tends to increase as a function of time. To use an often-quoted analogy, if we shuffle a pack of cards we will almost always make it more disordered: if we start with the cards arranged in "perfect order" (ace, king, queen of spades on the top, etc.) we will almost inevitably end up, after shuffling, at a less ordered distribution, but it is extremely unlikely that starting from a "random" pack we would end up, by the ordinary process of shuffling, at a perfectly ordered distribution, and indeed any player who achieved such a result would almost automatically be suspected of cheating. Such an increase of disorder (entropy) seems at first sight to be naturally asymmetric in time and hence to define a unique "direction".†

Let us assume for the moment that the ascription of more "order" to the unshuffled ("perfect") distribution than the "random" one is an intuitively transparent operation which involves no implicit anthropomorphic elements and moreover that it can be made satisfactorily quantitative (cf. previous footnote). Actually, the first of these assumptions at least is by no means unproblematical (what is a pack of cards *for*? Would Martians recognize the perfectly ordered pack as such? etc.) but it is not what I want to discuss here. Rather I want to focus on a difficulty which is well known and has been discussed very thoroughly in the literature,[1, 2] namely that in any purely physical process, the underlying dynamics of the system is time-reversible as mentioned earlier, and therefore that disorder will tend to increase in *both* directions: in fact, if we *knew* for certain that a given pack of cards had been shuffled "at random" and nevertheless found that at the time of observation it was completely ordered, we could legitimately conclude not only that a few minutes later, as a result of further shuffling, it would be more disordered, but also that a few minutes *before* the observation it had also been less ordered, and that we just happened to have caught it at the peak of a very unlikely statistical fluctuation. Now in practice, of course, if we found a perfectly ordered pack *of whose history we knew nothing*, we would conclude nothing of the sort, but rather that it had been deliberately prepared in order by purposeful human intervention. Thus, the apparent asymmetry of the increase of disorder with time in a case of this kind is a consequence of the fact that human beings can "prepare" highly ordered states, whereas they cannot "retropare" them: that is, they can set the initial conditions for a system at time $t < t_1$, and then let the laws of physics take their course undisturbed between, say, t_1 and t_2. If the initial conditions correspond to a highly ordered state, the degree of disorder is practically bound to increase in time despite the time-symmetry of the underlying laws. The inverse procedure is impossible, at least according to common sense: to determine the state of the system at the final instant, t_2, we would have to intervene *before* t_2, that is precisely during the interval when the laws of physics were supposed to be taking their course without outside intervention. So, in the end,

† The analogy as presented is, of course, very much oversimplified. In fact, if the "disorder" or "entropy" of a pack of cards is to correspond to the concept used in statistical mechanics, it should be a characteristic not of a particular arrangement of the cards ("microstate") but of a class of arrangements ("macrostate"), having some gross property or properties in common, e.g. a particular value of the number of pairs of neighbouring cards which are of the same colour. For details see, for example, ref. 4. However, for the limited purpose for which I need the analogy here it is unnecessary to go into these complications.

the apparent asymmetry implied by the second law of thermodynamics (increase of disorder) turns out to be intimately related to the fact that we can only affect the future – that is, to the second paradox indicated above, namely the fact that while we have a strong sense of the "direction" of causality in the macroscopic world, microscopic physics supplies no obvious basis for such an idea. Perhaps it is now becoming apparent in what way human consciousness, and human intervention, may at first sight at least be involved in the problem of time asymmetry even within the realm of physics.

At this stage it is as well to digress a moment to dispose of one matter which is not directly relevant to the present argument. It goes without saying that there are plenty of irreversible processes in nature where no human intervention is or could possibly have been involved. In such cases the asymmetry in time is generally and plausibly ascribed to the obvious asymmetry in the natural environment which arises from the fact that the Sun is radiating energy outwards rather than sucking it in "from infinity" – technically, it is a source rather than a sink of radiation. This fact is in turn usually related to the so-called "cosmological" time asymmetry – the fact that (according to most current theories) the Universe as a whole is expanding rather than contracting. This is a topic which itself involves some fairly deep conceptual problems,[(2)] but it does not directly affect the argument as presented above. However, we shall encounter it again later.

Turning back to the question of situations involving human agency, we have just seen that the apparent asymmetry of such situations in time is a consequence of the inability of human beings to affect the past. The question I now want to raise is: Does physics itself, directly or indirectly, via, say, biological considerations ultimately based on physics, provide any obvious reason for this inability? Or, indeed, for our somewhat related inability to "remember the future"? Such questions might seem at first sight absurd, and indeed may be made so by any of a number of quite natural misinterpretations: if, for instance, we were to interpret the second question as asking why it is the past we remember *rather than* the future, we risk inviting a reply in purely linguistic terms, namely that the past is, by linguistic convention, the "direction" which we can remember. To focus the discussion, therefore, it may be helpful to ask two more specific questions: Does physics by itself forbid: (a) the hypothesis that intelligent beings in a distant part of our Universe (Daleks for short) should have a sense of the "direction" of time which is reversed with respect to ours, (b) a limited degree of genuine precognition among us ordinary human beings?

Hypothesis (a) is only interesting (and indeed perhaps only meaningful) if, apart from their possibly inverted sense of time, the nature of Dalek "life" is sufficiently close to our own to be recognizable as such.† In that case we should have to ask, what are the conditions on our own environment which (for instance) allow an organism to differentiate in one direction in time rather than the other: if, as is commonly accepted, a necessary condition is the constant input of radiation energy from the Sun, then our "biological" and hence our (overall) "psychological" arrow of time is a consequence of the "cosmological" arrow. Since the Daleks are inhabitants of the same universe, they share this latter arrow and their sun is also presumably a source rather than a sink of radiation. It would then follow that hypothesis (a) is excluded – although it is by no means trivial to fill in the details of the above argument.[(6)]

However, even if we accept this conclusion it is by no means obvious (to me at least) that we should then exclude hypothesis (b) without further discussion. Indeed, to an unprejudiced observer, the evidence,[(7)] anecdotal as it inevitably is, for a strictly limited degree of

† The question of whether, and how, we could recognize "time-inverted" beings as intelligent[(5)] is a fascinating one, though unlikely to be of practical interest to space biologists!

precognition may be thought quite impressive. (Ability to "affect the past" would presumably be considerably more difficult to recognize!) And what I want to suggest is, that at least in the absence of a much more detailed understanding of the workings of the human brain than we at present possess, it is *not* entirely obvious that the laws of physics, even when combined with the given *overall* direction of biological process, exclude any possibility of genuine precognition over fairly small distances in time – or, by the same token, of a very limited ability to "affect the past". Needless to say, such a possibility, were it to exist, would have profound implications not only for philosophy but also for our view of physics itself.

Let us now turn to the second, at first sight unrelated, area of physics where human consciousness is sometimes believed to play a special role, namely the theory of the measurement process in quantum mechanics.[3] In the standard formulation of quantum mechanics,[8] one talks strictly speaking not about a single system, but about an "ensemble" (class) of identically prepared systems, and describes such an ensemble by a "wave function", knowledge of which enables us to predict, *purely statistically*, the probability of various outcomes of a given measurement. Consider, for instance, the measurement of some quantity A which can take one of a finite number of discrete values† $a_1, a_2, \ldots, a_i \ldots$. The wave function then allows us to predict unambiguously the *probability* p_i of getting a specified outcome a_i of the measurement on a system taken at random from the ensemble. Yet the wave-function description certainly cannot be replaced by a description in which a fraction p_i of the systems forming the ensemble are said to be *in* the state corresponding to a value a_i of the quantity A: the two descriptions would in fact give in general quite different predictions of the outcome of an experiment in which a quantity different from A is measured, and to date at least the predictions of the wave function description seems to be in good argument with experiment while the predictions of the rival description are not.

One is therefore apparently forced to say that until the quantity A is measured it *does not have a definite value* for any individual system in the ensemble. (It should be emphasized, again, that it is not just a question of us not knowing the value: if this were the case, the description rejected above would be adequate.) On the other hand, according to the standard prescriptions of quantum mechanics, the moment that the quantity A is measured the description of the system undergoes a discontinuous change: in fact the wave function changes abruptly so as to accommodate the information that A now "has" the value we have just measured (the so-called "reduction of the wave packet"). One is then faced with a dilemma: *either* (a) the wave function characterizes some physical property (which we at present find difficult to interpret intuitively) of each individual system: *or* (b) it is merely a shorthand for the statistical properties of the ensemble (and thus, naturally, has to be rewritten as soon as we obtain additional information about any particular system).‡ The difficulty with interpretation (a) is that in all but the very simplest cases it is hard to see how the assumed physical property can suddenly and discontinuously change when a measurement is made;[9] for instance, if a single electron (or photon) is diffracted through a narrow slit, its wave function spreads out over a wide area – and so, presumably, there is actually some kind of (presently unknown) physical disturbance over the whole of this area. Yet the moment that the position of the particle is "measured" (e.g. by observing the highly localized flash it makes on a scintillating screen) the

† Such a quantity might be, for example, the projection of the particle's intrinsic angular momentum on a given axis. Of course, the very fact that only discrete values of such a quantity are allowed is a consequence of the quantum theory and would not occur in classical mechanics.

‡ Technically speaking, by obtaining more information about a given system we assign it to a new ensemble.

wave function is supposed to contract into a very small volume near the position of the flash; and so, presumably, does the associated physical disturbance. Indeed, if we take the prescription literally, the contraction should actually take place at a rate exceeding the speed of light, thereby violating the canons of special relativity. With interpretation (b) above, in which the wave function represents nothing physical but is simply a statistical device resulting from our ignorance, the difficulty is precisely to see why the statistical description rejected above does not work. Faced with this dilemma, the majority of physicists have embraced, at least implicitly, the so-called Copenhagen interpretation[(10)] (more correctly, non-interpretation) of quantum mechanics according to which it is meaningless to ask questions, awkward or otherwise, about the meaning of the wave function, since it is simply and solely a mathematical interpolation enabling us to infer from a recipe for the preparation of a particular ensemble to the probability of various outcomes of possible experiments on it.[(11)]

While most practising physicists find this (non)-interpretation quite satisfactory in the context of their everyday use of quantum mechanics, it has become increasingly recognized over the last 20 years or so that the awkward conceptual problems it raises are not going to be exorcised so easily. Indeed, probably the single issue which divides physicists most deeply at present is the extent to which these problems cast doubt on the claims of quantum mechanics to constitute in some sense the ultimate description of physical reality. Let me sketch very briefly the nature of just one of these problems.[(3)] The notion of "measurement" clearly plays a special role in quantum mechanics, since the wave function is supposed to change continuously and causally between measurements but to jump (collapse) discontinuously and acausally as soon as a measurement is made.[(12)] This prescription, however, is ambiguous, since quantum mechanics contains no instructions for deciding exactly when a "measurement" has been made. In fact, it is possible to argue that what we normally call a measurement is nothing more than interaction with a (usually man-made) device such as a Geiger counter which itself consists of atoms and molecules, in which case it should itself be described by a quantum mechanical wave function; but if so, the device itself will in general not possess a definite value of all its physical quantities (including the readings of dials, etc.) until these are themselves "measured", and so on in an infinite regress. Indeed, some authors [(13)] have argued that the only satisfactory way to terminate this regress is to allow the "measurement" to be made only when the reading of the dials (say) is registered by a human mind, thus introducing human consciousness into the theory as an extra-physical ingredient. Clearly the paradox here is somewhat reminiscent of the one we encountered earlier in the context of time-reversibility: to enable thermodynamics to provide a unique "direction of time" we were forced in the end to assume that human beings can affect the future but not the past, an assumption which (apparently) cannot be justified within physics itself: to formulate quantum mechanics at all, we have to introduce the idea of measurement, a notion which (apparently) cannot be defined without internal inconsistency in physical terms alone.

In the face of this and other difficulties, some physicists have speculated that quantum mechanics may actually only be an approximate description of reality, the true description involving a consideration of a sub-quantum level characterized by some variables which at present we are unable to detect experimentally (the so-called "hidden variables"). Such a description, it is hoped, could be completely causal in character, and would if suitably constructed reproduce the quantum-mechanical results for existing types of experiment while avoiding the conceptual difficulties concerned with the problem of measurement; moreover, it might in principle predict results different from quantum mechanics under conditions more stringent than those hitherto attained.[(14)] A model of such a theory has been explicitly constructed[15] for a simple situation involving spin measurements on a single particle (or ensemble of such), thus finally disproving the widespread but erroneous belief[16] that no hidden-variable theory could reproduce all the results of quantum mechanics even for such a simple system. At first sight, therefore, it looks as if there are no very fundamental *a priori* objections to the idea of such a "hidden-variable" theory.

However, one of the most interesting and surprising developments in fundamental physics in the last decade or so has been the demonstration that the position is quite otherwise as soon as one considers a slightly more complicated situation, namely spin measurement (or the equivalent) on two systems which have interacted in the past but are now very far apart in space (so that, within the framework of presently accepted ideas, they should not influence one another). Let us assume (A) that the hidden-variable theory is such that after the two systems separate, each separately has a state described by its own hidden variables (which we do not, of course, know and which may be strongly correlated, in the usual statistical sense, with those of the other, remote, system); and moreover, (B) that the outcome of any measurement on (say) system 1 is determined only by the state of that system and by the properties of the apparatus set up to perform measurements on it, but not (for instance) by the variables of system 2 or the properties of *its* associated apparatus. (Such a hypothesis seems entirely natural since the two systems, and the two associated pieces of apparatus, are very distant from one another in space.) A theory having these properties is referred to as a "local" hidden-variable theory. There is now a remarkable theorem[17] which states that *no local hidden-variable theory can reproduce all the results of quantum mechanics.* Moreover, with some trivial modifications, the theorem can be strengthened to apply more generally to *any* "local" theory, that is any theory for which assumption (B) above is true and in which (A) is replaced by the more general assumption that there exists a description of the state of each system separately after they have ceased to interact. (In quantum mechanics no such description exists, which itself leads to a well-known paradox.[18] This theoretical conclusion has now been complemented by an experiment[19] which deliberately provided a situation in which the quantum mechanical predictions could not be reproduced by any "local" theory: the quantum predictions were, nevertheless, found experimentally, thus demonstrating fairly conclusively that no "local" description of nature can be correct.

It should be stressed that this conclusion at first sight runs violently counter to common sense, since it says that (at least in certain circumstances) a system cannot even be *described* individually and in isolation, even though it may be spatially separated from all other matter in the Universe. Indeed, if we were to take the argument to its logical conclusion it would seem to say that we can *never* describe any system in isolation, since it must have interacted with something in the past, however long ago!

It is possible to avoid this somewhat unpalatable conclusion if one is prepared to modify one or two of the "common-sense" assumptions embodied, perhaps implicitly, in (A) and (B) above in a sufficiently radical way. For instance, a possible hypothesis[20] is that by setting up the apparatus designed to measure the properties of system 1 one may, in some way not at present understood, affect the physical conditions prevailing in the region where system 2 is and thus the results of measurements made on it. Such a hypothesis, if it is not to violate the special theory of relativity, could in principle be tested by setting up the apparatus "at the last minute", so that there is no time for any signal to be transmitted to system 2.

However, to my mind a more intriguing possibility, and one which at last makes the promised contact with the first part of this essay, is that the "direction of causality" might in some sense be violated in this type of experiment[21] (and, perhaps, more generally in quantum measurement processes). In other words, instead of regarding the initial state of the system (whether described by hidden variables or not) as determining the outcome of measurements made on it, we might regard the outcome of the measurements as determining, at least partly, the initial state. This is perhaps somewhat more plausible in a hidden-variable picture (or some other theory which seeks a "sub-quantum-mechanical" level of reality) since a (temporary) "backward" interpretation of causality at the sub-quantum level might not necessarily conflict with the usual "forward" interpretation of the level of quantum mechanics, nor produce results which are clearly incompatible with the known initial conditions. It should be noticed, by the way, that the typical times involved in experiments of this type are usually extremely small by macroscopic standards (usually of the order of 10^{-9} sec) and indeed are probably less than the shortest microscopic "relaxation time" for the irreversible processes likely to be relevant here. It is therefore perhaps not so unthinkable that phenomena of a type (at first sight) unknown on the time-scale appropriate to the macroscopic world might exist over such short intervals.[22]

Speculating even more wildly, one might hope that if anything remotely resembling this proposal were to be true, it would not only resolve the "measurement paradoxes" of quantum mechanics at the atomic level, but also provide the microscopic basis for those phenomena, if they really exist, which *do* violate the "sense of time" on a *macroscopic* time-scale – such as precognition. That the time scale involved here is so many orders of magnitude longer (perhaps minutes or hours) is perhaps not so strange if one considers that the human brain, regarded as a physical system, certainly possesses a degree of complexity very many orders of magnitude greater than any of the instruments used in physics. On the other hand, it could equally well be that the questions of microscopic and macroscopic violations of the sense of time may turn out to be essentially unrelated, as seems to be the case with the violations of left-right symmetry at the levels of elementary particles and of biology.[23]

The above speculations may seem to be (and no doubt are) vague to the point of irresponsibility. Nevertheless, I do strongly suspect that if in the year 2075 physicists look back on us poor quantum-mechanics-besotted idiots of the twentieth century with pity and head-shaking, an essential ingredient in their new picture of the Universe will be a quite new and to us unforeseeable approach to the concept of time: and that to them our current ideas about the asymmetry of nature with respect to time will appear as naïve as do to us the notions of nineteenth-century physics about simultaneity.†

ACKNOWLEDGEMENT

I am indebted to Dr. Paul Davies for a very helpful discussion and criticisms of the original manuscript.

† For a deeper discussion of many of the questions raised here, as well as many recent references, see O. Costa de Beauregard, *Foundations of Physics* (in press).

1. For general discussions of the problem of time asymmetry, see H. Reichenbach, *The Direction of Time*, University of California Press, Berkeley, 1971, and
2. P. C. W. Davies, *The Physics of Time Asymmetry*, Surrey University Press, London, 1974.
3. For a general introduction to the problem of measurement in quantum mechanics, see B. d'Espagnat, *Conceptual Foundations of Quantum Mechanics*, 2nd edition, Benjamin, New York, 1976.
4. R. Kubo, *Statistical Mechanics*, North-Holland, Amsterdam, 1967, chap. 1.
5. Cf. N. Wiener, *Cybernetics*, Wiley & Sons, New York, 1948, p. 34; *The Nature of Time*, edited by T. Gold, Cornell University Press, Ithaca, N.Y., 1967, pp. 140-2.
6. Cf. H. F. Blum, *Time's Arrow and Evolution*, Princeton University Press, Princeton, N.J., 1968.
7. See, for example, J. B. Rhine, *The Reach of the Mind*, Faber & Faber, London, 1956, chap. 5.
8. L. Eisenbud, *Conceptual Foundations of Quantum Mechanics*, van Nostrand Reinhold, New York, 1971.
9. This point is emphasized in (e.g.) L. E. Ballentine, *Rev. Mod. Phys.* **42**, 358 (1970).
10. N. Bohr, *Phys. Rev.* **48**, 696 (1935).
11. For a more sophisticated version of the "Copenhagen interpretation" see H. Reichenbach, *Philosophic Foundations of Quantum Mechanics*, University of California Press, Berkeley, 1944.
12. For an alternative interpretation which avoids this "collapse" (but involves other difficulties) see H. Everett III, *Rev. Mod. Phys.* **29**, 454 (1957).
13. F. London and E. Bauer, *La Théorie de la mésure en mécanique quantique*, Hermann et Cie, Paris, 1939; E. P. Wigner, *Am. J. Phys.* **31**, 1 (1963).
14. For a forceful statement of this point of view, see D. Bohm, *Causality and Chance in Modern Physics*, Routledge and Kegan Paul, London, 1957.
15. D. Bohm and J. Bub, *Rev. Mod. Phys.* **38**, 453 (1966).
16. This belief goes back to a result of J. von Neumann, *Mathematical Foundations of Quantum Mechanics*, Princeton University Press, Princeton, N.J., 1955.
17. J. S. Bell, *Physics,* **1**, 195 (1964-5).
18. A. Einstein, B. Podolsky and N. Rosen, *Phys. Rev.* **47**, 777 (1935).
19. S. J. Freedman and J. F. Clauser, *Phys. Rev. Letters* **28**, 938 (1972).
20. D. Bohm in D. R. Bates (ed.) *Quantum Theory*, vol. III, Academic Press, New York, 1962, p. 385.
21. To the best of my knowledge this possibility was first pointed out in the present context by O. Costa de Beauregard, *Revue Internationale de Philosophie*, no. 61-62, 1 (1962); *Dialectica* **19**, 280 (1965).
22. Cf. the problem of "pre-acceleration" in electrodynamics (Ref. 2, p. 125), where the relevant time is of order 10^{-23} sec.
23. But see T. L. V. Ulbricht, *Q. Rev. Chem. Soc.* **13**, 48 (1959).

C. J. S. Clarke

Lecturer in Mathematics at the University of York.

Studied mathematics and general relativity at Christ's College, Cambridge, and was a research fellow at Jesus College, Cambridge, until 1974.

Recent work has been on the structure of singularities in general relativity and on the foundations of quantum theory. This extends into his more fundamental interest in the rediscovery of the intellectual content of contemplative Christianity, an enterprise pursued in conjunction with "the Epiphany Philosophers" at Cambridge.

THE HINTERLAND BETWEEN LARGE AND SMALL

1. PROLOGUE

I shall discuss an area where there is less than ignorance: where knowledge is negative, in that what we think we know must necessarily be false: the region of overlap between microphysics and macrophysics. Likening the growth of knowledge to the progressive construction of an atlas of our world, imagine two cartographers setting out to map the Pacific, one starting from the West and the other from the East, each one ignorant of latitude and longitude but using his own empirical coordinate system centred on his homeland. The initial gulf of ignorance between them would dwindle away until, finally, the maps overlap. In the region of overlap the maps would look different because of the use of two different coordinate systems. But a geometric transformation could convert from one to the other, so that it could indeed be said that the entire Pacific had been charted.

Now imagine their confusion if, in their common region, no such conversion were possible. Suppose that one map held a certain island to be mountainous, the other map flat. Was it really the same island? Did it change according to who observed it? Or had the entire process of map-making been in error all the time? Worse than ignorance, all that had previously been thought knowledge would be thrown into question.

Such, I claim, is the situation with micro- and macrophysics. As the two disciplines extend their maps outward towards each other, each quite satisfactorily, the less likely it becomes that any reconcilable overlap will be achieved unless one or the other undergoes a radical change. And the situation is revealed as crucial in the realization that it is just in the domain of possible overlap, the domain of threatened contradiction, that all the basic processes of life as we know it take place.

The arguments leading to this conclusion are simple ones, based on the general structures of classical and quantum physics. References and some more technical points are postponed to the notes at the end.

2. THE SMALL

(a) In Support of Realism

The theory of the small means, in modern physics, quantum theory – about whose nature and content no two physicists agree. The poles of the conflict correspond to the terms of what is perhaps the most basic classification of the psychology of Western thought, into nominalist and realist.(1) The first type is characterized by the tendency to see all general terms, or sometimes all terms whatever, as mere shorthands for specific events or experiences; the second sees them as pointers to structures that have some kind of "real" existence independent of our observing them. From the nominalist tendency springs operationalism, utilitarianism(2) and the approach to physics which regards theoretical concepts as merely parts of algorithms enabling one to predict successfully the results of experiments. From the realist comes a Platonic view of the world and a use of theories in which mathematical terms are seen as standing for real parts of the furniture of the Universe.

If science is nothing but the prediction of future dial-readings from present ones, there is no terrain to be mapped and so no map. I intend to reject such an extreme form of the nominalist tendency, and shall presume that scientific theories are *about* something, other than the experiments set up to test them. Clearly, all manner of questions are begged in saying this: a large part of philosophy, starting with Kant, has been concerned with just this question. This philosophy warns us, if the history of science were not warning enough, against being seduced by cartographical metaphors into supposing that we can usefully assess a theory by comparing it with some hypothetical "real world". Yet without some measure of realism there is neither knowledge nor ignorance, only greater or less success in prediction. And without a commitment to something underlying the formulae, some belief in a terrain which the map depicts, few steps in science would be possible.

The resolution of the conflict between large and small is to be sought here. Complete nominalism dissolves the conflict, but with it the possibility of knowledge; naïve realism provides satisfying pictures of the world, whose definiteness, however, renders the opposing views irreconcilable.

(b) "States" and "Collapse"

For quantum theory, the key term to place in the nominalist-realist spectrum is "state". The theory then tells us how to express everything in terms of relations between states, starting from the basic assumption that the set of all states has the same structure as the set of all complex one-dimensional subspaces of a complex infinite-dimensional Hilbert space.[3]

A *state* (or, more specifically, a state of an atom or particle) might be specified by what one has done in the past in order to set up some microscopic system; or by what results one can get, and with what probabilities, if one performs certain subsequent operations on the system. If "state" means nothing more than this[4] (the nominalist position) then quantum mechanical theories are nothing more than statements about connections between setting-up-actions and observing-actions. We must go on to ask, is there anything more that "state" can signify?

We customarily think of a state *of something*: that is, we imagine a substratum which possesses certain properties which constitute the state at any given time. In the realm of particle physics this idea is probably untenable, since particles can appear and disappear leaving no fixed substratum.[5] Therefore we must weaken the concept to that of the *state of affairs* maintaining at a given time[6] in some region of space – a weaker concept, but not so weak that we renounce all explanatory power. In the last section I shall explore the possibility of yet further weakening.

This now leads us to the central distinction between the physics of the large and the physics of the small. Macroscopically, so long as a region continues to be designated unambiguously, then a state of affairs can always be ascribed to it, a state which changes more or less continuously and at any rate in accordance with laws which generally produce a fairly predictable continuous change. But the assumptions that a state always has been defined and always will be defined is superfluous to most microphysical experiments, which are essentially limited in time. The experiment is set up (e.g. an electron is injected into an experimental chamber) then after an assigned lapse of time an observation is made which concludes that particular instance of the experiment. Outside this time-span there is no need to speak of a state at all. If, however, we do suppose that some state is definable after the observation, and that we are then quite free to choose some other observation to make, then the observational evidence forces us to assume that the state can change discontinuously and acausally every time it is observed.[7] The new state is determined by the outcome of the observation, an outcome which is quite unpredictable: the aim of quantum mechanical theories is to give ways of finding probability distributions for such outcomes.

This change is, from another point of view, described in the literature as the "collapse of the wave-packet".

3. THE CONFLICT

Two possibilities, which I shall describe in turn, now face us.

(a) The Two-world View[8]

We decide to limit the concept of "state" to the span of any particular experiment, making this a characteristic of descriptions of microphysics. Such a characterisation effectively divides the domain of science into two worlds: that in which clearly delimited experiments are carried out (e.g. particle physics) and that in which we participate in an ongoing state of affairs (e.g. astronomy). To the first is given a quantum description, while the second is described classically.

It is now crucial to realise that, while the general fields of the two studies may not be clearly distinct, the particular accounts which they give do not and cannot overlap (unless quantum mechanics be radically modified). This is because on any conventional quantum mechanical theory whatever, it is always possible to set up a quantum mechanical "state" which is a *superposition* of two other states which, on a macroscopic view (i.e. if they also had an interpretation in the other world), would be mutually exclusive. This, the heart of the conflict, has been given innumerable descriptions before, and I refer the reader who is unfamiliar with the idea to the very readable article by de Witt (1970).

A consequence of the inevitability of superpositions of "macroscopically opposed" states (that is states which, if they were part of the macroscopic description, would be exclusive) is the conclusion that the quantum and classical uses of the term "state" must differ: they are not talking about the same thing. Moreover, though the quantum description can be taken as an account of what happens in an atom, say, between its setting-up and its observation in any repetition of a given experiment, it cannot be taken as giving a complete account of any single instance of the experiment *as a whole* (including the final observation) unless an acausal link, analogous to "collapse", is postulated between the microscopic and macroscopic worlds. In itself, the quantum state provides information only about an ensemble of experiments, for whose results it provides a probability distribution. The classical description, on the other hand, refers in the first place to a single process; though it can then be applied to an ensemble of initial states to produce an ensemble of results with appropriate probabilities.

Thus the "maps" of the two regions interpenetrate and even touch but never overlap. The boundary of the quantum description can be joined approximately onto a part of the boundary of the classical one where probabilistic accounts of ensembles of experiments are concerned. It is known[9] that when the quantum description is carried to such large objects that the phase-information of the wave-function representing the state becomes small compared to the intrinsic thermodynamic uncertainty in the state, then, if the phase information is neglected, the quantum mechanical description can be translated into a classical probabilistic description. But where *individual processes* are concerned there can be no such translation.

One further consequence of this view needs noting: if the two worlds are distinct, the macroscopic world cannot be *made up of* microscopic entities. The table on which I am writing is not made up of atoms; an atom is only a state which is part of a description of some limited experiment to which I am quite external.

(b) A Unified View?

Unfortunately, this alternative does not really exist.[10] One would have to postulate that quantum mechanics and classical mechanics are in fact talking about the same things, but that the particular laws which govern large things are different from those which govern small ones. How, then, is the change-over to be made between the two? It might be possible to postulate some additional terms in the quantum mechanical equations of motion which only become significant on a large scale, but no way of doing this has found any general acceptance.

4. THE SCOPE OF OUR IGNORANCE

(a) Biology

This is the most striking example of a field in which there is a continuous transition from the microscopic to the macroscopic. Imagine, for instance, the processes involved in human vision: a single photon, which must be described quantum mechanically, falls on a rhodopsin molecule in a cone cell and is sufficient to trigger a nerve impulse which can affect the behaviour of the entire organism. Of course, the same is true of a Geiger-counter or of any particle detector: the difference is one of degree. In a particle-detector there is a fairly well-defined break between the microscopic processes and the macroscopic ones, to the extent that it would be unlikely to make any difference if one were to draw the dividing line in a different place. In the organism, on the other hand, there is every evidence of a finely tuned unity between the particle level and the molecular level, and between the molecular level and the structural. And it is precisely the links between these levels which are of interest biologically.

The two-worlds view must sever one of these links, regarding the rhodopsin molecule, for instance, as a quasi-classical observing apparatus when it receives the photon. A unified view, as far as one could envisage it now, could only postulate some unknown mechanism which inscrutably mediates the two worlds.

Although there is no evidence that the two-world view is false, a certain uneasiness attends it, particularly when one passes from biology to the psychological phenomenon of "free will". I would have liked to have catalogued this among areas of ignorance; but it is far from clear just what it is that we are ignorant of: despite millenia of attempts, no scientific language yet exists for turning one's vague feeling of mystery into sufficiently concrete questions. Whatever free will may be, it seems to have to do with organisms' unpredictability,[11] which may also be linked to the unpredictability that is peculiarly associated with the quantum domain (though there is no convincing proof of this).[12] At this point all one can do is to offer the speculative hope that a properly unified physics might dispel some of the obscurity which surrounds this most important topic.

(b) Astrophysics

So long as it is reasonably clear what is large and what is small (or, more precisely, where quantum interference effects are large and where they are negligible) then the two-worlds view will probably suffice and one could lay aside the desire for a unified physics as an outmoded relic of the nineteenth century. But in parts of astrophysics we seem obliged to speak of the large and the small in the same breath. Such is the case with theories in which the Universe has always expanded outwards and was arbitrarily small in the past. For the Universe of which we are now a part, must surely be classed among the large! The resolution of this paradox which seems to require us to apply quantum mechanics to the entire universe, is more than an academic exercise, since, on most current theories of cosmology, the fluctuations in density that must have been present at early times so as to form the galaxies and stars out of a homogeneous background cannot be explained classically, and appeal is usually made to the quantum realm of an even earlier epoch. Thus, without a resolution of the conflict, little progress can be made in explaining how everything around us came to be.

The problem has recently been presented in a time-reversed form. It is well known that most gravitational theories predict the formation of black holes. Now, if quantum processes are added onto a black hole background, then it has been shown by Hawking that the black hole must radiate energy, and in so doing radiate itself away to leave, in all probability, a "naked singularity": a point near which the length-scale of the gravitational field tends to zero and its strength tends to infinity (see Gibbons, 1976). Such an object can only be described quantum mechanically, and again we are faced with a conflict: the paradox of applying quantum theory to find the observable properties of an astrophysical object.

5. EPILOGUE

Many physicists and most biologists would, I surmise, argue that the problem I have described is (a) unimportant and (b) insoluble or inevitable, and it should therefore be ignored. Having given my counter-arguments to (a), I shall conclude with some hints as to (b). It may be that a position can be achieved (Clarke, 1974) which is less realist without relegating quantum mechanics to a computational algorithm. The result would be like the "guiding-wave" approach of the early interpreters of quantum mechanics,[13] with full reality granted to the observed states of the Universe, but a theoretical "reality" granted to a wave-mechanical model of a quantum universe which guided our observed states in a probabilistic way. This is, however, very far from a theory which will make knowledge possible in a region which seems condemned to ignorance.

NOTES

1. I use these terms in their scholastic sense, but generalize to psychological types; to the "Aristotelian" and "Platonist", if one wishes, though I would not take Aristotle as an eponym. The polarization is not unambiguous for philosophical theories (where do philosophical idealists stand, for instance?), but it is usefu to apply it to their authors.
2. In the sense of Stapp (1972), not Bentham!
3. The sort of mathematical approach I have in mind is that of Mackey (1963), which, though considerably extended by later work, remains a clear exposition of the ideas. The philosophical literature on the subject is immense: the best single work I know of is d'Espagnat (1971); while a good popular discussion of the central conflict is the article by de Witt (1970).
4. This position has been taken, for instance, by Stapp (1972).

5. It might be argued the space-time itself provides a substratum, and that the "state of affairs" is simply the set of properties of space-time. This is more than a mere playing with words, if one goes on to postulate that the properties of space-time are adequately described by Einstein's vacuum equations; this is the Geometrodynamics of Wheeler (1962). I believe this is mistaken, since Einstein's equations are essentially macroscopic, both as regards their conceptual background and their verification.
6. For the sake of simplicity I have used the Newtonian presupposition that space and time can be separated. Incorporation of relativistic ideas complicates the situation and makes the conflict a lot worse, since it becomes difficult to delineate the "beginning" and "end" of an experiment in many cases. A full discussion is in Clarke (1976).
7. There are theories, such as the Bohm and Bub (1966) hidden-variable theory in which causes are postulated for a quick but continuous change. But theories like this require one to "build-in" to the particle a series of hidden variables which will account for each of the experiments which it will subsequently encounter, as if all particles were created with a joint foreknowledge of what these were to be and a programme for what outcomes to produce. This is not causality in the scientific sense, but more a pre-established harmony. The alternative position, of collapse, is stated most uncompromisingly by von Neumann (1955).
8. It is only fair to add that this sort of view can be developed so as to become much more acceptable than the crude version that I have given in terms of maps. When thus developed, it becomes essentially the "complementarity" approach of Heisenberg (1974). While this may be the best that one can do, the defects to which I draw attention still remain.
9. This was most decisively demonstrated by Daneri, Loinger and Prosperi (1962, 1966), whose philosophical importance was discussed by Rosenfeld (1965). It will be seen that Rosenfeld's discussion is totally linked to the domain in which clearly delimited experiments are performed by outside observers. Further development has been carried out by Prigogine (1969), George (1972) and coworkers.
10. Which is not to deny that there are many attempts, some of which are successful, but all either *ad hoc* and contrived (including my own, Clarke (1971)), or else requiring such a radical revision of everything that they are a long way from physical utility. As a random selection, I would mention Bastin (1966, 1971) and Bohm and Vigier (1954). This latter paper has received negligible support because of the rather grossly classical framework in which it was presented; it has many formal features, however, which could be successful. A thorough-going attempt to construct a unified "atomism" has been given by H. P. Noyes (work to appear). A successful approach would have to start from foundations that differed radically from both classical and quantum physics as we know them, and then derive both as limiting cases. "Glueing together" classical and quantum physics is probably doomed to failure.
11. Without attempting to solve the problem of free will in a footnote, some expansion should perhaps be made here. First, it is not obvious that free will and determinism are incompatible. This has been extensively explored in the case of the "determinism" associated with predestination, where even Augustine (often erroneously thought to be a denigrator of free will) allowed our free choices to be part of the pattern of causation without their being thereby less free (*City of God*, Books XI, XII), and this view was explicitly espoused by Abelard. However, in an obviously deterministic Universe free will appears somehow gratuitous: one is forced to regard it as an epiphenomenon, and so as somehow fraudulent. A measure of indeterminism, whether it comes from thermodynamics or quantum theory, leaves the door wider open for free will, without thereby saying what it might be or what it does. And this happens more clearly with quantum theory, which, on most interpretations, has a kind of irreducible indeterminism built into it (Margenau, 1965). But then one must go on to say what free will is, if it is neither a quantum mechanical "hidden variable" (which, even if possible – see note 7 – would reduce everything to causality again), nor another name for mere randomness. I have attempted the first steps (Clarke, 1975) in constructing an intermediate status of "spontaneity", which exploits the mathematics of infinite vector spaces to allow the existence of a non-temporal causative factor which moulds the progress of events in time so that they acquire short-term significance but long-term randomness.
12. It is significant that very many critical processes in biology are associated with the cleavage of hydrogen bonds, whose energy lies in the range of thermodynamic, not quantum, energy fluctuations. The possible role of quantum processes could be in setting up co-operative effects that introduced a coordinated uncertainty in the timing of cellular events that could not be achieved by purely thermodynamic means.
13. Compare, for instance, the description of Born's views given in Jammer (1966).

REFERENCES

Ballentine, L. E. *et al.* (1971) *Physics Today*, April 1971.
Bastin, E. W. (1966) *Studia Philosophica Gandensia* **4**, 77.
Bastin, E. W. (1971) *Quantum Theory and Beyond*, Cambridge U.P., pp. 213-26.
Bohm, D. and Bub, J. (1966) *Rev. Mod. Phys.* **38**, 453-69.
Bohm, D. and Vigier, J. P. (1954) *Phys. Rev.* **96**, 208-16.
Clarke, C. J. S. (1971) *Int. J. Theoret. Phys.* **8**, 231-5.
Clarke, C. J. S. (1974) *Phil. Sci.* **41**, 317-32.
Clarke, C. J. S. (1974) "Eternal Life", *Theoria to Theory* **10**, 181-9.
Clarke, C. J. S. (1977) "Uncertain Cosmology", in *The Uncertainty Principle*, Ed. S. Chisick, Wiley, London/New York.
Daneri, A., Loinger, A. and Prosperi, G. M. (1962) *Nucl. Phys.* **33**, 297.
Daneri, A., Loinger, A. and Prosperi, G. M. (1966) *Nuov. Cim.* (B) **44**, 119-28.
d'Espagnat, B. (1971) *The Conceptual Foundations of Quantum Mechanics*, Benjamin, Menlo Park.
de Witt, B. S. (1970) *Physics Today*, Sept. 1970, p. 30.
George, C., Prigogine, J. and Rosenfeld, L. (1972) *Kong. Danske. Vidensk. Selsk. Mat.-Fys. Medd.* **38**, 3-44.
Gibbons, G. (1976) *New Scientist*, **69**, 54-56.
Heisenberg, W. (1974) *Double Dialogue, Theoria to Theory,* **8**, 11-34.
Jammer, M. (1966) *The Conceptual Development of Quantum Mechanics*, McGraw-Hill, New York.
Mackey, G. W. (1963) *The Mathematical Foundations of Quantum Mechanics*, Benjamin, New York/Amsterdam.
Margenau, H. (1965) *Phil. Sci.* **30**, 1.
Prigogine, I., George, C. and Henin, F. (1969) *Physica* **45**, 418.
Rosenfeld, L. (1965) *Suppl. Prog. Theoret. Phys;* Extra number, p. 222.
Stapp, L. P. (1972) *Am. J. Phys.* **40**, 11.
von Neumann, J. (1955) *Mathematical Foundations of Quantum Mechanics*, Princeton U.P.
Wheeler, J. A. (1962) *Geometrodynamics*, Academic Press, New York.

Ted Bastin

Strongly influenced as a student by Eddington's vision of the nature of the quantum. Studied at Queen Mary College, London, and the University of Cambridge. Formerly Research Fellow of King's College, Cambridge.

Research interests include foundations of quantum theory and the limitations of metrical concepts of the very large and the very small.

A CLASH OF PARADIGMS IN PHYSICS

To be asked to write on areas of our scientific ignorance is to be given an unusual and valuable invitation. Usually a scientist is asked to write about what he can demonstrate in detail, even though in deciding his orientation he will presumably have taken into account what he sees the important area of ignorance to be. With his title "The Art of the Soluble" Sir Peter Medawar has indicated a very important aspect of science which goes some way to justifying the usual approach. You do what you can do and you build on what has been done: you do not usually confront ignorance head-on.

It follows, however, that the very vision of the world that the science of a particular day gives you is heavily conditioned by what problems one can reasonably expect will turn out to be soluble given the existing methods and techniques (theoretical and mathematical), as well as experimental, that are available in that day. The areas of ignorance which are not penetrated, or at least not illuminated, by the existing methods and techniques tend to be overlooked. Sometimes it is held that they cannot exist, and we can surely expect better of scientists than that!

Still, however much one expects that scientific innovators should be sensitive to areas of ignorance, one would not expect them to *wallow* in ignorance. Some of the writers, for example, who advocate vitalism – the ones whom we may usefully call the "thus-far-and-no-further vitalists" – seem to rejoice in the possibility that there may be areas which are for some reason forever closed to rational investigations. Such a position seems really contrary to the scientific spirit. So we may well ask, under what circumstances could it be reasonable to allow one's scientific endeavour to be guided by a vague but overriding conviction that the existing approaches are fundamentally wrong even though no technique or organised way of thinking exists which could conceivably provide a better alternative? What I am seeking, therefore, is a way in which a survey of our ignorance might serve a scientifically creative end (*pace* Wittgenstein's aphorism "the limits of my language are the limits of my world" which in general sums up so much about the scientific vision).

I think that the current situation in the foundations of physics is one where a survey of ignorance can be valuable. I shall argue that there are two paradigms[(1)] discernible in this current situation which are incompatible. One, which I call "the classical paradigm", is so familiar in its application that it practically constitutes physicists' thinking as we have it at the moment. The other – which I shall call "the sequential paradigm" – has been forced into existence by our experimental knowledge but has no background of thinking out of which it naturally emerges. The latter is the one we need in our present situation; the former is the one which alone we can think with. The resulting conceptual confusion is what I shall mainly explore as my contribution to this "encyclopaedia of ignorance". To wind up, I shall probe into areas of even darker ignorance by suggesting that if we face this confusion we may find the strange phenomena of psychokinesis less perverse.

The "classical paradigm" has its basis in the idea of a continuous background of space and time, which are imagined as perfectly smooth, perfectly homogeneous, infinitely divisible continua, mathematically modelled by the continuum of all real numbers as it was finally formalized by Dedekind and Cantor between 1870 and 1880. Physical entities are located within these continua of space and time. The basic entities are particles – idealised as single points whose position in space changes continuously and smoothly with time – and fields – distributed through space and at each point of space having a certain intensity which again varies smoothly both with time and with changes in the special point at which the field is considered. The elegance of the classical paradigm as it appears in, say, Thomson and Tait's *Treatise on Natural Philosophy* (1887) compelled a feeling of universality; one was in possession of the way to describe reality direct. Indeed the classical paradigm has become for the physicist a mathematical elaboration of common sense and the automatic vehicle of his thought. And the quantum theory has produced a formalism which has strikingly different implications from what is implied by the classical paradigm without providing the new way of thinking that one would demand of a true change of paradigm. Indeed Niels Bohr, the profoundest thinker among the originators of the quantum theory, knew that there was a reality other than that described by the classical paradigm, and yet thought it inconceivable that we could learn to think about the world in terms other than those provided by the classical paradigm. He is reported to have said, "you might as well say that you and I are not sitting here drinking tea" (which is what they were doing) when this possibility was put to him. Bohr thought that one could consider phenomena under different aspects using the spectacles constituted by the classical paradigm, and that the limitations inherent in that form of description and understanding would show themselves as incompatibilities between these aspects as they were described. Bohr's complementarity philosophy for the foundations of quantum theory was based on the existence of these incompatibilities, and he was happy with this philosophy with its neo-Kantian flavour. Not merely did he think it impossible to change the spectacles; he did not want to.

Physicists in general have not the drive for consistency of thought of Bohr, and do at present think about the fundamental particles of physics in ways that I believe to be incompatible with the classical paradigm, while continuing to assume the general validity of the familiar concepts – field, momentum, mass, force and so on – that derive their currency within the classical paradigm.

If I could give a clear description of the new ways of thinking which I see to be intermittently emerging in the study of the fundamental particles – and to which I have referred as the "sequential paradigm" – I should not be writing this essay using it as an example of our ignorance. However, we get a clue from the fact that our direct knowledge of the fundamental particles is derived entirely from the processes of disintegration and association which they undergo. These we imagine to take place in sequences which we, as experimenters, interrupt. We call the point of our experimental interruption an *observation*. From such an observation or from a very small number of coupled observations we make inferences about the progressively more remote events in the sequences, but these inferences become rapidly more conjectural and dependent upon theoretical extrapolation as they become removed successive stages from the point at which the experimenter's interruption takes place.

Most of the time one thinks of a background to the experimentation on particles which is permanent, objective, and accessible to observations of the kind that we are familiar with from everyday, large-scale physics. A background that fits the classical paradigm, in fact, and into which the sequences of atomic events fit. Indeed one usually thinks of the macroscopic world as being *made up of* or *constituted of* the entities which take part in these sequences of events.

I conformed to this way of thinking myself in writing just now of the "interruption" by the experimenter when he makes an observation. Strictly speaking – within the way of thinking that I envisage – one knows only about such a background of events which are interrupted, through the information obtained from just such interruptions, and therefore the assumption that there is something going on in the background which can be described independently of the interruptions, is one that has to be demonstrated to be valid whenever it is used. What seems to be needed is a way of thinking in which apologies like these have not continually to be made for our inappropriate upbringing. Such a way of thinking would automatically steer us along lines which are realistic because they correspond to the operational situation we are actually in, by forcing us to see the world form the aspect of a point in a branching sequence of atomic events. This would be our sequential paradigm.

The first clear intimations to the world of physics of the facts which I claim to demand a new "sequential" paradigm resulted in the growth of the quantum theory. The quantum theory was elaborated over quite a long period, to take into account a certain class of experimental facts – namely, those facts which forced on our attention *discrete* attributes of the physical world which cannot be incorporated within an essentially continuous classical theory. It is reasonable to ask how far quantum theory has succeeded in the task of extending or replacing the classical continuum picture so as to incorporate these attributes.

The early forms of the quantum theory never attempted to *explain* discreteness in the sense that they could be said to have incorporated both the discrete and the continuous within one theoretical structure. They simply imposed discreteness as a mathematical constraint upon the ranges of values that were allowed to be interpreted as the results of measurements of certain physical quantities. As the quantum theory developed, however, physicists' attitudes to the problem of explaining the discrete values changed, and it came to be accepted that the modern (1926-1930) form of the theory was an intellectual structure within which discrete and continuous quantities could cohere.

If this were true, my concern with a sequential paradigm could be no more than a search for a prop to the imagination in solving problems for which a complete solution already existed in principle. But is it true? It is not now as easy as it was a quarter of a century ago to ignore the disquiet which many physicists have continued to voice about basic inconsistencies in the foundations of the quantum theory. This disquiet has been focused largely on what is called "the measurement problem" though this expression conceals muddled thought. To speak as though there were this one remaining problem to solve before quantum theory could be made properly consistent is a bit like saying that the project of finding an overland route to Australia was waiting only for the solution of a last remaining problem – the "closure problem" they might call it – constituted by the closed perimeter of that continent.

Physicists have felt justified in ignoring logical flaws simply because of the vast successes of the quantum theory. How could a theory be wrong which works so well and with such precision in so many detailed problems? Answer does indeed have to be found to this question if one proposes to question the universal applicability of the quantum theory, and the one I propose is to say that the successes of the quantum theory are without exception *combinatorial* in character, and therefore their experimental accuracy – however great – does not argue for the thesis that the quantum theoretical formalism constitutes a new dynamics to replace or generalise the classical one. I believe that my point of view can be justified though even to demonstrate its validity as a logical possibility would take great space, while to work it out in detail as an established point of view in theoretical physics would be to take on several major investigations. All I shall do to convey what I mean by "combinatorial" is to draw an analogy. Consider the proposition that the back wheel of a particular bicycle rotates exactly 2.4 times every time the pedals are turned once. I say that this proposition is *combinatorial* because it can be deduced from the observation that the numbers of teeth on the two sprockets are in the ratio 5:12. It is also true that this combinatorial relationship exists in a physical mechanism whose dynamics we understand in a fair degree of detail: such things as the strengths of the various parts of the structure of the frame, and the behaviour of the material of the tyres as well as the dynamical relationships of the parts – gyroscopic effects and so on – which contribute to the balance of the bicycle are all described to a high degree of approximation by classical mechanics. The value for the relative rates of rotation of the wheels, however, is not a part of this corpus of knowledge, even though one could measure it to a high degree of approximation experimentally. For no one would suppose that the high accuracy with which such experiments might give the value 5:12 would constitute an argument for the correctness of any of the branches of classical physics which had contributed to that corpus of knowledge.

The sequential paradigm that I have been arguing the need for must be combinatorial in character. The events it describes must happen as do the moves in a game of chess, and as with the moves in a game of chess the shape of the game depends upon the different possibilities of play that are allowed – again all combinatorial relationships.

The form of the quantum theory as we have it affords many clues to the way to go to work to set up a combinatorial form of that theory. For example, the well-known spin vectors – being two-valued – fit far more naturally into a combinatorial picture than they do into any conventional dynamics. The fact that physical quantities like the spin vectors (the old "quantum numbers", in fact, and their more modern counterparts) seem to need to be fitted into a combinatorial picture takes us only half-way to a sequential paradigm and beyond this half-way point our way seems blocked. There is good reason for this block.

A great deal of effort (one might say the central effort) of the quantum physicists of the past and present generations has gone into a search for a combinatorial scheme from which the quantum numbers and discrete attributes in general of the wide range of elementary particles that are now known could be deduced with as much consistency as, for example, was obtained for the energy levels in the low energy quantum theory. It is very difficult to say whether this search has been successful in any significant way. It is universally agreed that definitive success has not been achieved, in spite of the length and intensity of the search.

One therefore, naturally, looks at the raw materials that have been used for such combinatorial schemes – the origin of the bare numerical relationships from which they could be built up. One then discovers that in all the current approaches these numerical relationships have been sought in an essentially geometrical setting. The mathematical techniques used have often been very abstract but they have remained abstractions from what one could call, in the widest sense, geometry. Thus, for example, a simple combinatorial relationship could be derived from the connectivity of the Möbius strip. If a strip of paper is bent round and stuck onto itself like a link of an old-fashioned paper chain, and then cut with scissors along its middle line, then it will fall into two rings. If, however, one end of the strip is twisted through 180° before being stuck, then the scissors will produce *one* twisted link of twice the length. The half rotation has been responsible for a change from one object to two, and the whole phenomenon might well be taken for a model of the origin of how combinatorial relationship might be linked to something happening in the world, and hence the basis for the numerical value of, perhaps, a mass ratio in a fundamental particle. The process, however, which is imagined to be physical in character – namely the twisting and cutting of the strip – is essentially geometrical (or, more accurately, topological).

There is a small number of theorists who acknowledge the profound difficulties in assimilating any theory of the quantum phenomena to a continuum theory and who are convinced that the right course of action is to build the continua of space and of time from discrete processes of some kind. Even they, however, seem to fall back upon a geometrical origin for the combinatorial relationships that shall characterize the discrete processes, in spite of the obvious danger of circular argument that is entailed. In the "twistor" theory of Penrose and his associates, for example, the attributes of the different types of particle are sought by considering the possible associations and combinations of certain abstract group structures. One would suppose that the amount of freedom and power given by this approach would be exploited by using the very abstractness of the groups to import new kinds of connectivity and to get away from the familiar physical picture. What we find, on the contrary, is that the groups used are the familiar groups that describe the translations and rotations of bodies in 3- and 4-space. There seems a failure of the imagination at this crucial point.

Of course, I have already supplied my own answer to the question "how should the situation be exploited: what is the imaginative leap that is needed?" in talking about a sequential paradigm. We need a mathematics in which the dynamics of objects moving in space is replaced by a dynamics of branching sequences of events with the decisions at each branch point of which branch to take, itself contained in the system. In fact the dynamics is that of a computer programme.

It is not that the details of computer techniques, let alone computer engineering, are particularly relevant to the situation. It is rather that computer people work all day with relationships which are those required for a sequential paradigm whereas physicists work with ideas which are foreign to it. Not surprisingly the appropriate expertise is to be sought among the former. There one finds certain broad principles which are common to computer practice even though they have not yet been made the subject of an elegant formal mathematical system. They do not depend upon the particular ways the computers are designed, and certainly not upon what materials they are made of. These principles are the bare bones of a non-existent logic of sequential relationships.

They are:

1. That there exists a set of initial elements which have a structure represented most simply by an ordered set of 0s and 1, which fill positions in a string.
2. That these elements have labels or names expressed by numbers, which enable the elements to be retrieved. This provision is equivalent to providing a "store" or "memory".
3. That there exists a processing unit to which elements can be brought in pairs and in which logical operations can be performed.
4. There exists a process for creating new elements (and destroying them).
5. Each process or operation contains or generates instructions for bringing a "next" process into play, so that the working of the system proceeds automatically.

If we now consider the elements of our working system as corresponding to the "points" of the physicists' space, we see the strangeness of the new ideas. Instead of the freedom which the physicist (and the classical mathematician) assume as their right, to imagine themselves selecting one point or other as they wish, for consideration, secure in the knowledge that there is an intuitively obvious spatial relationship to guide them from one point to the next, these relationships have to be constructed before they have a meaning.

My proposal is not entirely without precedent. As a programme it has recently been suggested by Finkelstein.[2] With John Anson, C. W. Kilmister, Pierre Noyes and A. F. Parker-Rhodes and others I have been working on such ideas for some time and have produced principles relating especially to point (4) above in which the relative scales of the basic physical forces can be deduced from a very fundamental process of generating sets of points of increasing size by incorporating the operations on the sets into the sets. Noyes[4] is working from the point of view of high-energy theoretical practice to get an approach which can incorporate such principles. Looking back into history, I see the nearest thing to an origin for such ideas in Whitehead's efforts to consider spatial relationships as logical relationships.[3]

At this point in my argument I bring in a quite new kind of evidence – from psychokinesis. It happens that I have myself had extensive and variegated first-hand experience of experimentation with two well-known subjects who are able to influence physical objects in a paranormal way (that is to say, to exercise psychokinesis). These are Uri Geller and Suzanne Padfield. I have seen odd examples of psychokinesis produced by several other subjects. I mention the personal origin of my evidence only because the subject of psychokinesis has been much in the public eye and opinions are at the moment balanced between acceptance of it as a real phenomenon and rejection of it as too implausible to be considered. Little can be done by one person's experimental evidence to alter this balance (though evidence does have a cumulative effect) because the credibility of witnesses and experimenters is being called in question – and this is the case whichever side their evidence supports. Moreover, physicists with a lot of first-hand experience of psychokinetic effects are still not common.

Psychokinetic effects show an effect of "thought forms" directly on physical matter. By this I mean that the way a subject thinks about an object strongly influences what happens to the object. Physical effects which are influenced by thought forms are quite unlike the effects of physical forces of the familiar kind, and this dissimilarity is crucial.

It is very difficult to convey a sense of what I mean in a short space, but I can give some idea of the strangeness of what goes on if I point out that the effect a subject will have will depend upon the way the subject divides up the matter with which he is surrounded in his mind into "objects" with individuality. If a subject succeeds in having a psycho-kinetic effect at a distance on a spoon, then that spoon, which may be one among a heap of assorted objects of variegated materials, will be singled out: perhaps bent, perhaps apported,† perhaps both.

No ordinary physical force will do anything like this. A powerful neutron beam may have very odd effects on a heap of objects, but it will not know that one piece of metal is part of, say, a watch – and an adjacent piece part of the tray on which the objects are placed. Until one has faced up to this central fact about psychokinesis one has not faced up to psychokinesis. Many scientists who have set out with a serious intent of studying psychokinesis have been so put off by the obstinate intrusion of what I have called "thought forms" of this kind that – contrary to their own expectations – they have joined the ranks of those who did not want to know.

I do not believe myself that the fact that one has to come to terms with "thought forms" which have the power over matter of physical forces means one has to abandon rational inquiry. One does have to look for a minimal change in basic theory though, which will incorporate the thought forms without trivializing them. This is where my discussion of psychokinesis connects with the sequential paradigm, for I believe that the "minimal change" for which we have to look is just the inclusion of memory into physics which we have seen the sequential paradigm to demand.

I described earlier how violent a change the sequential paradigm would demand in what we take as "common sense" or "just what is to be expected". The classical laws of mechanics would no longer be the point from which investigations begin but would become the point at which some special deductive schemes with highly restrictive conditions might ultimately finish. Getting there at all would be a struggle, and the main difficulty would be to establish the persistence of macroscopic objects and the massive impression of objectivity they provide. However, those are the *difficulties* of the new theory: let us turn to what it would cope with rather naturally as soon as our understanding of the rudimentary interaction processes had led the way to something a bit more complex.

It seems plausible that the example we know best of a memory system – namely the human

† An object is said to be "apported" when it is transferred from one place to another by psychic means.

being – may play the same part in our sequential physics as the complicated classical physical object plays in the old paradigm: namely, that of the measuring instrument with which we get our direct information about the elementary processes. Of course, we should not expect to get a deductive build-up from the elementary sequences to the complexity of a human being any more than we ever get a deductive build-up from the atomic building bricks of classical (let alone quantum) physics to a thing like an armchair or a steam engine (the argument that these deductions are there "in principle" is just an act of faith which we can appeal to with our complex system no less – and no more).

The different position we would be in with our sequential paradigm would therefore be that whereas we should find it very difficult to give an account of a stable objective inorganic universe, we should find it not such a great step as to defeat the imagination to think of specific physical objects which incorporate histories which make them accessible to particular sensitives — persons who are the good psychokinetic subjects.

When this very unpredictable and special rapport exists between these structures (namely the history of the object and an atomically specified memory trace in the human subject) then changes of the kind we call psychokinesis, and involving thought forms produced by the subject in his efforts to elicit these memory traces, will take place.

Let us look at the changed situation another way. Our new paradigm frees us from the preconceptions of spatial and temporal and what it gives us instead to play about with un-selfconsciously (as we may put it) is *similarity of pattern.* In so far as two physical entities have a pattern in common which is specified at the microscopic or quantum level, there ceases to be any problem of how they interact. They do so automatically; for to the extent that they possess similarity of pattern they are to that extent the same entity, even though, as we should normally consider them, they are separated in space or in time.

If now, we suppose that the human being – presumably through the operation of his brain tissue – is endowed with extreme sensitivity to the patterns which exist in physical objects as a result of the particular circumstances and situations through which they have gone in their past, then we have no need of any further hypothesis to explain how interaction takes place in spite of the absence of mechanical connection. In the case where one entity is a physical object, therefore, and the other a human being, one should imagine that the human being is capable of reproducing a part of the pattern which is constituted by a temporal sequence of events in the history of the object. If we want to think temporally we can imagine the brain of the person running along the paths into the past of the object. But we need not do so; we can equally well think of a whole path of this kind run together to form a more complex pattern which is directly comparable to one in the brain. In the context of this speculation it is very relevant that sensitives are usually very much more easily able to affect objects which have played an interesting part in the lives of some other person or persons (like watches and ornaments) than they are able to affect mass-produced artefacts.

My account of psychokinesis is no more than a sketch. I have not even suggested how, for example, a physical object might include in its structure coded reference to the events in which it took part, and there are several equally glaring gaps. On the other hand, there are gaps in our understanding of the relation of brain tissue to the workings of the imagination which amount to almost total ignorance anyway. Which takes me back to my beginning: familiarity may make us see a reasonable coherence where in fact there are great areas of ignorance while denying any coherence to unfamiliar ideas which may be no worse in their incoherence. Therefore to juxtapose an unfamiliar picture with the familiar may be a very good way to construct our encyclopaedia of ignorance.

NOTES

1. "Paradigm" is the word Thomas Kuhn has associated with the view that scientific change characteristically takes place by a revolution in ways of thinking rather than by continuous change. The paradigm is the essential kernel of the way of thinking. I subscribe to the revolutionary view and use the term "paradigm" in the general sense that I have just defined, and not necessarily in any of Kuhn's detailed senses.
2. See, for example, David Finkelstein, "Space-time Code IV". *Phys. Rev.* D.9,8 (15 April 1974). Also, contribution to *Conference on Quantum Theory and Structures of Space and Time, Feldafing*, July 1974.
3. Pierre Noyes, S.L.A.C., Stanford University, California. The only account published as yet is "A Democritean Phenomenology for Quantum Scattering Theory". *Found. Phys.* 6, 83 (1976).
4. A. N. Whitehead, "On Mathematical Concepts of the Material World". *Phil. Trans. Roy. Soc.* (1906). Also reprinted in *A. N. Whitehead, an Anthology*, edited by Northrop and Gross, Cambridge, 1953.

Sir Alan Cottrell, F.R.S.

Master of Jesus College, Cambridge.

Deputy Chief Scientific Adviser to H.M. Government 1968-71 and Chief Scientific Adviser to H.M. Government 1971-4.

Awarded numerous honorary degrees and prizes including the Harvey Prize, Technion, Israel, 1974, and the Rumford Medal of the Royal Society, London, 1974.

Main interest has been in the atomic theory of the properties of matter, especially those of metals, with emphasis on the theory of the strength, ductility and brittleness of steel, and also on problems of nuclear radiation in solids in connection with the development of nuclear power. More recently interested in the roles of science and technology in national affairs and also in industrial policy. Another recent interest is in the presentation of science to the general public.

EMERGENT PROPERTIES OF COMPLEX SYSTEMS

WHOLES AND PARTS

How does quantity become quality? How do the distinctive properties of bulk matter emerge out of those of its constituent particles? How do the characteristic properties of, say, a plasma, a superconductor, or an insulator, emerge from those of the charged particles in them; or those of diamond or graphite emerge from those of the carbon atom; or the self-reproduction of DNA or the enzymatic action of protein, from organic molecules; or the self-awareness of mind, from neurons; or the significance of a newspaper picture, from a set of dots? Such questions bring up the general problem of the origin and nature of *emergent properties*, i.e. properties of a whole system not possessed by its parts.

Even by setting out the problem in this way we have already taken up a definite point of view, which is the orthodox scientific one that wholes are in principle – if not yet always in practice – entirely explicable in terms of their parts. In the light of so many triumphs today in the application of atomic theory and quantum mechanics to the understanding of the structure and properties of bulk matter, no great courage or originality is demanded by such an attitude, certainly within the physical sciences and perhaps also in molecular biology. But when we come to the behavioural, psychological and social sciences, too little is understood, yet, for there to exist a scientifically objective basis for rejecting the old vitalist belief that some of the emergent properties of life cannot be completely reduced to physics and chemistry. However, irrespective of whether one believes in vitalism or reductionism, it remains a sound research tactic to proceed on the working assumption that all wholes are in principle understandable entirely in terms of their parts, since this has been so consistently successful in the natural sciences and since it may be the means of bringing science right up to the edge of the supposed gap between the material and mental worlds, a gap which may then be seen to be either illusory or profound.

In a sense, most of the ordinary properties of bulk matter are emergent properties since the only fundamental properties in physical systems are the kinematic properties of elementary particles. Exact science begins with the laws of *motion* and motion is kinematics. Even forces have no fundamental significance, being only auxiliary concepts introduced because they enable the correlations in the motions of two different particles to be neatly epitomised. All other features of the physical world, as we experience them, derive from the kinematic properties of elementary particles *and from our imperfect knowledge of them*. When we stand in front of a fire, the movements of electrons in the flames bring about corresponding movements of other electrons in our eyes and skin; and we see light and feel heat. From the movements of molecules in a gas emerge the properties of pressure, conduction and convection. The frictional properties of matter or, more generally, the properties associated with thermodynamic irreversibility, do not exist at the level of elementary particles. They arise entirely from the fact that we can recognise and describe *certain* overall kinematic states of large groups of particles – those in which the individuals share a common motion or some other uniform kinematic feature – *more simply* than we can recognise and describe any members of the much larger class of disordered states which lack such distinctive features. We thus meet here a most important feature of complex systems: *some new properties of matter emerging out of our ignorance of the individual motions of its constituent particles.* Moreover, it is not merely a lack of knowledge but also a lack of *interest.* For, if we had a science of bulk matter which told us just where every particle is and how it is moving (which is, of course, quantum-mechanically impossible), delivered as a vast computer print-out of billions of positions and velocities, we would then *know* everything about the piece of matter in question but *understand* nothing about it. It is not merely the imperfections of our senses which cause new bulk properties of matter to emerge from large assemblies of particles; it is also the nature of our minds, which crave understanding and seek only such general knowledge as is necessary for that understanding. It is our own subjective qualities, even when we are practising rigorously as natural scientists, that open a Pandora's box of emergent properties of complex systems.

ORDER, DISORDER AND ORGANISATION

Since a complex system contains many particles which constrain one another's positions and motions – and thereby develop a *structure* between themselves – we have to consider structure as well as properties. For systems such as crystals it has been useful to distinguish between *simple* and *complex* distributions of the particles in a structure. Consider from this point of view a large, regular checkerboard of $2N$ squares, each of which is given a single nought or cross selected from N noughts and N crosses. Regard the board as two interpenetrating square grids, labelled A and B respectively. Then an example of a simple distribution is that described by the statement "all the noughts are on grid A, and the crosses on grid B". The geometrical and conceptual simplicity of this distribution are displayed by the fact that it is *completely* describable in a sentence of a *few* words. By contrast, a complex distribution lacks this special feature and leaves us with no option but to describe it in a long cataloguing sentence which separately reports the particular nought or cross state of each individual square.

A *simple* distribution is, of course, an *ordered* one. In physical systems such as solids, liquids and gases it has been quite sound to infer also the opposite to this, i.e. that a *complex* distribution is a *disordered* one. But this is not generally true in biological systems, because here some complex distributions are specifically produced from detailed instructions and have spectacularly different properties from the usual type of disordered distribution.

Such systems, i.e. DNA and protein molecules, are not ordered, at least in the sense by which they might be regarded as structurally simple. They require long sentences for their description and, considered purely structurally, appear to be disordered. But they are not disordered, either. They have unique properties which are critically dependent on their precise structures. We may call them *organised* systems and regard them as a second and distinct class of complex systems which are produced by biological action.[(1)] The contrast between *order* and *organisation* is made clear by comparing, say, a crystal with an amoeba. The first is highly ordered. The amoeba is not at all well ordered, since it consists of a shapeless bag containing a sticky fluid in which float irregularly shaped long-chain molecules. But it is organised to a sophisticated extent that leaves the crystal far behind and it has some spectacular emergent properties: it can feed and keep itself alive, adapt to different circumstances, and make replicas. Another example of an organised system is the haemoglobin molecule. Its special ability to capture or release oxygen depends on a configurational change of the whole molecule, so that the complexity of the structure is essential, for the special chemical properties of the molecule, and in fact is specifically organised to provide just those properties.

PHYSICO-CHEMICAL ANALOGUES OF BIOLOGICAL PROCESSES

The link between purposive organisms, such as the amoeba, and ordered crystals, is provided by molecular biology. The properties of the DNA molecule, though spectacular, are nevertheless still recognisably "physico-chemical". Its ability to reproduce itself – by each of the arms of its double spiral, when separated from the other, serving as a template for the construction of a replica of the other – is structurally an extreme elaboration of the ability of one crystal to seed others, in a saturated solution, and is kinetically an example of the auto-catalysis which occurs in many chemical reactions. An important difference, however, is that DNA can exist in many different distributions, according to the sequence of nucleic acid units along its arms, and these differences profoundly influence the relative abilities of the various forms of the molecule to compete for survival by reproduction. Of course, some crystals are better fitted for competitive growth than others, for example by the presence in them of dislocations which provide growth spirals on their surfaces, but this is not a useful analogy. Moreover, by means of mutations, which alter the sequences of nucleic acids along the arms, new varieties of DNA emerge from the general population, some of which may be even more competitive for survival than their predecessors. Out of this emerges the great biological property of self-improvement by evolution.

As well as seeking physico-chemical foundations for biological processes, people have constructed mathematical models of these processes, by generalising the kinds of expressions familiar in theories of chemical kinetics and of quantum-mechanical transitions. One of the most interesting of these is due to Eigen,[(2)] who constructed equations for the concentrations of nucleic acid polymers for given rates of multiplication (auto-catalytic), mutation and decomposition, subject to an overall conservation condition which leads to competition and hence "natural selection" between the different molecular varieties. In this way he was able to derive solutions which display the basic features of evolutionary biology. In this mathematical model the mutations occur indeterministically (stochastically) but, once started, the populations of new strains develop deterministically. Eigen's system is thus *unstable* against small spontaneous fluctuations (i.e. mutations) and it is from this instability that its ability to evolve "biologically" emerges.

EMERGENCE OF MACROSCOPIC STRUCTURES AND PROPERTIES

A more general analysis of the emergence of macroscopic structures and properties of unstable systems has been made by Glansdorff and Prigogine.[(3)] Living organisms are examples of *open systems*, which receive highly ordered energy (e.g. sunlight) from their surroundings and use some of this order to maintain and construct themselves. In this respect they are like heat engines and heat pumps which also concentrate energy by operating on an energy flow. There is a well-developed subject of irreversible thermodynamics which deals with energy flows and derives relations such as Ohm's law, Fourier's law, etc., but this is limited to systems very near to thermodynamic equilibrium, systems that attain stable steady states by attempting to relax to equilibrium in the face of the small and continually perturbing effect of their imposed non-equilibrium boundary conditions (e.g. different temperatures across a heat exchanger).

Glansdorff and Prigogine have shown, however, that when the deviation from equilibrium and the ensuing flows go beyond a certain threshold, the system is often no longer stable in a steady state. As a manifestation of this instability a *macroscopic structure*, such as a convection cell in a heated fluid, may appear spontaneously in the system. Below this threshold, the energy all goes into the individual thermal motions of the particles, but above it some is channelled into energy of macroscopic structures and stream patterns. The most interesting examples occur in chemically reactive systems, in which various reactants move through the system by diffusion, as well as being created or annihilated by chemical reaction. Depending on the chemical conditions, in such systems above a certain size, there can emerge spontaneously rhythmic oscillations in the concentrations of the reactants (analogous to the predator-prey cycles in ecological systems, or the depression-boom cycles in industrial economies); and also in some cases there emerge spatial patterns which break the originally homogeneous symmetry of the system. Such spontaneously formed oscillations and macroscopic patterns – which exemplify Herbert Spencer's idea of the "instability of the homogeneous" as the foundation of natural evolution – have been studied experimentally by Zhabotinsky and Zaikin[(4)] and by Winfree,[(5)] and were earlier predicted theoretically in a remarkable paper by Turing.[(6)] The instability of the homogeneous may also be significant in the theory of the early evolution of the Universe.

These extensions of classical thermodynamics and kinetic theory to conditions far from equilibrium are an important development which is enabling a new bridge to be built from physics and chemistry to biology, as well as providing a theoretical basis for the discussion of emergent properties of complex systems. In the words of Glansdorff and Prigogine: ". . . there is only one type of physical law, but different thermodynamic situations: near and far from equilibrium. Broadly speaking *destruction of structures* is the situation which occurs in the neighbourhood of thermodynamic equilibrium. On the contrary, *creation of structures* may occur . . . beyond the stability limit. . . ."

QUANTUM EFFECTS IN BIOLOGICAL PROCESSES

In so far as a biological cell is a factory where many chemical reactions and diffusion flows take place, its behaviour is expected to be analysable in these physico-chemical terms, perhaps even to the extent of explaining the formation of membranes and other cell structures. Nevertheless, biological systems are, at heart, *atomistic and quantised.* Their energy comes in quanta – as in the action of a photon on a chlorophyll molecule, or that of an X-ray quantum in causing a mutation in a DNA molecule – or equivalently, it arrives embodied in certain molecules (ATP) in well-defined excited states. Similarly, the elementary biological processes are operated by individual molecules whose effectiveness depends on a precise composition and structure; even the exact way the molecular chain is folded up is critical. One of the great emergent properties of living systems – the ability of thermodynamically unstable molecules to resist decomposition – may thus be a consequence of quantization, as Schrödinger supposed.[(7)]

SELF-CONSTRUCTING SYSTEMS

Another great property of living systems is that they are *self-constructing* (as when a chick is formed from an egg, by internal actions directed by self-contained instructions) and *self-reproducing.*[(8)] It has often been thought that no automaton could have these properties, i.e. that a machine could construct only things simpler than itself. This, however, is known to be false, at least in principle. Turing[(9)] proved in 1937 that a "universal digital computer", composed of a finite number of parts, was possible which, by scanning and acting on information fed to it bit by bit, from an arbitrarily long tape, was unlimited in its ability to process mathematically expressible information. Later, von Neumann[10] applied this theorem to computer-controlled constructional machines. A Turing computer is made of a *finite* number of parts and so needs a finite amount of information to describe its construction. This information, set down on a tape, could thus be processed by another such computer and if this were designed like a numerically controlled machine tool, so as to *act* on its processed information instead of merely recording its output on paper, then one such computer could construct its fellow. It could even make one more complicated than itself! Needless to say, a computer with such properties would be a very complex system indeed (200 pages of von Neumann's book were needed to describe it!). But the key to its properties lies in a simple consideration: since a system of N parts can in principle have of the order of N^2 distinct binary cross-connections, versatility can increase rapidly with complexity. Thus, von Neumann concluded that there is "a minimum number of parts below which complication is degenerative, in the sense that if one automaton makes another, the second is less complex than the first, but above which it is possible for an automaton to construct other automata of equal or higher complexity". In other words, self-constructibility is an emergent property of a complex system.

SUBJECTIVE ASPECTS OF EMERGENT PROPERTIES

We have already seen, in the case of friction and thermodynamical irreversibility, that some emergent properties result from our own ignorance of or disinterest in detailed information about complex systems. They should not be dismissed lightly on this account, for some of their roots reach down into the deepest foundations of physics. Gold[11] has emphasised the subjective component in many scientific concepts, as follows:

> The basic concepts of physics have evolved from primitive subjective notions about the external world – space, time, force, velocity. They were singled out from other subjective notions because 'objective' measurements were possible . . . certain 'laws of physics' could then be defined that described these regularities with very high precision. No such demonstratable success attended the use of any other concepts . . . which did not allow themselves to be turned into 'objective' ones through the process of measurement. It is obviously a serious question whether the criterion that 'all observers have to agree' really is sufficient to define objectivity. What if all observers bring the same subjective notion into the measurement? . . . The concept of the passage of time is one that does not seem quite to fit into this 'objective' world, and that is not really needed in a definition of the laws that characterise the behaviour of matter. Yet, that time passes is so apparently self-evident that it hardly allows of further discussion . . . every object, it is thought, experiences the *flow* of time. . . . But which way does the clockmaker make the hands go round? There is nothing objective about that: he fixes the gears so that in *his* appreciation of the flow of time the hands progress from the low numbers to the high. . . . If the big pattern of world lines, which contained everything there is, had no indication of any flow, why do we insist on taking this concept into physics, when otherwise we try and free the discussion of physics from subjective 'impressions'?

Thus the flow of time – and we may add to it the related notion of "the present" or "now" – appears to be an emergent property of ourselves, a subjective embellishment of the more austere physical concept of time as no more than a coordinate along which events exist tenselessly. Recent investigations have supported the view, against all intuition, that there is no role for the "flow of time" or "now" in the physical world (for a summary, see Davies[12]). Yet in the mental world these have the unshakeable validity of direct experience. Here is one place, then, where science appears to have brought us right up to the edge between the material and mental worlds and it does look as if there might be an unbridgeable gap between them. In the face of such dilemmas as this, some physical scientists and modern philosophers tend to dismiss the subjective aspects of nature as mere illusions, thus turning their professional backs on clearly dominating aspects of their ordinary lives. Why do they not tell us now the winner of next year's Derby? That would be a quick way to dispel our illusions about the flow of time.

The property *par excellence* of this kind is *consciousness* or *self-awareness*. We have no idea of how consciousness might be explained from physics and chemistry; and, not surprisingly, some people have scorned it as a primitive belief in a "ghost in the machine". But there is perhaps one small constructive step we might take, in the direction of this problem. It is to remind ourselves that our *impressions* of the collective properties of complex matter are often not at all like the elementary processes which constitute them. When we touch a hot surface, we detect a molecular motion but our experience of this sensation is nothing like kinematics. The feeling of *hotness* is a different kind of sensation altogether. Again, when we look at a newspaper picture we see, not an array of dots, but the image of a well-known face. In both cases, we subjectively dredge collective properties out of the complex of our sensations, and these collective properties belong to entirely different categories of experience than their physical causes. A complex system may thus appear to us as something transcendentally different from the individual elements which constitute it. Looked at in this way the disjunction between the singleness of the mind and the complexity of the brain, and with it the great problem of the relation of mind to matter, may become a little less formidable.

ACKNOWLEDGEMENTS

I am grateful to Professor W. H. Thorpe and Dr. C. J. Adkins for useful comments on the draft of this article.

REFERENCES

1. K. G. Denbigh, *An Inventive Universe*, Hutchinson, 1975.
2. M. Eigen, *Naturwiss.* **58**, 465 (1971); *Quart. Rev. Biophys.* **4**, 149 (1971).
3. P. Glansdorff and I. Prigogine, *Structure, Stability and Fluctuations*, Wiley-Interscience, 1971.
4. A. M. Zhabotinsky and A. N. Zaikin, *J. Theor. Biol.* **40**, 45 (1973).
5. A. T. Winfree, *Science*, **181**, 937 (1973).
6. A. M. Turing, *Phil. Trans. Roy. Soc. Lond.* B, **237**, 37 (1952).
7. E. Schrödinger, *What is Life?*, Cambridge U.P., 1951.
8. J. Monod, *Chance and Necessity*, Collins, 1972.
9. A. M. Turing, *Proc. London Math. Soc.* **42**, 230 (1937).
10. J. von Neumann, *Cerebral Mechanisms in Behaviour*, Wiley, New York, 1951.
11. T. Gold, in *Modern Developments in Thermodynamics*, edited by B. Gal-Or, p. 63, Wiley, 1974.
12. P. C. W. Davies, *The Physics of Time Asymmetry*, Surrey U.P., 1974.

R. W. Cahn

Professor of Materials Science and Dean of the School of Engineering and Applied Sciences, University of Sussex.

Recent research has been on the recrystallisation of metals, ordering processes in alloys, formation of metastable phases by ultrarapid cooling of molten alloys. Has built up in the last 4 years a substantial research group concerned with "splat-quenching", and is now increasingly concentrating on glassy alloys made by this process.

TRANSFORMATIONS

Circe, the enchantress, turned men into swine: but the beasts, as they snuffled among the acorns, wept for their lost human forms, for they preserved the minds of men. The change of outer form while some inner essence is maintained intact – the process of *transformation* – is a recurrent theme in literature, mathematics and science alike. In science, the notion of transformation is linked with that of *structure*, for structure determines appearance. The bones of the skull fix the features, the sequence of amino acids specifies the gene that codes for eye colour, the arrangement of atoms or molecules in a crystal determines its shape. Note, however, a crucial distinction between the gene and the crystal. It is of the essence of a gene that it is almost always invariant and replicates precisely true to type; when it does undergo a minute mutation, that in turn becomes invariant, and the resultant biological change with it. Not so with a crystal: the array of atoms is labile and may change reversibly from one pattern to another.

Thus a crystal represents something special in nature, for it is at the same time a single substance and two or more substances. When iron is heated, a well-defined temperature (910°C) is reached when the stacking of iron atoms all at once changes; in crystallographic language, it changes from body-centred cubic to face-centred cubic. On cooling, the structure changes back, and this change on cooling brings with it changes in properties (Fig. 1). In pure iron these changes are trivial, but dissolve a small amount of carbon in the hot iron and the resultant alloy – a simple form of steel – transforms with a dramatic change of properties. The previously soft and pliable steel becomes very hard and brittle.

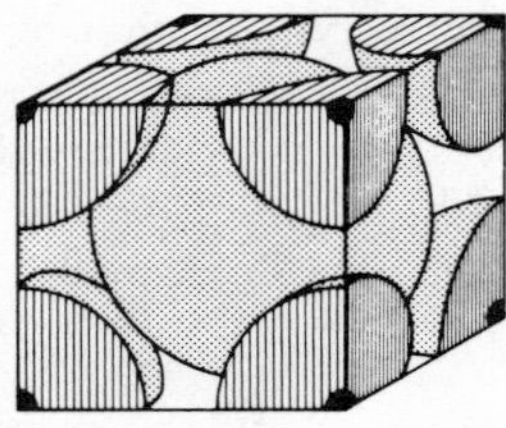

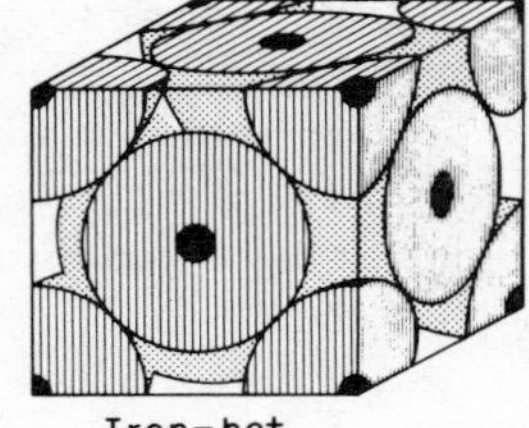

The crystal structures of iron below and above 910°C. Each sphere represents an identical atom.

Not all crystals behave in this fashion. Some transform, sharply, at one or more well-defined temperatures. Some (including steel) transform over a range of temperatures. Some transform suddenly and others take their time. Some crystals can transform only if the pressure to which they are subjected changes: the most celebrated instance of this is the conversion of almost pure carbon in the form of graphite to the form of diamond when it is heated under a very large confining pressure. Still others cannot transform at all; melting, that is, the loss of all crystalline order, preempts any change of crystalline pattern. The generic name for transformations in crystals is "polymorphic transitions"; the Greek word implies the existence of many *forms*. The concept of polymorphism, however, implies that some feature is common to all the forms: there is a unifying essence, just as there is between the grub, the pupa and the moth. What is it?

It is easy to say that the chemical identity of a polymorphic crystal is invariant: but what does the term mean? A chemical substance is defined by the nature and proportions of its constituent atoms and the way they are assembled. Two organic chemicals can be made up of the same atoms in the same proportions in each molecule, yet be put together differently and have quite disparate properties. Such *isomers* are not in general mutually convertible. So the whole molecule is not likely to be the invariant feature in a polymorphic crystal and indeed most such crystals do not contain recognisable molecules at all. The archetypal polymorphic crystal is either an element – such as iron, cobalt, phosphorus, sulphur, uranium – or a simple inorganic compound or solid solution, such as $CaCO_3$, ZnS, CuZn, $Cu(Al)_x$. Any such substance forms a crystal that is in effect a single giant molecule: one cannot pick out a single zinc and a single sulphur atom and call that pair a molecule. The only constant feature in a polymorphic crystal is the heap of atoms, the elementary chemical building blocks. What varies when such a crystal is heated or compressed is the nature of the chemical binding; the strength, length and mutual inclination of the chemical bonds changes and the constituent atoms may cluster and rearrange themselves so that the local composition varies from point to point.

The understanding and control of polymorphic transitions is the central concern of the science of metallurgy. This is inevitable, for metallic artefacts have to be formed – and so must be soft and pliable – and they also have to be strong and hard to withstand the shocks and stresses of service. That paradox can only be resolved by transforming the structure of the artefact *after* it has been shaped and put together.

The technological importance of transformations in metals and alloys, then, is evident enough and the scientific problems are subtle and varied in the extreme. To appreciate why this is so, it is necessary to invoke another dimension, that of *microstructure*. Most metallic objects consist not of a single crystal but of an assembly of small irregularly shaped crystal grains, which can only be seen with the aid of a microscope. Generally more than one species of crystal is present and useful alloys are most often composites of several distinct crystalline *phases*. The sizes, shapes, proportions, compositions and mutual disposition of these phases – collectively, the microstructure – are all variable and subject to control. Heat-treatment, designed to alter the microstructure, is the metallurgist's central skill.

The behaviour of a simple carbon steel – the most important industrial alloy – will serve to exemplify the range of phenomena which are covered by that deceptively simple term, *transformation.* When the high-temperature form of steel (right-hand sketch, Fig. 1) is slowly cooled, it breaks up on transformation into two crystalline forms: almost pure iron in the form shown on the left, together with a compound of iron and carbon (Fe_3C) with a more elaborate crystal structure. (The high-temperature form of iron can readily dissolve carbon; the low-temperature form cannot.) Small crystallites of Fe_3C are independently nucleated in many sites. Part of the microstructure consists of thin plates of iron and of Fe_3C in alternation, and the whole assembly is fairly soft. If the high-temperature form is instead cooled suddenly, the transformation process is entirely different. The dissolved carbon is pinned in its existing sites and cannot segregate, for lack of time, and the alloy transforms by an ordered shift of millions of atoms into a new pattern. In the slow process, atom movements are at first random and uncoordinated; in the fast process, they are disciplined and simultaneous. The terms "civilian" and "military" have been applied to the two categories of transformation.

The product of the military transformation, containing as it does a great deal of carbon in enforced solution, is extremely distorted and therefore hard but also incapable of resisting intense shocks. If now this product is tempered by slow progressive heating, a new civilian transformation begins and a succession of iron-carbon compounds is formed in sequence, in the form of minute crystallites. Any desired compromise between hardness and shock resistance can be achieved, and different parts of the same object can be made to have quite different microstructures. The classical Japanese sword represents the most sophisticated application of these skills.

Military and civilian transformations each exist in rich variety, with many subtle distinctions of mechanism. Steels in particular form a large metallurgical family because of the variety of alloying elements which can be added to the basic iron/carbon constituents. Some of these elaborate steels are sensitive to imposed distortion. In such steels, transformations are induced when a hot sheet or rod is forced into a changed shape, as happens when it is passed through a pair of rollers. An early form of this was the use of a special steel – Hadfield's manganese steel – for the construction of railway points; every time a wheel crashed against a crossover point, the point became a little harder because of the stress-induced transformation. The study of such *thermomechanical treatments* is a new chapter in metallurgical research and is at a scientifically most intriguing stage. The strangest variant is the *shape memory effect*. Certain alloys – the alloy NiTi is the best known – can be extensively deformed and then, on heating, will return to their pristine shapes. This behaviour is quite different from that of an elastic spring: a spring obstinately returns to its original shape when it is let go, whereas an SME alloy humbly accepts its pummelling and stays put in its new shape. Only when it is heated does it home to its original form, even against a strong mechanical force seeking to prevent it. This mode of behaviour is always based on a stress-induced polymorphic transformation of the military type, followed by a reverse transformation when heat is applied. It is as though the natural pugnacity of an army forced into precipitate retreat could only be regenerated by an exposure to sunshine!

The shape-memory effect is extremely intriguing, both for its engineering applications and because of the difficulty of understanding the long-range forces which powerfully drive the transformed alloy back towards its original shape. A most detailed investigation of the microstructural changes is necessary in order to come to grips with this phenomenon, which is part of a much larger complex of questions concerned with the interplay of temperature, stress and transformation.

Quite apart from the technological justification, many metallurgists, physicists and chemists have long found this field of study irresistible for its purely scientific attractions. It is a satisfyingly hydra-headed creature: two questions raise their heads for each question that is resolved. For instance, the field of transformations in liquid crystals has arrived in the past few years, as a new branch of physics (stolen while no one was looking from the chemists who opened it up). Liquid crystals are half-crystalline, half-disordered substances that respond sensitively to heat and electric fields: the transformations in liquid crystals from greater to lesser degrees of order have close family resemblances to ferromagnetic and "atomic-order" transformations, and are proving amenable to interpretation in terms of a form of "catastrophe theory" which was applied to ferromagnetism and atomic order long before it was taken up and generalised by mathematicians.

The attractiveness of transformations as a subject for scientific investigation may have something to do with a universal, prescientific human obsession, attested by much ancient legend and folklore. The Greeks told tales of Proteus, a sea-god. Men would seek and seize hold of him as he sat on the rocks, in order to force him to grant a wish. The evasive god would transmute through all the varieties of living appearance, many repellent or terrifying, in the hope of frightening off his captor so that he might escape beneath the waves. Only the brave man who dared to keep his grip on the god till he had run through his entire gamut of forms and returned to his own godlike shape was assured of his wish. Each creature has its proper form and any departure from it is an affront to the natural order of things. A man's sense of identity is indissolubly linked to his own physical body: those who destroy this link, like Circe, like the jealous fairy who turned the prince into a toad, have always been seen as wicked and destructive. To retain the burden of identity while suffering a mutation in appearance is one of man's enduring terrors.

Underlying this is the philosophical dogma that in nature, building-blocks determine structure and structure defines appearance. Man can use a pile of stones to build a cathedral or a bank, but a pile of iron sulphide "molecules" is not subject to man's will and is bound to form a particular kind of crystal, always, with a particular yellowish colour and cubic external form. Snowflakes grow in different patterns according to the change of temperature and humidity, but they are all variations on a strictly constant theme, set by the hexagonal structure of ice. The structure of a simple crystal is implicit in its building blocks – atoms or molecules – and the predetermined forces which bind them together. Complex biological molecules, proteins, are half-way between the determinate simple crystal and the building whose form is subject to man's free will. The same basinful of amino acids can form a multitude of different proteins according to the template provided. It is still a matter of atomic fit: Monod showed how one amino acid selects another out of a copious mixture and by this means the acids assemble into a preordained pattern. As Monod puts it: "The epigenetic building of a structure is not a *creation*; it is a *revelation*." This is as true of an element as it is of a protein.

Polymorphic transformations (and indeed the melting of crystals) thus conflict with a deep philosophical sense of fitness, of match between form and essence. Those crystals which transform easily are unstable, not sure of their own proper nature, full of doubt, like man tormented by his Freudian unconscious. It is ironic that the most protean of elements is also the most unstable and dangerous: plutonium has four polymorphic forms. By gaining an understanding of transformations, man masters his fears, keeps hold of Proteus, gains his desires, fulfils his needs. At this level, the philosophical, scientific and technological desires of man all impel him to the same study. The mastery of transformations reassures man, the magician, that Jekyll and Hyde are under control.

Sir Denys Haigh Wilkinson, F.R.S.

Vice-Chancellor of Sussex University from September 1976. Formerly Professor of Nuclear Physics and Experimental Physics, University of Oxford.

Hughes Medallist of the Royal Society, London, 1965. Knight Bachelor 1974.

Founder Member of the Governing Board of the National Institute for Research in Nuclear Science 1957-65; Chairman of the Nuclear Physics Board of the Science Research Council 1965-8; Chairman, Physics III Committee and Member of the Scientific Policy Committee, CERN, Geneva, 1971-5.

THE UNKNOWN ATOMIC NUCLEUS *

Since the discovery of the neutron in 1932 we have said that the atomic nucleus contains neutrons and protons. More cautiously we should have said that when neutrons and protons (nucleons) are brought together they coalesce to form atomic nuclei. The two statements are not necessarily the same: the nucleons, on coalescing, might lose their identity and form some sort of undifferentiated nuclear "black hole" matter, without granular structure, characterised just by a few overall quantum numbers such as electrical charge and angular momentum. The fact that when nuclei are struck violently together nucleons come out does not prove that nucleons were inside in the first place – barks come out of dogs but that does not prove that dogs are made of barks.

It is, indeed, very difficult to get anything like direct, as opposed to inferential, evidence that there *are* nucleons inside nuclei. Perhaps the nearest approach to direct evidence comes from bombarding heavy nuclei such as lead with very energetic protons, of about 25,000 MeV in the best (CERN) experiments, measuring the energy distribution of the outgoing protons closely in the forward direction and comparing that distribution with what would be expected if the nucleus were indeed composed of so many protons and so many neutrons that interacted individually with the bombarding proton just as they would in the free state. The theory, put like that, sounds easy but it is extremely complicated to work out in full numerical detail because you have to allow for the possibility that the bombarding proton makes not just one but many successive collisions as it struggles through the nucleus and also for the possibility that it creates secondary particles (mesons) as it goes. However, when theory (the so-called Glauber method as worked out by O. Kofoed-Hansen) and experiment are compared they agree very well (better than 30 per cent) over the whole investigated energy range which is about as good as could be expected in view of the inaccuracies in our knowledge of the input data themselves, viz. the free-space nucleon-nucleon collision probabilities.

Convinced that nuclei contain nucleons pretty much as in the free state our next question is: what are those nucleons *doing*? The answer is a surprisingly simple one, namely that the nucleons are just cruising around essentially independently of one another as if any given nucleon were moving in some central field of force that must obviously, in the last analysis, be due to the fellow nucleons of the one in question but into whose origins and detailed nature we do not need to inquire for an initial discussion of the motion of the given neutron or proton. Of course, the motions of the several nucleons must be correlated to some degree because neutrons and protons are fermions and so obey the Pauli exclusion principle – each nucleon must occupy a separate quantum state: just as with electrons in an atom or in a metal every quantum state from the energetically lowest is occupied until all the nucleons of the nucleus are accommodated; the ensemble forms the nuclear Fermi sea but the associated correlations between the nucleons of the sea are due to the laws of quantum mechanics, the Pauli principle, the anti-symmetrisation of the overall wave function, and do not in the slightest imply that any forces explicitly operate between the particles even though we know that such forces must exist because there is nothing else available to bind the nucleons together into the nucleus; cf. the situation in an atom where we similarly build up our zero-order picture of atomic structure by ignoring the forces between the electrons but where the overall field of force within which the electrons move is primarily due to an external agency – the central atomic nucleus.

We can easily determine the depth of the nuclear Fermi sea because it must be equal to the depth of the effective potential within which the nucleons are moving (and that we can infer from the way in which bombarding nucleons are refracted as they enter and leave nuclei) minus the energy with which the topmost nucleon of the Fermi sea is bound into the nucleus (which is just the least energy that we must pay into a nucleus to persuade it to release a nucleon); the answer is about 100 MeV.

We know that this picture of nucleon motion in independent non-interacting orbitals – the zero-order nuclear shell model – is reasonably correct because we can knock, say, protons out of a nucleus by bombardment with, say, energetic electrons (500 MeV or so in the best (Saclay) experiments) – an $(e,e'p)$ reaction – and thereby – through the e vs. e',p *energy* imbalance – map out the energy distribution of the protons within their Fermi sea and also – through the e vs. e',p *momentum* imbalance – map out the associated momentum distribution of the protons which is linked by general laws of quantum mechanics, having nothing in particular to do with the nucleus, to the *angular* momentum distribution of the protons, this latter being itself specified by our independence, shell model, hypothesis. Energy and (angular) momentum distributions tally most impressively with expectation in light nuclei such as ^{12}C.

How can we understand this remarkable simplicity, this atomic-like structure, of the nucleus? The simplicity of the atom occasions no surprise because the electron-electron interaction, that perturbs the simple orbital motions of the individual electrons, is feeble compared with the interaction of each electron with the central nucleus that produces the simple orbital motions. But in the nucleus the nucleon-nucleon forces that tend to *break down* the simple orbital motions (nucleons "bumping into each other") are the very same that must *give rise to* the overall field of force that somehow generates those simple orbital motions. So how can the motions be so simple? The answer to the conundrum lies in the Pauli exclusion principle that gives rise to the notion of the Fermi sea in the first place. Imagine that the nucleon-nucleon interaction is zero and that the nucleons' orbital motions are established by some external field of force so that every quantum state up to some maximum is occupied by a single nucleon. Now switch on the nucleon-nucleon interaction that will try to make nucleons bump into each other and so break down the simple motions of the Fermi sea; the "bump", however, will not be effective unless it is sufficiently violent to lift *both* participant nucleons right out of the Fermi sea into unoccupied quantum states above its surface since smaller excitations would put the bumping nucleons into already-occupied states and this is forbidden – one state, one nucleon. So whether or not the simple Fermi-sea motions survive in a real nucleus is a question of the depth of the Fermi sea (100 MeV or so as we have seen) in relation to the interaction energy between two nucleons at their usual spacing inside the nucleus. It is not too easy to give a simple answer for this interaction energy because it depends on the details of the nucleon motions and on the effective spectroscopic states in which the "bump" takes place but it is certainly only of the order of 10 MeV. It is therefore not surprising that the Fermi sea survives except in the immediate vicinity of its surface where excitations into a few higher states may take place: the "configuration mixing" that distinguishes the modern shell model from its primitive forebear of 1950.

Turning to the quantification of these ideas, we find that detailed shell-model calculations of nuclear structure, starting from nucleons whose orbitals are adjusted in size to fit actual nuclear dimensions, but *including* the experimentally determined nucleon-nucleon interactions operating *between* the nucleons, give quite good agreement with experiment not only in the level schemes but in static properties such as magnetic moments and in dynamical properties such as radiative transition probabilities. Technically much-more-elaborate unrestricted self-consistent calculations, similar in principle to Hartree-Fock calculations of atomic structure, in which the nucleons are left to work out their own salvation on the basis of their mutual interactions, without the presupposition of any particular dimensions for the resulting orbitals, yield theoretical nuclear dimensions, surface thickness and binding energies that agree quite well with nature. An additional feature of these Hartree-Fock calculations is that they predict that many nuclei should be not roughly spherical but rather strongly deformed – half as long again as they are wide in some cases – and this is in striking accord with experiment.

Many nuclear properties can be simply represented by speaking in classical-sounding terms of the bulk properties of the nucleus: it rotates as a whole or vibrates as a whole. Such collective descriptions of nuclear behaviour are completely consistent with the microscopic shell model that we have just sketched and may be directly derived from it – in full quantitative detail in sufficiently simple cases and in convincing outline in more complicated ones where we believe we are held back from a full description only by lack of adequate computer power.

All this we shall call the "conventional" approach and it makes it sound as though nuclear structure is a closed book – we measure empirical forces between nucleons, we calculate, on the basis of those forces, the properties of complex nuclei and the answer comes out right. This is true, but only up to a point.

There are two problems. The first is that our detailed accounts of nuclear structure really involve only a few "valence" nucleons near the top of the Fermi sea – just a few per cent of the total nuclear contents. True it is that these few nucleons turn out experimentally to have the quantum numbers we should expect of them if all those below them in the Fermi sea were doing what the shell-model would have them do: but we have no evidence that the lower nucleons are in fact behaving in this "conventional" manner. It is unlikely that the quantum numbers of the valence nucleons would come out right if the under-pinning nucleons were behaving radically differently from the "conventional" prescription, e.g. were set into a crystalline lattice. On the other hand, it is quite possible that, to continue to use "conventional" language, the appropriate admixture of configurations lying *above* the valence nucleons into those lying *below* could transform the simple shell-model orbital motions of some at least of the under-pinning nucleons into quite tight self-centred clusters such as alpha-particles. In this case a gentle probing of the nucleus, involving only rearrangement of the valence nucleons among themselves, would give the "conventional" answer but a more vigorous probing may reveal a predilection of the nucleus to fall apart into what we should then probably have to take to be literally pre-formed clusters: we should begin to think of the nucleus as a raspberry; there are some signs that this may be so; we must keep an open mind. The second problem is that when the "conventional" approach is pressed really quantitatively in a place where it should be able to withstand that pressure, i.e. in a place where, in its own terms, its predictions are unambiguous and not qualified by uncertainties in its own formulation or its own relevant parameters, it fails.

A single immediate illustration of this quantitative failure of the "conventional" approach will suffice. The simplest nuclear reaction is the radiative capture of slow neutrons by protons to form deuterons: $n + p \rightarrow d + \gamma$. The known nucleon-nucleon forces permit an accurate and unequivocal prediction of the cross-section for this process: $(0.303 \pm 0.004) \times 10^{-24}\,\mathrm{cm}^2$. The experimental number is also accurately known: $(0.332 \pm 0.002) \times 10^{-24}\,\mathrm{cm}^2$. The discrepancy between theory and experiment is therefore (10 ± 1.5) per cent. We find similar discrepancies between theory and experiment wherever we look in systems of the type where the "conventional" approach has no escape: the β-decay of ^{3}H and other light nuclei; magnetic moments of nuclei of supposedly well-known structure; the density of nuclear matter; the binding energy of nuclear matter. Something almost always goes wrong at the few per cent level: what is it and can we put it right? To see what is wrong in the "conventional" approach we go back to the beginning: we have spoken of empirically determined forces acting between nucleons as the basis of our theories of complex nuclear structure; but where do those forces come from and should we not have regard to their origins in working out the consequences of their actions? In other words the properties of nuclei may depend not only on the forces between the constituent nucleons but also on the means by which those forces are engendered.

Before we look at the forces between nucleons let us look at the individual nucleons. Nucleons may or may not have an explicit sub-structure of quarks, and that is not for our present consideration. But, in any case, the nucleon must be thought of as a source of virtual mesons: just as the electron is surrounded by its electric field, the quantization of which is virtual photons, a tenuous cloud of which must therefore be thought of as surrounding any electric charge, so a nucleon is surrounded by its nuclear field, the quantization of which is virtual mesons, a cloud of which surrounds the nucleon. But since the meson-nucleon coupling is much stronger, by a factor of 100 or so, than the photon-charge coupling, the meson cloud around a nucleon must be pictured as dense rather than tenuous, with several virtual mesons around at any moment, so that the meson cloud dressing a nucleon is to be thought of as part of the normal essential substance of the nucleon and not, as in the case of the virtual photon dressing of an electric charge, an insubstantial frill that may be disregarded to a first approximation.

Indeed, when the internal charge distribution of a proton is probed by the scattering of energetic electrons from it, it is found that the proton, unlike the electron, is not a point charge but an extended structure of $\langle r^2 \rangle^{1/2} \simeq 8 \times 10^{-14}$ cm. Now mesons are particles of finite mass m so that to bring them into virtual being costs energy of at least $\Delta E = mc^2$ and therefore the time Δt for which they can enjoy their virtual existence is limited by Heisenberg's uncertainty principle to $\Delta t{\cdot}\Delta E \simeq \hbar$, viz. $\Delta t \simeq \hbar/mc^2$: they will therefore not be able to stray to distances greater than about $c{\cdot}\Delta t \simeq \hbar/mc = 1/\mu$ which distance sets a natural spatial scale for the nucleon through a factor of form $e^{-\mu r}$. Now for the lightest meson, the pion, $mc^2 \approx 140$ MeV, $1/\mu \approx 1.4 \times 10^{-13}$ cm; heavier mesons, with correspondingly smaller values of $1/\mu$, are legion (for example, the strongly coupled ρ- and ω-mesons have $1/\mu \simeq 3 \times 10^{-14}$ cm) and it is not surprising that the whole mesonic dressing should add up to the observed $\langle r^2 \rangle^{1/2} \simeq 8 \times 10^{-14}$ cm.

So nucleons may be thought of as fuzzy balls of size about 10^{-13} cm; when two nucleons approach, each has no means, apart from the relatively weak electric force that we disregard at this stage, of knowing that the other is there until their mesonic clouds begin to interpenetrate; when this happens we can imagine that occasionally a meson that has emerged from one nucleon will not go back in again but will rather pass across to the second nucleon and thereby carry word of the first one's presence, i.e. establish an interaction, a force. Again, we should expect this force to display the factor $e^{-\mu r}$.

It is therefore this exchange of mesons, pions at the greater distances, heavier mesons, pairs of pions and so on at smaller distances, that constitutes the force between nucleons that we can empirically measure and that, then forgetting the mesonic origins of the force, we use as the starting point of our "conventional" theories of nuclear structure – that typically get things wrong by about 10 per cent. However, the overall properties of a complex nucleus must depend on the totality of what is going on inside it and not just on what its nucleons are doing; for example, the flow of charged pions between the nucleons, establishing the force between them, constitutes an electric current in addition to the currents due to the motions of the protons themselves and so will contribute to all current-linked phenomena, for example the magnetic moment. Another possibility is that a pion "in the air" between two nucleons may disappear to give an electron-neutrino pair, $\pi \rightarrow e + \nu$, thereby contributing to the weak (β-decay) properties of the nucleus. Such mesonic effects must show up everywhere to some degree and when we calculate that degree to the best of our present ability, it turns out to be typically a few per cent. We must obviously "mesonate" our "conventional" nucleons-only approach and let in the mesons explicitly.

"Mesonation" has further consequences: when real pions bombard nucleons in the laboratory we find that the nucleons can be thereby raised into a multitude of excited states, generically called resonances or isobars, just as the quanta of the electromagnetic field, photons, can raise systems held together by that force, e.g. atoms or molecules, into excited states. So when nucleons interact, which they do by bombarding each other with pions and other mesons, in the virtual state to be sure but that makes little essential difference, there must be a certain chance that that bombardment will result in the excitation of isobars. This has two important consequences for our discussion of complex nuclei:

1. If an isobar is formed in a free nucleon-nucleon (N-N) low-energy collision it cannot persist because it is more massive than the nucleon and so, by the Heisenberg uncertainty relation, must disembarrass itself of its excitation energy ΔE within a time Δt given by $\Delta E \cdot \Delta t \simeq \hbar$; the lightest isobar is the Δ for which $\Delta E \simeq 290$ MeV so the longest Δt is about 2×10^{-24} sec; this de-excitation must be effected by, for example, the re-emission of a pion such as gave rise to the excitation and its reabsorption by the other nucleon; this merely constitutes part of the N-N force such as we use in the "conventional" shell model computation. If, however, the two nucleons in question are part of a complex nucleus another option is open: the pion that de-excites the isobar may be absorbed by a third nucleon thereby constituting a force of an essentially 3-body nature that comes into play only when three nucleons are close enough together and that is distinct from the sum of the various possible 2-body forces that operate between the three. This 3-body N-N-N force must therefore be introduced into our computation of nuclear structure additionally to the 2-body N-N forces of the "conventional" approach. (Such genuine N-N-N forces can arise in other ways: for example, the exchange of a ρ-meson between two nucleons is part of the N-N force; but the ρ-meson, after emission by one nucleon of a complex nucleus, may decay, $\rho \rightarrow \pi + \pi$, and not be absorbed as a ρ-meson by a second nucleon at all; if both decay pions are then in fact absorbed by the second nucleon this is again just a piece of the N-N force but if one decay pion is absorbed by the second nucleon and the other by a third we have an N-N-N force.)

2. The isobar, having been formed, may de-excite to a nucleon not by emission of a pion but by emission of, for example, a photon, real or virtual, or an electron-neutrino pair, which will therefore constitute an electromagnetic or weak interaction of the nuclear system additional to that involving nucleons alone (or nucleons plus mesons only).

A further possibility is that a pion, in flight between two nucleons A and B, may dissociate into a nucleon and an anti-nucleon; the anti-nucleon then annihilates with nucleon B giving a pion that is absorbed by nucleon A so that the final state, as the initial one, consists just of two nucleons and this process is just part of the N-N force. However, if nucleons A and B are part of a complex nucleus, the pion coming from the annihilation of the anti-nucleon with nucleon B could be absorbed not by nucleon A but by a third nucleon of the initial system thereby constituting another component of the N-N-N force. Similarly, the annihilation could yield a photon or an electron-neutrino pair rather than a pion so that this pion-dissociation mechanism contributes also to the electromagnetic and weak properties of the nucleus.

What is the quantitative importance of these processes whose qualitative role is undeniable? We examine the example of radiative neutron-proton capture that we saw above to have a (10 ± 1.5) per cent discrepancy with "conventional" nucleons-only theory. We now see that the "conventional" theory in which the gamma-ray "comes out of the nucleons" must be supplemented by at least three additional mechanisms:

(i) a pion is emitted by one nucleon but, while it is "in the air", before being absorbed by the other nucleon, emits the gamma-ray;

(ii) a pion is emitted by one nucleon and on absorption by the other raises it into an isobaric state that de-excites emitting the gamma-ray;

(iii) a pion is emitted by one nucleon and dissociates "in the air" into a nucleon and an anti-nucleon; the anti-nucleon annihilates with the second initial nucleon giving the gamma-ray.

The sum of these three mechanisms in fact completely removes the 10 per cent discrepancy. Other discrepancies among the properties of the lightest nuclei, for which we think we have quantitatively-reliable "conventional" predictions, are similarly removed, or significantly reduced, by similar mesonic interventions.

It is therefore quite wrong just to think of the mesons as the generators of the N-N force then to use that force for nuclear structure computations and forget about the mesons that gave rise to it: they must be brought in *explicitly* if we want to get the right answer.

So far we have spoken chiefly of pion-exchange because the pion is the lightest meson and so has the longest "reach": $1/\mu = \hbar/mc \simeq 1.4$ fm. But the heavier mesons can also be exchanged and give important contributions to the N-N force at smaller distances; for example, the ρ-meson's coupling to nucleons is very strong and although its reach is only about 3×10^{-14} cm it is an important contributor to the N-N force. However, the fact that a meson is heavy does not exclude it from certain long-range effects. For example, a ρ-meson could emerge from a nucleon, quickly convert to a π-meson by emission of a photon or interaction with the electromagnetic field, $\rho \rightarrow \pi + \gamma$, giving a "long-reach" pion to be absorbed by another nucleon. This process contributes to certain nuclear properties, specifically to the sum of the magnetic moments of nuclei having complementary numbers of neutrons and protons, e.g. ^{3}He and ^{3}H, to which the process $\pi \rightarrow \pi + \gamma$ does not, and so it can be, and has been, picked out and identified.

Complex nuclei must similarly contain "strange" mesons and hyperons by virtue of processes such as $\pi \rightarrow \Sigma + \bar{\Sigma} \rightarrow \pi$ and $N \rightarrow \Sigma + K \rightarrow N$ but it is more difficult to put their undoubted presence into evidence.

Although it is now clear that explicitly mesonic effects are important at least at the 10 per cent level this does not mean that the procedures by which we have introduced them are adequate or even right. We have taken the line that we start from an empirical N-N force that is presumably generated by meson exchange in a manner that we do not understand in detail; the mesons are then forgotten for the time being, the empirical N-N force being used to calculate the structure of the "conventional" system of nucleons only, moving under the influence of that N-N force; such mesons as we know how to treat adequately are then let back into that nucleonic structure without changing it and their explicit additional effects such as their electromagnetic and weak properties and their generation of isobars and N-N-N forces are essentially treated as perturbations that do not fundamentally change either the initial structure or the nucleons that constitute it. We will now look at a few difficulties.

The first difficulty is that this "mesonated conventional" approach is essentially ambiguous, for example we quickly find that we are running the danger of double counting: when long-lived nucleon states, such as the zero-order shell model orbitals of the Fermi sea, are involved it is fairly clear that the mesonic exchanges between them, of time-order $\hbar/mc^2 \simeq 5 \times 10^{-24}$ sec for a pion and shorter for other mesons, are relatively fleeting additions that can be explicitly separated out and discussed; when, however, we are considering more complicated nucleon motions such as are associated with the successful N-N "bumps" that *do* occasionally lift nucleons out of the otherwise tranquil Fermi sea and that may involve excitations of $\Delta E \simeq$ 100 MeV or much more, we see that such nucleonic excitations are themselves fleeting ($\Delta t \simeq \hbar/\Delta E \simeq 7 \times 10^{-24}$ sec when $\Delta E = 100$ MeV) and are comparable in time scale with the exchange times of the mesons themselves that gave rise to the "bump". The meson exchange and the associated nucleon motion can no longer be cleanly separated from each other and we clearly must not simply add their separately computed effects because they must to some degree just be different representations of each other.

In principle the way out of this first difficulty is to abandon the "mesonated conventional" approach, in which we take the empirical N-N interaction as the starting point and add explicit mesons and isobars later, in favour of a true many-body relativistic field theory in which nucleons, mesons and isobars partake with equal right in an overall description of the resultant system whose properties, such as its interaction with the electromagnetic field, would then be given unambiguously; there would be no forces as such in the input data but only the identity of the various particles and the associated coupling constants appropriate to the various particle combinations.

This counsel of perfection cannot yet be followed but we might hope for progress along the following lines: take a very simple model, soluble in relativistic many-body field theory, a system containing perhaps one species each of nucleon, isobar and meson then: (i) compute the N-N force that would arise from the meson exchanges, use this force in the "conventional" way to calculate "nucleon-only" wave functions, "mesonate" this "conventional" structure by adding the mesons and isobars perturbatively as sketched above and then compute whatever properties you are interested in using various recipes for resolving such ambiguities as you can spot; (ii) compute the same properties from the proper full solution.

In this way we should get a feel for the right way to handle the ambiguities and for the degree to which the "mesonated conventional" approach can be relied upon as a stand-in for the real thing. Although it may not be possible to treat field theoretically a realistic assembly of mesons, nucleons and isobars it may be possible to treat separately several sub-sets, viz. different limited selections of mesons and isobars, and so get a feel for whether the ambiguities can be systematically resolved or whether different recipes are going to be necessary for different meson exchanges, etc.

The second difficulty of the "mesonated conventional" approach is related to the first and is that it may miss entire new modes of nuclear behaviour essentially because it closes its eyes to the possibilities. For example, our "mesonation" considers only meson exchanges between pairs of nucleons, or threes when isobar excitation is included. But it is entirely possible that mesonic circuits involving more than three nucleons may get excited with the very strong $\pi + \mathrm{N} \rightarrow \Delta(\mathrm{N}\Delta\pi)$ interaction, involving the Δ-isobar, as a kind of catalyst for the generation of pions. Indeed, so strong is the $\mathrm{N}\Delta\pi$ interaction that it is quite on the cards that the *entire nucleus* may participate in such pionic circuits with the pions then pursuing orbits referred to the nucleus as a whole like those of the nucleons themselves. The total energy of the system could then minimize at a very high pion density: the so-called "pion condensation" phenomenon, associated particularly with A. B. Migdal, that has been much discussed recently. Although it seems unlikely that nuclei of normal density already contain such condensed pions (a point made especially by G. E. Brown) it also seems likely that this phenomenon might set in at densities elevated by only a factor of two or three above normal, so that real nuclei may already contain the beginnings of such condensations, on a short time scale, associated with density fluctuations; these mini-condensations or multi-nucleon pion circuits may already have some importance for the overall energy balance and, specifically, might lower the energy of certain types of nucleon cluster states, viz. promote nucleon clustering into more-or-less well-defined nuclear sub-structures the possibility of which we noted earlier; these sub-structures would not necessarily be long-lived but might be numerous and even fairly well defined if seen in a snapshot.

If quasi-stationary sub-nuclear many-$N\Delta\pi$ clusters exist then the Δs themselves in such clusters will be modified from their free-space properties just as the properties of a coupled system of identical resonators of any kind differ from the properties of the individual identical resonators: the effective energy and width of the Δ will depend on the type of cluster of which it is part. This we might hope to see by exciting such clusters through the bombardment of nuclei by real pions: in principle the Δ-resonance might shift or split and we should probably stand our best chance of seeing this by looking specifically in final-state channels that might be preferentially fed in the ultimate decay of such clusters, viz. involving high-energy complex fragments such as α-particles, etc., as has been emphasised by E. Vogt.

Even more radical "whole-nucleus" consequences of mesonic exchanges might be envisaged. Although the $N\Delta\pi$ interaction that may make pion condensation possible is very strong it is a p-wave interaction that peaks for free pions bombarding stationary free nucleons at $E_\pi \simeq$ 200 MeV so that the individual condensed pions tend to be "energetic" and this tends to raise the energy of the condensate. However, consider the case of the hypothetical $J^\pi = 0^+$ σ-meson that would have a strong *s*-wave interaction with nucleons; this could promote condensation for "zero-energy" σ-mesons and it has been suggested by T. D. Lee and G. C. Wick, on the basis of a σ-N many-body field-theory, that the associated nucleon density could be very high – perhaps 10 times normal – and the system more stable than the normal nuclear ground state. It is therefore conceivable that there are some such collapsed nuclei, "abnormal nuclear states", already around in nature but if so they are very few. Recent work by M. Rho suggests that although such abnormal states of high density can probably be made they are probably also high in excitation above the normal ground state and so will be difficult to identify. However, their transitory role, riding upon density fluctuations in ordinary nuclei as discussed above for pions, may not necessarily be totally negligible. Note that even if an identifiable σ-meson does not exist as such (and a broad structure of the correct quantum numbers has been seen in the $\pi\pi$ system at a total energy of about 600 MeV) states of appropriate properties could be synthesised out of pairs of pions, $K\bar{K}$, etc.

An interesting question of practical importance is raised by these field-theoretical and other discussions of the density of mesonic fields as a function of density of nucleons: the time scale. If the nucleon density is suddenly changed, say, in a shock-wave generated by the collision of two fast heavy ions or in a spontaneous nuclear fluctuation, *how long* does it take, say, the pion field to establish its equilibrium value, *how long* does the phase change to the condensed state, if there is to be one, take to come about? The calculations of, say, the pion field density are equilibrium calculations in which, as a function of nucleon density, pions have themselves a right to a certain equilibrium density by virtue of the magnitudes of the various coupling constants; but these equilibrium calculations do not tell us how in detail the pions *get there*: obviously, in a sense they must "come out of the nucleons" but the calculations do not tell us how and so they do not tell us how long equilibrium will take to reach. The situation is analogous to that of the Planck spectrum of black-body radiation: if I suddenly heat up a cavity how long does it take for the Planck spectrum to establish itself? The spectrum is an equilibrium spectrum to which the cavity has a right determined only by its temperature and not dependent on the material out of which its walls are made and yet the photons must "come out of the walls" and take a time to do it that will be determined by the nature of the walls.

This question of time scale is obviously of importance in the nuclear case because the only means we have available, at least on Earth, for changing the nuclear density are themselves of short time scale; it is quite possible that critical densities for condensation might be exceeded in a heavy ion collision but the nucleus pulls itself apart before condensation has time to establish and manifest itself: that would be a pity.

Another radical possibility is that the many-body forces that are effectively induced by the more generalised sorts of meson exchange that we have just been considering may be such that large nuclei actually become *crystalline*, at least in their interior. This would be an extreme form of cluster structure and there is nothing specific that suggests that it may be happening but so long as we remain ignorant about the organisation of the depths of large nuclei we must keep an open mind. (It may well be that this possibility could be eliminated, theoretically, by arguments based on the shell-model quantum numbers of the valence nucleons – see our earlier remarks about cluster models – or, experimentally, by the Glauber analysis of high-energy p-nucleus scattering which, as we saw, gave good agreement with experiment for an essentially uncorrelated neutron-proton nuclear gas; but neither of these exercises has been carried through.)

Returning to the difficulties of the "mesonated conventional" approach we find a third substantial problem that takes us right back to our discussion of *nucleon* structure. The free nucleon is a dynamical structure containing pion and other mesonic currents. Nucleons interact with each other only by virtue of the exchange of mesons such as contribute to the build up of their individual structures; in saying this we must additionally bear in mind that mesons are not conserved so that the exchange of mesons *between* nucleons will affect the flow of mesons *within* the individual nucleons but not in any simple way. The situation is very much like that of the near-zone of a radio antenna; energy flows out of the antenna and back in again (circulation of virtual photons); when a second, identical, antenna is brought near the first energy now flows out of one and into the other but the original flow of energy out of each and back into the same antenna again is also changed. Thus those properties of the individual nucleons that are affected by their internal mesonic constitution (the *intra*-nucleonic meson currents) such as their magnetic moments and intrinsic β-decay rate will be changed when they are brought into each other's vicinity, in a manner additional to the contribution to the overall nuclear property in question that is due to the *inter*-nucleonic meson currents. Indeed just as in the analogy of the antennae one cannot separate the "intra" from the "inter" effects in any unambiguous way; we find ourselves, at this deeper level, facing the same kind of problem as arose in our discussion of the exchange-current/configuration-mixing ambiguity; in the present case we clearly need a model of *nucleon* structure before the resolution can be effected and this is one stage beyond the relativistic many-body field theory that constituted our in-principle hope for resolving the earlier problem. It is impossible to guess how important such "nucleon modification" effects might be but they will clearly be greater for processes that require nucleons to be in strong interaction and we might hope to get some indication from, for example, high-energy photodisintegration that will depend on, among other things, the magnetic dipole moments of the "individual nucleons".

We have seen how essential it is to take explicitly into account the mesonic structure of the nucleus if we want to get "ordinary" questions answered correctly to better than 10 per cent or so and how far we are from an unambiguous clearing-up of that 10 per cent, a satisfactory treatment of which will have to await at least a semi-realistic many-body field theory. We have also seen how mesonic effects of a type as yet only dimly envisioned may have far-reaching consequences for nuclear structure by, for example, promoting clustering through incipient condensation phenomena. The two messages are distinct but equally clear: (i) even conventional nuclear structure physics already demands "mesonation" for an acceptable understanding; (ii) the new types of phenomena that mesons may engender are so radically different from the expectation of conventional nuclear structure physics that we must keep a very open eye and mind indeed.

Abdus Salam, F.R.S.

Professor of Theoretical Physics at Imperial College, University of London and Director of the International Centre for Theoretical Physics, Trieste.

Awarded the Hopkins Prize (Cambridge University) for the most outstanding contribution to Physics during 1957-8, the Adams Prize (Cambridge University), 1958, first recipient of the Maxwell Medal and Award of the Physical Society, London, 1961 and recipient of the Hughes Medal of the Royal Society, London, 1964. Awarded the Guthrie Prize and Medal, 1976.

Member of the Institute of Advanced Study, Princeton, 1951. Elected Fellow of the Royal Society, 1959; Fellow of the Royal Swedish Academy of Sciences, 1970; Foreign Member of the USSR Academy of Sciences, 1971 and Honorary Fellow of St. John's College, Cambridge, 1971. Chairman of the United Nations Advisory Committee on Science and Technology, 1971-2.

Published widely on aspects of the physics of elementary particles and the scientific and educational policy for Pakistan and developing countries.

PROBING THE HEART OF MATTER

What is Matter? What are the laws which govern its behaviour? Do there exist fundamental entities of which all matter and all energy is composed? These are the quests of Particle Physics – the frontier discipline of Physical Science, an area of intense search, profound ignorance, and also of some of the deepest of man's insights into the workings of nature.

The historical approach to the subject centred on the search for the most elementary constituents of matter – a search whose tenor was so remarkably predicted by Isaac Newton, "Now the smallest of particles of matter may cohere by the strongest attractions, and compose bigger particles of weaker virtue; and many of these may cohere and compose bigger particles whose virtue is still weaker and so on for diverse successions, until the pregression ends in the biggest particles on which the operations in Chemistry and the colours of natural bodies depend, and which by cohering compose bodies of a sensible magnitude.

"There are therefore agents in nature able to make the particles of bodies stick together by very strong interactions. And it is the business of experimental philosophy to find them out."

In Newton's day, no "small particles" were known, and the only "virtue" bulk matter was definitely known to possess was the gravitational force of attraction, a long-range force between two objects of masses m_1 and m_2, at a distance r from each other, which acted according to the law

$$F = \frac{G_N m_1 m_2}{r^2} \text{ (with } G_N \text{ the Newtonian constant).}$$

The first of "small particles" to be discovered was the electron in late nineteenth century, followed by the proton, the neutron and the neutrino – all discovered by the year 1934. In 1934, these four appeared to be the ultimate constituents of which all matter might be composed. They appeared to possess three further "virtues" besides the gravitational. To characterise these "virtues" quantitatively one ascribes "charges" to these particles; the electron and the proton carry fixed (equal and opposite quantities (e and $-e$) of) electric charge; the proton and the neutron carry equal "nuclear charges", while all four (the electron, the neutrino, the proton and the neutron) carry a "weak" charge in addition to the "gravitational charge" (this latter is better known as "mass"). The "nuclear" and "weak" forces are short range (of the form

$$g_1 g_2 \frac{e^{-r/r_0}}{r^2}$$

where g_1 and g_2 are the magnitudes of charges carried by the relevant particles) in contrast to the electric and the gravitational forces, which are both long range (of the form $e_1 e_2/r^2$ and $m_1 m_2/r^2$). The effect of nuclear and weak forces is felt predominantly when the participating particles are closer than $r_0 = 10^{-13}$ and $r_0 = 10^{-16}$ cm respectively.

To complete the story of these charges, in 1934, Dirac added to the subject the concept of "anti-particles"; every particle has as a counterpart an anti-particle which has the same mass (gravitational charge) but an electric, weak or nuclear charge which is equal in magnitude but opposite in sign to that of the particle.

The subsequent history of particle physics is the history of proliferation and discovery of other objects equally as small in size – or in some cases smaller still – than the four particles mentioned above, so that the suspicion has arisen that possibly these four – or at least some of these four – may themselves be composite of some more basic constituents – "quarks" – not so far experimentally isolated as free particles. In fact the whole concept of "elementarity" – and the notion of "basic" constituents – may have become untenable. A more profitable way of summing up the new discoveries appears to be the statement that as one has probed deeper into the structure of matter (and the standard method for this is collisions of higher and higher energy beams of protons, neutrons, electrons and neutrinos with each other and the study of their reaction products) – one has discovered new "charges", besides the ones mentioned above.

At present some nine charges may be distinguished; there are three so-called "colours", two "isotopic" charges, "strangeness", "charm", "baryon" and finally "lepton" charge. The weak, the electric and the nuclear charges are made up from these nine basic charges. And further – and this is where the biggest excitement of 1973-4 has centred around – we have discovered that there exists a basic symmetry principle, which states as a unifying principle *that all these charges are on par with each other*; that (barring the gravitational charge) all other charges are aspects, are manifestations of one basic charge. The result seems experimentally borne out for weak and electromagnetic charges. The sorts of questions which arise are the following. Do there exist nine basic objects each carrying one single "charge"? Is there anything sacred about the number *nine* for the "basic" charges? Or indeed, is there really a finite number of charges – or shall we go on discovering newer and newer charges as we probe deeper and deeper into the heart of the particles we create in high-energy collisions?

To summarise then, the basic dilemma of our subject is the "charge" concept and its nature. There is one charge – the gravitational (mass) – which we believe is rooted within the concepts of space-time. Einstein identified gravitational force with the curvature of a continuous space-time manifold. Are the other charges, weak, electromagnetic, and nuclear, equally rooted within space-time, perhaps giving an indication of its topology in the small? When looked at from a distance, the surface of the ocean may appear unruffled, as a continuum; but looked at closer one may distinguish all manner of granularities, all manner of topological structures. Do the various "charges" correspond to these? Or is it that these charges are telling us of new dimensions besides the four dimensions of space and time; dimensions we have not yet apprehended? And what is the role of Planck's fundamental constant of action and Pauli's exclusion principle in this search for the meaning of charge concomitantly with the structure of space-time? We are very far yet from discovering the heart of the matter.

Roger Penrose, F.R.S.

Rouse Ball Professor of Mathematics and Fellow of Wadham College, Oxford.

Awarded the Dannie Heineman Prize of the American Physical Society and the American Institute of Physics, 1971, the Eddington Medal of the Royal Astronomical Society of London, 1975 and the Adams Prize, Cambridge University 1967.

IS NATURE COMPLEX?

1. NATURE AND NUMBER

Though Nature is undoubtedly subtle, she is surely not malicious. This, at least, we have on the authority of Einstein.† [1] But is she complex? The word has more than one meaning: the first, nearer to "complicated" than "subtle"; and the second, a mathematical meaning concerning the nature of number. It is this second meaning that forms the main subject of this essay, though there will be relevance also to the first.

That Nature can be usefully described, at least to a considerable degree, according to the laws of number, has been in evidence for many centuries. But what is not so familiar to those without a mathematical background is that there are several *different kinds* of number, many of which are nevertheless subject to the same arithmetical laws.

Most primitive are the so-called *natural numbers*:

$$0, 1, 2, 3, 4, \ldots$$

and just about anything in our world can be quantified with their aid. We may speak of 3 apples, 17 eggs, 0 chickens, 500 people, or hydrogen atoms, or eclipses of the Moon, or lightning flashes. Next in order of abstraction comes the system *integers*:

$$\ldots, -3, -2, -1, 0, 1, 2, 3, \ldots$$

Initially these were defined as a convenience, introduced in order to make the laws of arithmetic more systematic and manageable. But there are many things in the world which cannot be quantified using them. We cannot, for example, accurately speak of −3 people in a room! On the other hand, there are various slightly more abstract ideas for which a description in terms of integers *is* appropriate, a bank balance being, perhaps, the example which springs most readily to mind. But in basic physics, also, the integers have their role to play. The clearest case of a physical attribute which seems to be accurately quantified by integers is electric charge. As far as can be told from accurate experiment, there is one basic unit of electric charge, namely that of a proton, and all other systems in nature have a charge which is an exact integer, positive, negative, or zero, when described in terms of this unit. (If the hypothetical quarks are, after all, eventually discovered existing as free particles – and intensive searches have so far failed to reveal them – then the values of this unit would have to be divided by three, but physical systems could again be described using integer charge values.) There are also other less familiar physical quantities which are described by integers, such as baryon number, the various leptonic numbers (presumably) and the quantum mechanical spin of a physical system about some axis.

While the arithmetical operation of subtraction is simplified by the introduction of negative integers, the operation of division is not. For that, the fractions are needed:

$$0, 1, \tfrac{1}{2}, -1, -\tfrac{1}{2}, \tfrac{1}{3}, \tfrac{2}{3}, 1\tfrac{1}{2}, 2, -\tfrac{1}{3}, \ldots$$

Any two such numbers can be added, subtracted, multiplied or divided (except by 0). But, while convenient mathematically, the system of fractions has seemed not to play any clear role

† "Raffiniert ist der Herrgott aber boshaft ist er nicht"; see, for example, Banesh Hoffman, *Albert Einstein, Creator and Rebel* p. 146, Hart-Davis, MacGibbon Ltd., London, 1972.

in quantifying features of nature. Instead, we must apparently leap further in abstraction and include all the so-called *real* numbers

$$0, 1, \tfrac{1}{2}, -1, \sqrt{2}, 2^{\sqrt{2}}, \pi, -6.1974302\ldots, \ \ldots$$

each of which can be represented in terms of a (signed) infinite decimal expansion. The rules of finite arithmetic are precisely the same as for fractions, but the system of reals is more complete in that certain *infinite* operations can now be performed. In particular, there are such infinite sums as:

$$\frac{\pi^2}{6} = 1 + \frac{1}{2^2} + \frac{1}{3^2} + \frac{1}{4^2} + \frac{1}{5^2} + \ldots$$

$$\sqrt{2} = 1 + \frac{1}{8}\left(\frac{2}{1}\right) + \frac{1}{8^2}\left(\frac{3\times 4}{1\times 2}\right) + \frac{1}{8^3}\left(\frac{4\times 5\times 6}{1\times 2\times 3}\right) + \frac{1}{8^4}\left(\frac{5\times 6\times 7\times 8}{1\times 2\times 3\times 4}\right) + \ldots$$

According to normally accepted views, measurements of time and distance are to be described in terms of real numbers. Indeed, it would seem that such physical notions supplied the original motivation for their invention.

The famous paradoxes of Zeno were conceived before the mathematical notion of a real number had been formally introduced. These paradoxes expressed a puzzlement about the continuous nature of time and space which had long been felt, and which have seemed to be resolved now that the rigorous mathematical concept of the real number system is at hand. In fact, such has been the success enjoyed by this number concept, that it is now hard to contemplate that time and space could be described in any other way. One gets a strong impression that time and space *are* continuous – with just the same sort of continuity that finds rigorous expression in the real number system (bearing in mind, of course, that space is three-dimensional rather than one-dimensional). But perhaps we are being blinded by appearances and by our long familiarity with the real numbers. It is this familiarity which makes *appear* natural the extrapolation to the infinitely small of a seeming continuity which is present on ordinary scales of time and space. But it could turn out that time and space do not, after all, have the kind of continuity on a very small scale that has been assigned to them. In my view – and in the views of some others[(1)] – such a radical change in our understanding may well be in the offing. This is a question to which I shall return later in this essay.

But if we accept the picture presented by the physics of today, the real number system provides an unending supply of elements which is called forth again and again for the quantification of physical concepts. For example, there is velocity, energy, momentum, frequency, mass, temperature, density, force, and so on. The use of real numbers for their description can, however, be traced back to the real-number continuity of space and time, so there is, in essence, nothing new to be gained from the observation that such numbers are being employed here also.

But mathematics does not stop here. Whereas subtraction, division and infinite summation have been made systematic concepts within the real numbers, the operation of solving equations has not. For example, the simple-looking equation

$$x^2 + 2 = 0$$

has no solution among the reals, whereas the apparently similar one

$$x^2 - 2 = 0$$

has the two real solutions $x = \sqrt{2}$ and $x = -\sqrt{2}$. However, for several centuries mathematicians have become used to the idea that the first equation, also, can be considered to have two solutions, written $x = \sqrt{-2}$ and $x = -\sqrt{-2}$. These are certainly not real numbers (since no negative number can have a real square root), but they are good numbers, nevertheless. Several abstractions have, in any case, had to be made in the passage from the natural numbers to the real numbers. The one further abstraction that is needed is that an extra number, denoted i, be adjoined to the system of reals, together with all those numbers formed from i by multiplying and adding real numbers to it. The number i is to be a square-root of -1:

$$i^2 = -1,$$

and the general element of the extended number system so produced, called the system of *complex numbers*, has the form

$$a + bi$$

where a and b are real numbers. In particular, the solutions to the first equation considered above are given when $a = 0$ and $b = \pm\sqrt{2}$.

At first, the notion of a complex number may seem a little bewildering. One has become accustomed to the idea that negative numbers do not "really" have square roots. It may seem, perhaps, that although these "complex numbers" are doubtless logically consistent, they serve merely as a formal device introduced for purposes of mathematical convenience. And mathematically convenient they certainly are! Not only do they enable square roots to be taken with impunity, but also cube roots, fifth roots or, indeed, any other complex power of any complex number (except perhaps zero) whatsoever. Furthermore, polynomial equations of any degree can now be solved in a completely systematic fashion. And new previously unexpected properties arise of extraordinary power and beauty. That the system of complex numbers occupies a regal place within the realm of abstract mathematical ideas, there can be little doubt. Moreover, they are not simply a "convenience"; they take on a life of their own. Yet one still may have the lingering feeling that they are not part of "reality", but merely creations of human thought.

But we must ask why it is that real numbers themselves give an impression of having this "reality" that the complex numbers seem not to have. Partly it is a question of familiarity. We are used to calculating with real numbers (or at least with finite decimal expansions) and come across them at an early stage in life. Complex numbers we encounter only much later, if at all. But more than this, it is the feeling that physical measurements, notably those of space and time, employ real numbers, whereas no such obvious physical realisation of the complex number system makes itself felt.

This, at least, seems to be the case with classical physics. With the coming of quantum mechanics the situation changes. For complex numbers play a crucial role in one of the most basic axioms of quantum theory – as will be explained shortly. But important as this fact is for the physics of the submicroscopic world, its significance begins to get lost on the macroscopic scale. Yet the role of the complex can still be discerned in the nature of phenomena at any scale

– particularly where relativity is involved. The geometry of our world is, indeed, more "complex" than it seems to be at first.

2. THE GEOMETRY OF THE COMPLEX

To try to see how this can be so, let us first indicate the standard way in which complex numbers are represented in terms of ordinary geometry. We envisage a Euclidean plane, with Cartesian axes x, y. The complex number $\zeta = x + iy$ (x, y real) is represented in this plane (called the Argand plane of ζ) by the point with coordinates (x,y). There are simple geometrical rules for the sums, products, differences and quotients of complex numbers.[2]

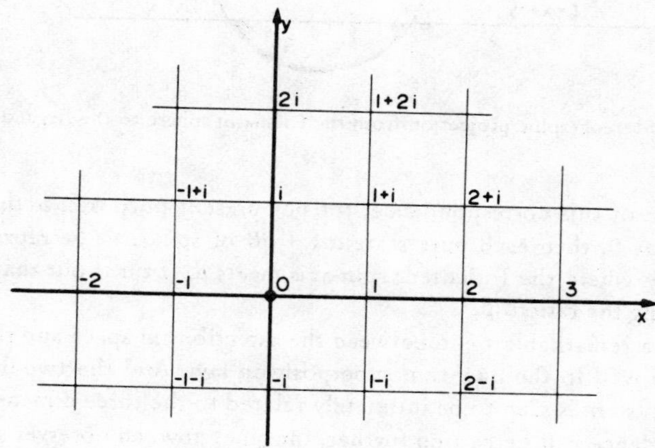

Fig. 1. The representation of complex numbers on the Argand plane.

So it seems that the world of the Flatlander could be described by the geometry of the complex number system. But what has this to do with *our* world, with its three dimensions of space and one of time? One answer is to be found in the basic rule of quantum mechanics known as *linear superposition*. If A and B represent quantum states, then for every pair of complex numbers α and β (not both zero) there exists a new state represented as $\alpha A + \beta B$, where for each distinct ratio $\alpha{:}\beta$ we get a physically distinct situation. Though rather formal and abstract, as stated, there is a link, apparently tenuous but in my view fundamental, between this rule and the geometry of space and time.

This may be seen at its most primitive level in the states of spin of an electron. Each such (pure) state of spin may be represented by a direction in space – where the electron is viewed as spinning in a right-handed sense about that direction. If A is the state where this direction is upwards and B where it is downwards, then any other spin direction can be represented as $\alpha A + \beta B$. There is therefore a one-to-one relation between the directions in space and the ratios $\alpha{:}\beta$ of pairs of complex numbers.

Now the only difference between such a ratio and just a *single* complex number is that with a ratio, *infinity* is included (i.e. when $\beta = 0$). The space of such ratios can be represented as the *Riemann sphere* S, which may be related to the Argand plane P as follows. Imagine a third axis (the z-axis) perpendicular to the plane P and through the centre 0. Take the sphere S to have unit radius and centre at 0. The north pole N of S is the place where the positive z-axis meets S. Consider a point R of P, representing the complex number ζ. The straight line RN (extended, if necessary) meets S in a unique point Q, other than N (stereographic projection). This point Q then represents the complex number ζ, on S, where ζ is now viewed as a ratio (say $\alpha:\beta$) and where N represents $\zeta = \infty$.

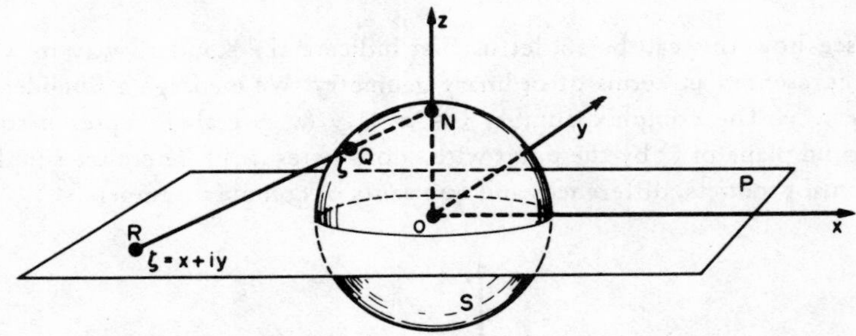

Fig. 2. Stereographic projection from the Riemann sphere to the Argand plane.

The significance of this correspondence, for our present purposes, is that if we imagine our electron situated at 0, then each pure state $\alpha A + \beta B$ of spin may be represented by a unique point of S, namely where the (oriented) spin-axis meets S. It turns out that this is precisely the point Q representing the ratio $\alpha:\beta$.

Thus, we have a remarkable tie-in between the directions in space and the fact that complex numbers are employed in the quantum superposition law. And the two-dimensionality of the complex number system is seen to be intimately related to the three-dimensionality of space.

This correspondence can be carried further. Imagine, now, an observer situated at the centre 0 of the sphere S. He looks momentarily out at the sky and represents each star that he sees by the point of S in that direction. In this way he can assign a complex number (ratio) ζ to each star as its coordinate. We now envisage a second observer in any state of motion whatever, who momentarily coincides with the first at the instant that the stars are observed. The second observer carries his own reference sphere S' and obtains a different complex number ζ' to label stars. The correspondence between ζ and ζ' is given by a remarkably simple formula:

$$\zeta' = \frac{\kappa\zeta + \lambda}{\mu\zeta + \nu},$$

where κ, λ, μ, ν are complex parameters defining the relative motions and orientations of the two observers. This formula takes into account the fact that the two observers may be rotated relative to one another, the relativistic transformation between fast-moving observers and the finiteness of the speed of light being also correctly incorporated.

The formula is especially remarkable in that it is what is called *holomorphic*. This means that it is expressed analytically entirely in terms of $\zeta = x + iy$, the so-called *complex conjugate* $\bar{\zeta} = x - iy$ being nowhere involved. This holomorphic property is important and arises only because the problem has been treated correctly according to relativity.[3] No such result arises within Newtonian theory. The full power and beauty of complex analysis really only arises when holomorphic properties are considered. What we are beginning to see here is the first step of a powerful correspondence between the space-time geometry of relativity and the holomorphic geometry of complex spaces.

3. TWISTORS

We have, in effect, been just studying an aspect of the geometry of the space of *photons*. Those photons which enter the eyes of our two observers are labelled holomorphically in terms of complex numbers, as a first indication of the fact that the *space of all possible photons* is really a *complex* space. That is, the different (classical) photon states can be described holomorphically by complex parameters. Photons are examples of massless particles. All such particles travel with the speed of light, but photons are distinguished from the others by the fact that they spin about their direction of motion with a magnitude $\hbar$. For the others the spin value is some different multiple of $\hbar$. Intuitively, the massless particle may be visualised in terms of a straight line drawn in space-time, namely its world-line. But this is not strictly accurate, since the particle is not quite localised when the spin is non-zero. More correctly, the state of a (classical) massless particle should be described by its energy and momentum, and its relativistic angular momentum about some origin. It turns out that all this information can be encoded as a certain complex quantity called a *twistor*[(4),(5)] which is (holomorphically) parameterised by four complex numbers.

These twistors may be taken as the primary elements in an approach to physics for which space-time points are superseded in their role. Instead of describing a particle in terms of a space-time locus, we describe it using twistors. We may think of a twistor as lying, conceptually, somewhere between a point and a particle. Points themselves are to be *constructed* from the twistors. And so also are the particles. In place of the wave function of a quantum mechanical particle, it turns out we require a holomorphic function of one or more twistors. The 1-twistor particles are all massless (photons, gravitons, . . .); the 2-twistor particles appear to be "leptons" (electrons, muons, . . .); and the 3-twistor particles, "hadrons" (protons, neutrons, pions, kayons, . . .) while possibly there are also 4-twistor particles, etc., forming further families (e.g. the newly discovered J/ψ particles). Moreover, the number of twistors used can apparently be increased when needed so that an n-twistor particle can also play a role as an (n+1)-twistor particle. The usual space-time descriptions are accordingly dispensed with altogether, although translations to space-time terms can be achieved when needed. The description of elementary particles which arises in this way must be viewed, so far, as rather speculative. Yet the mathematics seems compelling – and the classification scheme which arises appears to be compatible with what is known from standard elementary particle theory.

My contention is, then, that the twistor description of our world is more fundamental than space-time. On a classical macroscopic scale, and when general relativity is not involved, the two descriptions are *equivalent*. But for the micro-world of quantum particles (say at the level of 10^{-13} cm or less), my claim is that the twistor description will eventually give a *more* accurate picture. And the geometry of twistors is fundamentally complex, achieving an essential interweaving of quantum-mechanical principles into the ideas of geometry.

It should be emphasised, however, that twistor theory remains somewhat conjectural and incomplete at present. Although some surprisingly economical twistor expressions are already at hand for the description of much of physics, a great deal more needs to be done. The large-scale phenomena of general relativity have yet to be adequately incorporated, though there are some very promising indications. And a twistor theory of elementary particles or of quantum fields is only just beginning to emerge. Nevertheless, the increasing role of complex spaces, as opposed to real ones, seems clear. This much arises not only from twistor theory but from many other approaches to fundamental physics. But in the twistor approach the role of these complex spaces is all-embracing. If twistor theory does indeed provide a more satisfactory framework for the description of basic physics than does the normal space-time approach, then we shall truly be able to say that nature *is* complex – in the sense of the word intended here.

4. WHERE NEXT?

What of more complicated kinds of number than the complex? There are such mathematically defined objects as quaternions, octonions and others of various different kinds. But in each case it turns out that some law of arithmetic must be sacrificed. If we wish these laws to remain intact, then complex numbers are as far as we can go. Moreover, it appears that without retaining these laws of arithmetic, a suitable function theory cannot be obtained.

It may seem ironical that complex numbers, with their own brand of two-dimensional continuity, should have had first to be brought into physics in a basic way in order to give mathematical description to various observed physical phenomena of *discreteness*. As its name implies, quantum theory is a theory designed to handle the discrete, yet it is a theory based on the use of the complex *continuum*. However, this is not so much of a paradox as it might seem at first. There is a certain kinship – even a kind of duality, reflecting the wave-particle duality of quantum physics – between complex continuity and discreteness. There is a rigidity exhibited by holomorphic functions which is quite absent in the case of functions of a real variable. Already in standard quantum theory a discreteness emerges out of the seeming morass of complex continuity. With twistor theory, it appears that we may be able to say more. In particular, the fact that electric charge occurs in integral multiples of some fixed value finds explanation within the theory – as a consequence of the rigidity of holomorphic functions.

So perhaps we shall eventually come full circle. By generalising the concept of number as far as possible away from the natural numbers of which we have direct experience, while retaining the maximum of arithmetical properties, we are apparently led back to a discreteness which we had seemed to have left behind. Perhaps, after all, the physical laws are of a discrete basically simple combinatorial nature – while at our present level of knowledge, it seems to be the complex continuum which provides the clearest route to a deeper understanding of physics.

FURTHER READING

1. E. Schrödinger, *Science and Humanism*, Cambridge University Press, 1952.
2. E. T. Copson, *An Introduction to the Theory of Functions of a Complex Variable*, Clarendon Press, Oxford, 1935.
3. R. Penrose, "The apparent shape of a relativistically moving sphere". *Proc. Camb. Phil. Soc.* **55**, 137-9 (1959).

Concerning twistor theory:

4. R. Penrose, "Twistors and particles", in *Quantum Theory and the Structures of Time and Space*, edited by L. Castell, M. Drieschner and C. F. von Weizsäcker, Carl Hanser Verlag, München, 1975.
5. R. Penrose, "Twistor theory, its aims and achievements", in *Quantum Gravity*, edited by C. J. Isham, R. Penrose and D. W. Sciama, Clarendon Press, Oxford, 1975.

Hans J. Bremermann

Professor of Mathematics and Biophysics, University of California at Berkeley.

Programmed and operated John von Neumann's pioneering computer at Princeton in 1955. Early researches into the physical limitations of computer processes influenced theoretical cybernetics and emphasized the importance of the modern theory of computational complexity.

Recent research has been concerned to some extent with complexity theory, but is mostly concerned with understanding naturally occurring "computers" and efficient methods of solving cybernetic problems.

COMPLEXITY AND TRANSCOMPUTABILITY

In this essay I wish to point out serious limitations to the ability of computers to carry out enough computations to solve certain mathematical and logical problems. The same limitations also apply to data processing by nerve nets, and thus ultimately to human thought processes.

The present era has seen advances in computing that are no less spectacular than advances in space travel. Today a single computer can do more arithmetic operations in a year, than all of mankind has done from its beginnings till 1945 when the first electronic computer became operational. I will show in the following that this achievement is insignificant when measured against the vastness of what one might call the "mathematical universe".

Computers are physical devices and as such are subject to and limited by the laws of physics. For example, no signal can travel faster than light in vacuum (3×10^8 m/sec) and this restriction applies especially to signals inside a computer. Let τ_{switch} be the switching time of computer components. (In the fastest computers τ_{switch} is of the order of 10^{-9} to 10^{-8} sec (1 to 10 nanoseconds).) The travel time of signals between different parts of a computer is determined by the distance a signal has to travel. That means τ_{travel} = distance/velocity $\geqslant$ distance/light velocity.

The travel time of signals between different parts of a computer should not exceed the switching time. Otherwise travel time would be the limiting factor in the speed of the computer.

Hence $\tau_{switch} \geqslant \tau_{travel} \geqslant$ distance/light velocity. Hence, the distance between different parts of the computer is bounded by $\tau_{switch} \times 3 \times 10^8$ m/sec $\geqslant$ distance. Thus, if τ_{switch} is 10^{-9} sec, the distances in the computer are limited to 30 cm. In other words, the entire computer must be quite small. For $\tau_{switch} = 10^{-10}$ sec the maximum size would be 3 cm, etc.

Size and signal speeds are not the only limitations. More fundamental are the following: The different components of a computer communicate with each other by signals. The reception and interpretation of a signal constitutes a physical measurement. Physical measurements are governed by the uncertainty principle of quantum mechanics. One can show: the faster the measurement the larger is the energy that is required to make the signal readable with sufficiently small error probability. If the total signalling energy is limited, then there is a trade-off between the number of distinguishable signals that can be sent and the time required to identify them. It is customary to call the logarithm (base 2) of the number of distinguishable messages the *information content* (measured in bits (binary digits)). It turns out that for given energy the amount of information that can be sent is proportional to time. The proportionality factor is given by E/h [$\sec^{-1}$]. Where E is the energy, h is Planck's constant. *The amount of signal flow (in bits/sec) in a computer is thus limited by E/h, where E is the energy available for signalling.* (Bremermann, 1962, 1967, 1978; R. Thom, 1972, p. 143 (English translation Thom, 1975).)

This fundamental limit of data processing applies to computers, irrespective of the details of their construction. It can even be extended to computers other than digital machines (and thus becomes applicable to data processing by nerve nets). We will return to this question later.

Granted that computers are thus limited, what does it mean? Is the fundamental limit a serious barrier to computations of practical importance, or is it a subtlety without significant implications? To answer this question one must know something about the computational requirements of mathematical problems.

COMPLEXITY OF COMPUTATIONS

What exactly is the role of computation in mathematical problems? This question has been asked in earnest only in recent years. Very few results have been published prior to 1962, but since 1968 many papers have appeared that deal with this subject.

It will be useful to consider some specific examples before we attempt to examine the general question. Consider a system of two linear equations in two unknowns:

$$a_{11}x_1 + a_{12}x_2 = b_1,$$

$$a_{21}x_1 + a_{22}x_2 = b_2.$$

If $a_{11} \neq 0$ we multiply the first equation by a_{21}/a_{11} and subtract it from the second which gives

$$a_{11}x_1 + a_{12}x_2 = b_1,$$

$$(a_{22} - a_{12}a_{21}/a_{11})x_2 = b_2 - b_1 a_{21}/a_{11}.$$

Solving for x_2, substituting the result in the first equation and solving for x_1 requires altogether nine arithmetic operations.

If we have three linear equations in three unknowns we may proceed analogously. Multiply the first equation with a_{21}/a_{11} (provided $a_{11} \neq 0$) and substract it from the second. Multiply with a_{31}/a_{11} and subtract it from the third. This reduces the second and third equation to a system of two equations in two unknowns which we solve as before.

The analogous method can be applied to four, five, any number of linear equations. It is known as *Gaussian elimination* and it is an example of what mathematicians call an *algorithm*. An algorithm is a method that takes the data that come with the problem (in our case the coefficients a_{11}, a_{12} . . .) and transforms them step by step until the numbers are obtained that constitute the solution of the problem (in our example the values of $x_1, \ldots x_n$ if we have n linear equations). It can be shown that the number of arithmetic operations required to carry out the Gaussian elimination algorithm is $\frac{2}{3}n^3 + \frac{3}{2}n^2 - \frac{7}{6}n$.

We may call this number the *computational cost* of the algorithm. As the number of equations, n, increases, the cost goes up, and it goes up faster than n. According to our formula the dominant term is $\frac{2}{3}n^3$, thus the computational cost (as measured in terms of arithmetic operations) increases as the third power of the number of equations.

There are other methods to solve systems of linear equations. For example, Cramer's rule which computes $x_1, \ldots x_n$ by means of determinants, and in turn there are algorithms for computing determinants. The popular algorithms of computing determinants by developing them with respect to the elements of a row or column multiplied with smaller subdeterminants have a computational cost of the order of $n! = n(n-1)(n-2) \ldots 1$. The number $n!$ increases much faster than n^3.

When we implement an algorithm on a computer we must see to it that the computational cost of an algorithm does not exceed the number of arithmetic operations that a computer can carry out within a reasonable span of time. Before the advent of electronic computers, the computational capacity of existing machines and of hand computation by humans was quite limited.

Early electronic computers could perform 100 to 1000 arithmetic operations per second. Today's computers have reached performance rates of between 10 and 100 million arithmetic operations per second, or less than 10^{13} arithmetic operations per day. If we use Gaussian elimination, we thus can at most solve $\sqrt[3]{(3 \times 10^{13})} \approx 3 \times 10^4$ linear equations in a day. (The current monetary cost of a day of computer time may run as high as £10,000.) (In practice the maximum number of linear equations that can be solved is smaller than 3×10^4 because of round-off errors that are introduced, since numbers must be limited to a fixed number of digits.)

To increase the number of linear equations that can be solved we may explore the possibility of (a) algorithms that require fewer arithmetic operations and (b) computers of greater speed. These two questions are quite different; we will first discuss question (a).

As we have seen, there is an algorithm for solving linear equations that requires of the order of $n!$ arithmetic operations. This algorithm is worse than Gaussian elimination because $n!$ grows much faster than $\frac{2}{3}n^3$. In fact, by the mid-1960s Gaussian elimination had empirically proven itself as the best all-purpose algorithm for solving linear equations. Can this be proven rigorously?

This is not an easy task. For each n we must determine the minimum of the computational costs of all possible algorithms that solve systems of n linear equations. Since the computational cost of any algorithm is positive or at most zero, the computational costs of all algorithms are bounded below by zero. Therefore, for each n the minimum exists. It is some integer between 0 and $\frac{2}{3}n^3 + \frac{3}{2}n^2 - \frac{7}{6}n$.

In the mid-sixties some mathematicians conjectured that Gaussian elimination is indeed the best algorithm for solving linear equations and tried to prove that this is the case. However, in 1968 V. Strassen (then at Berkeley, now in Zürich) described an algorithm which for large n has a lower computational cost than Gaussian elimination. Its computational cost is less than $4.7\, n^{\log_2 7}$, and $\log_2 7 \approx 2.807$. For large n this number is less than $\frac{2}{3}n^3$. Strassen did not prove that his algorithm is the best of all possible algorithms. His result provides a better upper bound for the minimal cost. It can easily be shown that any algorithm requires at least n^2 arithmetic operations (in the worst case). Thus, we have the minimal computational cost bounded by n^2 below and by $4.7n^{2.807}$ above (Strassen, 1969).

So far we have considered a single mathematical task, namely solving systems of n linear equations in n unknowns. There are numerous other tasks of widespread interest in applications, such as finding the roots of a polynomial, solving systems of non-linear equations, optimizing a function of several variables, computing solutions of differential equations, etc. Operations research, which analyzes the efficiency of business operations and production processes, has its share of high cost computational problems. Some of these are known as "travelling salesman problem", "allocation problem", "shortest path problem", etc. It would take us too far to explain all of these problems in detail. The reader is referred to textbooks on operations research or to the paper of Karp (1972) which describes these and other problems and their interrelation in a concise way.

It may suffice to describe one of them, the travelling salesman problem, which is as follows: given n+1 cities $A, A_1, A_2, \ldots A_n$, any pair of cities has a distance between them. Suppose a salesman wants to visit each city exactly once, except A, the city from which he started and to which he wishes to return at the end of his trip. The problem is to find that routing of his trip which minimizes his total mileage; that is, the sum of the distances between the cities that he has visited. Each routing is given by a sequence: $A, A_{j_1}, \ldots A_{j_n}, A$ where $A_{j_i}, \ldots A_{j_n}$ is a permutation of the cities $A_1, \ldots A_n$, and A is the city where the trip starts and ends. There are $n!$ such permutations, and hence the problem can be solved by examining $n!$ mileage sums and picking the minimum. Picking the minimum of N numbers can be done with $N-1$ comparisons. Thus we could solve the problem at the cost of $n!$ comparisons between numbers (and an additional arithmetic cost of computing the mileage sums). This method, however, is not practical for large n. For example, for $n = 100$ we have (by a formula known as Stirling's formula):

$$n! = 100! > \sqrt{(200\pi)}\left(\frac{100}{e}\right)^{100}, \text{ where } e = 2.718.$$

This computational cost exceeds the capacity of any computing resource on earth.

Better algorithms than the one just described are known, but all known algorithms have a computational cost which increases faster than any finite power of n, where n is the number of cities. The problem of whether there exist algorithm whose computational cost is bounded by some polynomial in n, for all n, is unsolved. (If such an algorithm exists the problem is called *polynomial.* Whether the travelling salesman problem is polynomial is a famous unsolved problem, and lately Karp (1972) has shown that in this respect many of the problems of operations research are tied together. Either they are all polynomial or none of them is.)

The theory of the computational costs of mathematical numerical problems is known as *complexity theory*. Before 1959 it was virtually non-existent. In recent years it has made giant strides and it is developing into an entirely new branch of mathematics and theoretical computer science. For a sampling of recent results the reader is referred to Miller and Thatcher (1972). (Some further discussion of complexity problems is also contained in Bremermann (1974) and in the author's forthcoming lecture notes on biological algorithms (1978).)

For many practical mathematical, engineering and accounting tasks the computational costs of available algorithms and the capacities of available computers are satisfactory, but there are exceptions. We already mentioned operations research. Another area of difficulty is the numerical solution of large systems of differential equations, especially differential equations that are *stiff.* (That is, systems that combine processes of greatly differing speeds.) Partial differential equations pose a problem, as do the *ab initio* calculations of molecular configurations from Schrödinger's equation.

In another area, artificial intelligence, the excessive computational cost of known algorithms has been the main obstacle to having, for example, computers play perfect games of chess (or checkers, or Go). If at each move a player has k choices, then n moves comprise k^n possible move sequences. This number grows exponentially with n. For some games (like Nim) there are shortcuts that eliminate the need for searching through all the alternative move sequences. However, for chess (a game that is considered a true intellectual challenge and not mere child's play) such shortcuts have never been found. All known algorithms involve search through an exponentially growing number of alternatives and this number, when search is pursued to the end of the game, exceeds the power of any computing device.

A similar situation prevails in mathematical logic and in other branches of mathematics where there are formalized proof procedures. The search for the proof of a (conjectured) theorem always seems to involve search through an exponentially growing sequence of alternatives of formula transformations. Quite generally: most artificial intelligence problems require computing in amounts that grow exponentially with the depth of the search. The required search effort, in most cases, exceeds any available computing resource before it has reached sufficient depth to solve the problem (cf. Nilsson, 1971).

In summary: many mathematical, logical, and artificial intelligence problems cannot now be solved because the computational cost of known algorithms (and in some cases of all possible algorithms) exceeds the power of any existing computer.

THE FUNDAMENTAL LIMIT OF DATA PROCESSING

Granted that certain problems cannot be solved with existing computers, may we expect that eventually all problems will become solvable through advances in computer technology?

Computer performance has indeed increased dramatically since 1945, when ENIAC became operational. However, as I indicated in the introduction, computer performance cannot be improved indefinitely. The signal flow in a computer is limited by E/h [bits/sec], where E is the energy available for signalling. How serious is this limit?

It is easy to derive an ultimate upper bound. This bound is not meant to be realistic in the sense that it would seem practically possible to build computers that come close to this bound. Practical bounds would be much harder to derive. Thus our fundamental limit is merely a far out yardstick beyond which improvement cannot go. It is comparable to saying: astronauts cannot travel at speeds exceeding the light velocity, though in practice their speeds are much more limited. Estimates of realistic limits of the speed of space travel would be much harder to derive, having to take into consideration rocket technology, etc.

As Einstein first observed, there is an equivalence relation between mass and energy. Energy has mass, and mass, if converted, yields energy in the amount of $E = mc^2$, where c is the velocity of light in vacuum and m is the mass that is being converted. An atomic power plant is, in fact, a device for converting mass to energy.

Consider now a closed computing system, with its own power supply. Let m be the total mass of the system. This includes the mass equivalent of the energy of signals employed in the computer, while another share of the total mass is contained in the materials of which the computer and its power supply are made.

In existing computers the structural mass by far outweighs the mass equivalent of the signal energy. It would be difficult, however, to derive a realistic upper bound for this ratio. Thus, in order to avoid complicated arguments, we simply observe. *The total mass equivalent of the energy that is invested in signals cannot exceed m, where m is the total mass of the system.*

By combining this limit with the fundamental limit of the data processing we obtain:

No closed computer system, however constructed, can have an internal signal flow that exceeds mc^2/h bits per second. (Here m is the total mass of the system, c the velocity of light in vacuum and h is Planck's constant.)

This limit was derived by the author in 1961 (see Bledsoe, 1961; Bremermann, 1962). An improved argument was given by the author, Bremermann, 1967. A new discussion is to be contained in Bremermann, 1978.

The numerical value of c^2/h is 1.35×10^{47} (bits per second per gram). The number is large or small, depending upon the perspective. Existing computers process no more than $\approx 10^4$ to 10^5 bits per second per gram, and the fundamental limit appears much too large for practical purposes. When compared with the complexity of some algorithms, however, the limit appears small. The limited age and the limited size of the Universe constitute (far out) outer limits to computing. Again we choose these very unrealistic outer limits to avoid complicated arguments that could be made in order to establish more realistic and stringent bounds to the product of mass and time that could be considered as available for computing. Current estimates of the age of the physical universe run to about 20 billion years, that is 2×10^{10} years or 6.3×10^{17} sec. The total mass of the Universe is estimated as about 10^{55} gs. Thus we have 6.3×10^{72} gram seconds as an outer limit for the mass time product.

TRANSCOMPUTABLE ALGORITHMS

We call an algorithm *transcomputable* if its computational cost exceeds all bounds that govern the physical implementation of algorithms.

It can be shown that the exhaustive search algorithm for chess is transcomputable. The same is true for many algorithms of artificial intelligence and operations research. In fact, any algorithm whose computational cost grows exponentially with a size parameter n is transcomputational for all but the first few integers n.

This is a rather disturbing thought and many people have chosen to ignore it. (Analogously many people have for a long time chosen to ignore the fact that earthly resources of space, air, fossil energy and raw materials are limited.) One exception to this trend has been Ross Ashby, who more than any person in the world has emphasized the consequences of this limit in many of his writings between 1962 and his death in 1972 (cf. Ross Ashby, 1967, 1968, 1972).*

Another kind of limitation to computation is *thermodynamic*. R. Laundauer (1961) has pointed out that when in the course of computation information is discarded entropy is generated which must be dissipated as heat. How much information must be discarded when computations are carried out? Initially this question was not well understood. Recently Bennett (1973) has shown that any computation can be carried out essentially in a logically reversible way which implies that it can be done with little or no entropy generation (cf. also Landauer, 1976). In Landauer and Woo (1973) both thermodynamic and quantum limitations are discussed and an extensive bibliography is given. There are many open questions.

So far we have stated the fundamental limit for *signal flow* in a computer. Readers may wonder whether there would be an escape from the limit if we consider larger classes of computers – analogue computers, special-purpose circuitry (hardwave simulations), etc. This is not the case. In a forthcoming article I am trying to show that the limit applies to any physical implementation of any kind of algorithm. In essence, the fundamental limit is identical with the uncertainty principle of quantum mechanics.

In particular, the limit applies to nerve nets, and thus, ultimately, it imposes limits on human intelligence. This statement presupposes that the human brain is subject to the laws of physics and that it cannot solve logical and mathematical problems without implementing algorithms.

We may compare the phenomenon of transcomputability with limitations that apply to space travel. In order to reach a distant point in space the traveller has to perform a motion which requires time and energy. Since both are in limited supply the accessible portion of the Universe is limited. Analogously, in order to reach knowledge of mathematical theorems, optimal moves in a (mathematical) game, or in order to explore the trajectories of differential equations, etc., computations must be performed which require time and energy. Since both are in limited supply the accessible portion of the mathematical universe is limited.

The fundamental limit has epistemological consequences, for example the following: many systems (biological or physical) are composed of *parts*. The interactions between parts obey certain laws (e.g. gravitational interaction between mass points, electromagnetic, weak, strong interactions between elementary particles, chemical interactions between molecules, etc.). The *reductionist approach* tries to derive the total system behavior (the trajectory of the system in its state space) from the laws that govern the interactions of the component parts and from the initial state and inputs to the system. For complex systems knowledge of the systems

* *Note added in print*: Recently Knuth (1976) has written an article in *Science* that clearly explains the dual problems of complexity and "Ultimate Limitations" of computing (which he derives in a different way).

trajectories can be transcomputational with respect to sequential digital computation. In that case, if an analog of the system can be obtained, put in the proper initial state and if the state of the system can be observed, then the system trajectories are predictable, provided that the analog system runs faster than the original. If no such analog system is obtainable, then prediction becomes impossible, even if all the parts and the laws governing their interactions are known.

REFERENCES

Ashby, R. (1967) "The place of the brain in the natural world", *Currents in Modern Biology* 1, 95-104.

Ashby, R. (1968) "Some consequences of Bremermann's limit for information processing systems", in *Cybernetic Problems in Bionics* (Bionics Symposium, 1966), edited by H. L. Oestreicher and D. R. Moore, Gordon & Breach, New York.

Ashby, R. (1973) Editorial, *Behavioral Science* 18, 2-6.

Bennett, C. H. (1973) "Logical reversibility of computation", *IBM J. Res. Devel.* 17, 525-32.

Bledsoe, W. W. (1961) "A basic limitation of the speed of digital computers", *IRE Trans. Electr. Comp.* **EC-10**, 530.

Bremermann, H. J. (1962) "Part I: Limitations on data processing arising from quantum theory", in "Optimization through evolution and recombination" in *Self-organizing Systems*, edited by M. C. Yovits, G. T. Jacobi and G. D. Goldstein, Spartan Books, Washington, D.C.

Bremermann, H. J. (1967) "Quantum noise and information", *Proc. Fifth Berkeley Sympos. Math. Stat. a. Prob.*, Univ. Calif. Press, Berkeley, Cal.

Bremermann, H. J. (1974) "Complexity of automata, brains and behavior", in *Physics and Mathematics of the Nervous System*, edited by M. Conrad, W. Güttinger and M. Dal Cin, *Biomathematics Lecture Notes*, Vol. 4, Springer Verlag, Heidelberg.

Bremermann, H. J. (1978) "Evolution and Optimization", planned for publication in the *Biomathematics Lecture Notes* series, Springer Verlag, Heidelberg.

Karp, R. M. (1972) "Reducibility among combinatorial problems", in Miller and Thatcher (1972).

Landauer, R. (1961) "Irreversibility and heat generation in the computing process", *IBM J. Res. Devel.* 5, 183-91.

Landauer, R. (1976) "Fundamental limitations in the computational process", *Bericht Bunsengesellschaft Physikal. Chem.* **80** (to appear).

Landauer, R. and Woo, J. W. F. (1973) "Cooperative phenomena in data processing", in *Synergetics*, edited by H. Haken, B. G. Teubner, Stuttgart.

Miller, R. E. and Thatcher, J. W. (Eds.) (1972) *Complexity of Computer Computations*, Plenum Press, New York.

Nilsson, N. J. (1971) *Problem-solving Methods in Artificial Intelligence*, McGraw-Hill, New York.

Strassen, V. (1969) "Gaussian elimination is not optimal", *Numerische Math.* 13, 354-6.

Thom, R. (1972) *Morphogénèse et stabilité structurelle*, Benjamin, Reading, Mass.

Thom, R. (1975) *Morphogenesis and structural stability*, English translation of Thom (1972) by D. Fowler, Benjamin, Reading, Mass.

Knuth, D. E. (1976) "Mathematics and Computer Science: Coping with Finiteness", *Science* **194** , 1235-1242.

C. W. Kilmister

Professor of Mathematics at King's College, University of London.

Early interests in theoretical physics included general relativity and the work of Eddington. In recent years his interests have changed somewhat to two other fields, logic and the applications of mathematics in the social sciences. The essay in this volume represents his first considered statement in the latter field.

MATHEMATICS IN THE SOCIAL SCIENCES

1. THE PROBLEM OF THEORISING

I want to deal here with the following fundamental puzzle: mathematics has proved itself a valuable tool for carrying on sustained argument and for unifying points of view, both within its own domain and in many applications in the physical sciences. How is it that it has had no similar success in the social sciences? As a matter of fact a book appeared more than 30 years ago, which set out clearly the basic problems confronting anyone bold enough to apply mathematics in the social sciences. This was von Neumann and Morgenstern's *Theory of Games and Economic Behaviour*(1) (numbered references are to the combined list of references and notes on the text at the end of the chapter), and as I shall have cause to refer to this frequently, I denote it by vNM. Yet in the field of particular interest in vNM, economics, it is almost as if it had never been written, and the only substantial discussion of the problem of theorising – in sociology – largely misses the point of vNM. In 1943 vNM could state correctly: "Mathematics has actually been used in economic theory, perhaps even in an exaggerated manner. In any case its use has not been highly successful." The only real change today is a greater success in the provision of quantitative data, a change which obviously has far to go, so the discussion of reasons in vNM is still of interest. I can indeed make my statement of the puzzle above more precise: to what extent is the criticism of vNM on the mark, and why has it not had more effect?

von Neumann and Morgenstern dismiss the argument that mathematics is inappropriate for sciences involving the "human element" or where important factors arise which, though quantitative, defy measurement by pointing to the history of physics and chemistry and, in particular, the theory of heat. Here there was originally complete confusion between the concepts of heat and of temperature, so that theory was a prerequisite of correct measurement.

Not that von Neumann and Morgenstern would deny the inappropriateness of attempts to apply mathematics to economic problems unclearly formulated and with deficient empirical background. But the need then is to clarify the problems and improve the background. Instead, vNM utilises "only some commonplace experience concerning human behaviour which lends itself to mathematical treatment and which is of economic importance". This refers to the *theory of games* (see Section 4, p. 181), a rather simple model (to use modern jargon) of the market-place.

The value of this model is not in its empirical application, and this, I think, brings us to the important part of the argument. "Newton's creation of a rational discipline of mechanics" involved at a crucial stage the discovery of the calculus.(2) It was no straightforward application of a mathematical tool ready-made beforehand, but the development of mechanics and the calculus hand-in-hand. But, vNM argues, social phenomena are at least as complex as those of physics, so it is "to be expected – or feared – that mathematical discoveries of a stature comparable to that of calculus will be needed in order to progress in that field". The authors modestly discount the importance of the theory of games, but it does provide an example of the need for mathematical techniques not used in physics.

2. A MODEL APPLICATION

Since the call to arms in vNM has produced so little action, I shall begin by analysing it more closely. In later sections I will then review some more recent developments. To fix ideas I shall confine myself (except in Section 6) to the same field as vNM, i.e. economics. I begin the analysis by studying a very simple application which could have been made at the end of the last century. In this way I hope to concentrate on the way the mathematics is used, with all (economic) passion spent. Marshall(3) used supply and demand diagrams to determine prices in a (stable) free market. The demand diagram (D in Fig. 1) shows the amount D which would be bought for any price P measured along the bottom axis, so long as this amount is available. The supply curve, S, shows the amount S which would be brought to market when any given price is reigning. When the curves are superimposed, as in Fig. 1, they intersect at the equilibrium price P_0.

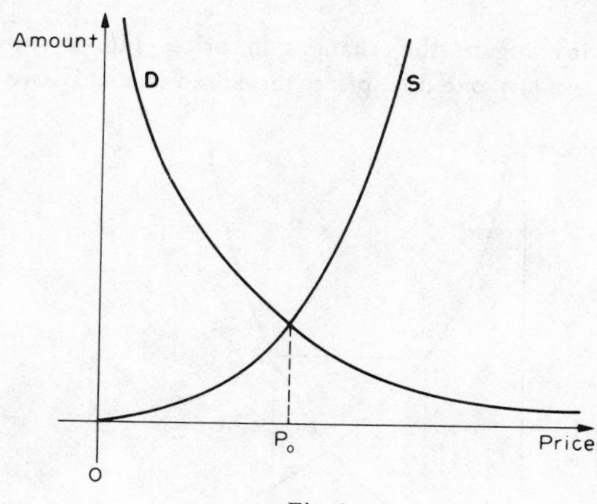

Fig. 1

Now suppose, as followers of Marshall, we try to import time into the theory so considering a non-stable system. Supply takes time to adjust to demand; let us make a simple model in which today's supply is determined by yesterday's prices, and in turn determines today's prices. And let us make the model precise by supposing that Marshall's curves are described by the equations

$$S = aP + bP^2, \quad D = c/P, \tag{1}$$

where a,b,c are some fixed positive numbers.(4) The equilibrium price is then determined by the equation

$$bP^3 + aP^2 - c = 0.^{(5)} \tag{2}$$

In the dynamic model, if P_n is the price on the nth day, (1) is replaced by

$$S = aP_{n-1} + bP_{n-1}^2, D = c/P_n,$$

so that, instead of (2) we have

$$\frac{c}{P_n} = aP_{n-1} + bP_{n-1}^2. \tag{3}$$

It is easier to choose new units of money so that prices are measured by x_n rather than P_n, where $P_n = x_n(c/b)^{1/3}$, which means that (3) reduces to

$$\frac{1}{x_n} = Kx_{n-1} + x_{n-1}^2, \tag{4}$$

where K stands for $(a/c)(c/b)^{2/3}$.
The equilibrium price x_0 now satisfies[6]

$$x^3 + Kx^2 - 1 = 0$$

but when we come to investigate the changes in price, Fig. 2 shows rather unwelcome conclusions. Suppose that, on any one day, prices measured in x-units are x and P on the supply

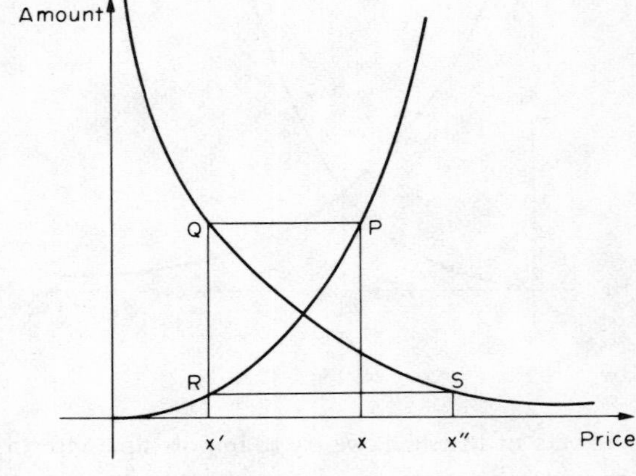

Fig. 2

curve represents the next day's supply (which is high, since prices have been taken as higher than equilibrium). Such a supply gluts the market and demand can only match supply at a price-level x' corresponding to Q on **D**. On the third day supply is therefore cut to R on the supply curve, producing demand T and price x'', greater than x, and so farther from the equilibrium instead, as one might hope, of approaching equilibrium. The market prices show increasingly severe fluctuations[7] as Table 1 shows (where the special case $K = 3.5$, $x_0 = 0.5$, $x = 1$ is chosen).

TABLE 1

Day	1	2	3	4	5	6	7	8	9	...
Price	1.0	0.22	1.21	0.18	1.55	0.13	2.16	0.08	3.40	...

The model is a bad one for a changing market.[8] We therefore want to know the extent to which its inadequacies can be remedied. Do they lie in the simple form (1) chosen for S and D, or are they inherent in any delayed supply-demand model? As it happens we can now say much more about the applicability of models than was possible 10 years ago and this will be dealt with in Section 7. But it will be useful first to look at possible changes. We could alter the supply curve; we might well argue that $a = 0$ (the stimulus to supply from unit change in price might be lacking altogether for very small P). More important, supply cannot continue to shoot upwards as prices increase, for limitations of "overheating" (labour shortages, raw material shortages) produce a cut-off. All in all, a supply curve like S in Fig. 3 seems more reasonable.(9)

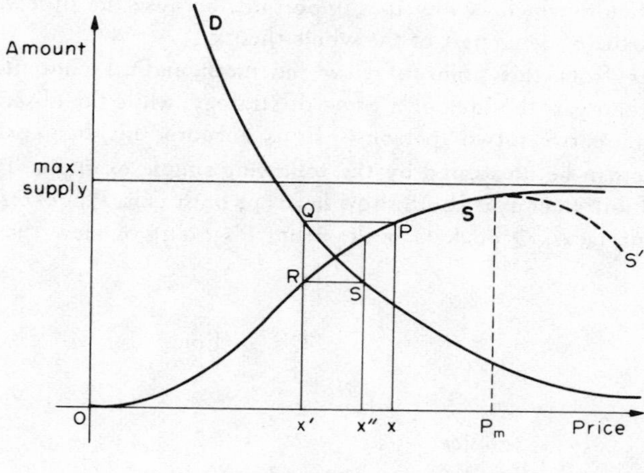

Fig. 3

And it is obvious that a suitable choice of constants can mean that the Marshall critical point gets near enough to P_m for Fig. 3 to show quite different behaviour from Fig. 2 in the dynamic case, to show, in fact, prices settling down towards the equilibrium (x'' in Fig. 3 is less than r!). Again it would be plausible to suppose that the demand curve did not fall to zero even for very high prices, but there was a constant residual "subsistence level" of demand.(10)

There is clearly no end to the variations that can be played; and each complication involves a great deal of work as soon as we get beyond the merely descriptive level. It is this fact that makes the critique of models sketched in Section 7 so important. But in any case there is a strong presumption that we are only playing at the serious task of describing reality and it is time to turn back to see what lessons are to be learnt from this analysis.

von Neumann and Morgenstern were critical, by implication, of the attempted application of calculus, the tool fashioned in the image of mechanics, to quite different phenomena. But it is surely not the calculus *per se* which is at issue; it was not used in the first model above, it proved useful in the later ones (in the proofs in the notes). Yet all models shared equally the strong air of ineffectiveness condemned in vNM. The distinction, and it is a crucial one for my argument, is between criticism of *method* and *ontology*. We do not wish to deny the usefulness of calculus as a tool for solving some particular mathematical problem, as it was used in note 10, for example, in discussing the existence or otherwise of equilibrium prices. But it is quite another matter to assume that the concepts used in calculus would be equally appropriate to incorporate into the basic formulation of an economic rather than a mechanical theory. This is what I mean by the ontological criticism.

3. THE PROBLEM OF THEORISING AGAIN

I would paraphrase an argument in physics of about 20 years ago[11] by saying that an economic theory is usefully thought of as consisting of thoughts, sentences written or spoken, observations of the economy or manipulations of it. That is, an economic theory is a union of all these things and not a pure calculating tool which exists in our minds and which we then apply to the economy. We can try to ensure that we do think in this way by substituting for the "thought" in the theory the observations of the economy which gave rise to it. This substitution is difficult, which is why it is important, because the observational thing derives meaning only by virtue of being part of the whole theory.

Let us consider from this point of view the mathematical content of vNM. Here the "thought" in the theory is the idea of a *game* of strategy, while the observational thing is that of a conflict situation between two "persons" (firms, corporations) in an exchange economy.[12] The idea of a game can be illustrated by the following simple example. Two players, P and Q, simultaneously exhibit a penny; if both show heads or both tails, P pockets both pennies, but if they show different faces, Q pockets both. From P's point of view the pay-off, in obvious notation, is

		Q's choice	
		H	T
P's	H	2	−2
choice	T	−2	2

and it is quite clear that (as the game has really no structure) it is best for each player to exhibit H or T at random. The game is fair, in the sense that, in a long run of plays, neither player will gain or lose. If, however, the rules are changed and in the new game P's pay-off is

		Q's choice	
		H	T
P's	H	2	−3
choice	T	−1	2

the new game has still an air of fairness about it. But suppose that P, instead of exhibiting H in ½ the occasions in a long run of plays, does so in p of the occasions (so $0 \leqslant p \leqslant 1$). The average pay-off to P, if Q shows H, will then be $2p - (1 - p) = 3p - 1$ (since P exhibits T in $(1 - p)$ of the occasions), whilst if Q exhibits T, the average pay-off to P will be $-3p + 2(1 - p) = 2 - 5p$. So long as $3p - 1 < 2 - 5p$ (i.e. $p < 3/8$) it will be best for Q (who wants to minimise the pay-off) to exhibit H; if $p > 3/8$ he should show T. This suggests a likely hint for P; by choosing $p = 3/8$, he puts Q in the position where he cannot gain from either H or T, and the pay-off is then $1/8$ to P. The full theory substantiates this surprising result.[13]

In the economy the situation is that of two persons in an exchange economy, for which the results for each depend not only on his own actions, but on those of the other. Such a problem of conflicting maxima and minima was nowhere dealt with in classical mathematics. If P,Q have more possibilities open to them (e.g. if each exhibits one face of a six-sided die) this means only that the analysis is more tedious to work out. But if more than two players are involved there are fundamental changes. Consider, for example, a particular three-person game, as devised in vNM with the intention of having no extraneous structure to confuse the argument. The three players, named 1, 2, 3, each choose the name of one of the two others. If a chooses b and b chooses a, then a and b together constitute a *couple* and they receive and share equally between them a stake (agreed before the game starts) paid by the remaining player. In such a game a couple may form, or it may not; but there is absolutely nothing more that can be said about it. So the whole structure of the game now lies outside the actual play and depends on which two players agree, by discussion beforehand, to form a coalition against the third.

If each player's agreed stake is the same the game (as far as its rules are concerned) is obviously fair, and if the stakes are different it can be made fair by giving each player a suitable (positive or negative) bonus payment for each play. But four-person games can have an intrinsic unfairness, which cannot be removed by such bonus payments.

We return to the lesson we can learn for vNM's argument in the next section.

4. THE REAL NUMBERS

We can now make better sense of the ontological aspect of vNM's argument. The mathematical objects in the theory are to have corresponding observational things. One might, for instance, make a detailed criticism (in line with the Marshallian analysis of Section 2) of the neo-classical notion of *utility*, which formed the basis of neo-classical price theory. Utility started as the property of commodities which cause them to be bought by a purchaser who seeks to enjoy their utility by consumption. In the neo-classical picture of the housewife maximising utility in filling her shopping basket it is clear that utilities are some sort of numbers. The concept lives on; it is fundamental to two-person games in vNM, and throughout they insist on a numerical estimate of advantage and disadvantage, i.e. a numerical utility for two-person games. But also vNM analyses other games, by coalitions, in terms of two-person games; though in the final sections the authors sketch a way of by-passing the two-person game so as to leave the n-person game in a position where numerical utilities are not required. (This sharp distinction between two-person games and others seems natural when we compare the actual methods of solution. For it is a remarkable fact that, although the two-person game exhibits new features, its solution can be reduced to a mathematical problem (linear programming) which – though admittedly not part of mathematics when vNM was written – is of a sufficiently classical flavour to turn the vNM argument applied to two-person games into a mere debating point. It is quite otherwise with other games, the theory of which is left in an incomplete form in vNM.)

I do not wish to argue against the concept of utility, only against its necessarily numerical character (a point raised before in many places), and I would want to extend this to argue against unthinking assumptions of numerical quantities in general. Such unthinking assumptions are not confined to economics, but mirror a development in mathematics itself, which is worth considering. So far we have freely used the real number field, that is, the set of numbers represented by all, terminating or non-terminating, decimals. So ubiquitous is this field that we tend to regard it as part of nature, instead of as an artefact. But, like so much else, it is part of the Greek contribution to our culture (though painstakingly reconstructed in the nineteenth century). When the Pythagoreans discovered the theorem now universally given their leader's name, it was probably by means of a similar triangle proof (Fig. 4). Triangles ABC, DBA, DAC are all similar, so that

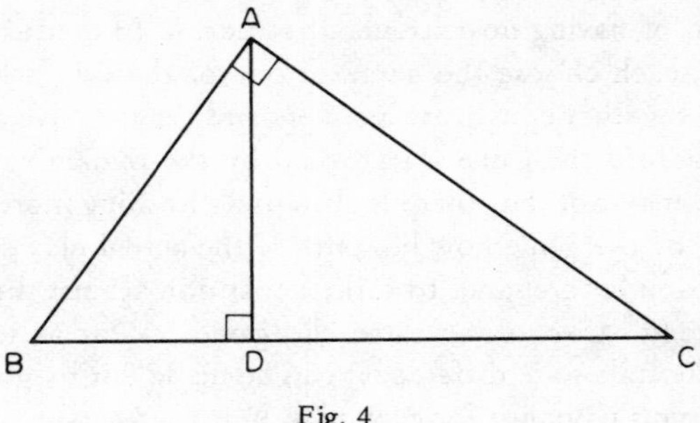

Fig. 4

$$\frac{BD}{AB} = \frac{AB}{BC}, \qquad \frac{DC}{AC} = \frac{AC}{BC}$$

and so

$$AB^2 + AC^2 = (BD + DC)\, BC = BC^2.$$

And their theory of similar triangles was, in turn, based on an argument in which two proportional sides were divided, one into p parts, on into q, so that the parts were equal. In short, the length of one side in terms of the other as unit, was p/q, a ratio of integers or *rational number*. Yet the Pythagorean theorem itself gives for the diagonal of the square of unit side $(\text{diagonal})^2 = 1 + 1 = 2$. There cannot be a fraction p/q whose square is 2, for (assuming first that p/q is in its lowest terms) this means $p^2 = 2q^2$, so p^2 is even. But the square of an odd number is odd, so p is even, $p = 2r$ (say). Hence $4r^2 = 2q^2$, $q^2 = 2r^2$ and so q is even and this is in contradiction to our assumption that p, q have no common factor.

The Greeks reconstructed their theory of magnitude – one version is in Euclid's *Elements* – and did much more. The original rational numbers formed a totally ordered field (for any two rationals r, s, exactly one of $r < s$, $r = s$, $r > s$, holds) and this ordering satisfies the axiom of Archimedes (for *any* r, s, we can take a suitable number N of copies of r so that $Nr > s$). The measure of the Greeks' achievement was the construction of the real numbers as a new, Archimedean ordered field and, as we know now, the largest (in the sense that any Archimedean-ordered field has a copy inside the reals). The uniqueness of this determination of the reals has had a great psychological effect; but, in the last decade, non-standard number systems have radically altered the picture. A non-standard system can be thought of as the reals augmented and filled in by infinite numbers and their reciprocals, the infinitesimals. The ordering can no longer be Archimedean, for the characteristic of an infinitesimal is that no number of copies of it will make it finite, but it is a total ordering. Even in the physical (geometrical) world such quantities can appear. A familiar example is the comparison of the "size" of the "horn-shaped angles" between a family of touching circles (Fig. 5). In the sense of

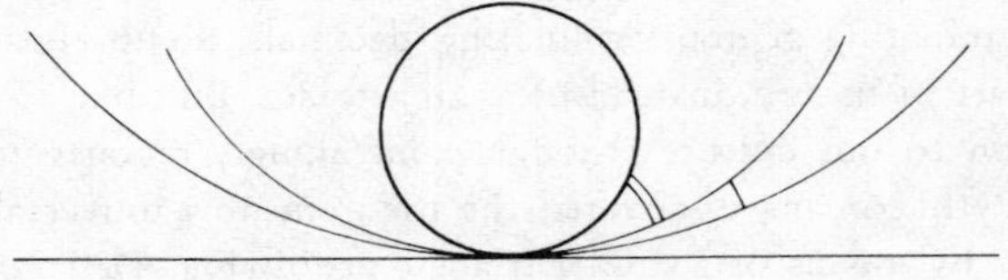

Fig. 5

the real numbers, all such angles are zero.

Now my aim here is not to promote non-standard analysis in economics, though it might just conceivably be useful in some restricted regions. (For example, the vNM argument for real numerical utilities rests on probabilities being real numbers. The axioms in vNM could be satisfied if both utilities and probabilities were non-standard; non-standard probabilities have much to commend them, for dealing with very unlikely situations.) Rather I aim to break the psychological barrier produced by the long hegemony of the real number field in mathematics, because my interpretation of the ontological aspect of the vNM *critique* is that new developments are likely to take place by use of mathematical structures different from the real numbers.

5. NEW DEVELOPMENTS – 1. AGGREGATION

It is one thing to sketch out a bold but vague scenario, as in the last sentence; it is another to carry it out. In these last three sections I wish to describe three developments which seem to be pointing in the same direction, from inside the subject, as that to which I, as armchair critic, have pointed from outside. The first of these can be seen (it was not so intended by its author) as an attempt to compromise with the ontological criticism by beginning with any amount of illicit furniture and then removing it by ingenuity and hard-work. The second (in Section 6) takes the ontological *critique* absolutely seriously and undertakes to work with only the mathematical furniture which is fully legal. Finally, in Section 7, is described a compromise which essays to show that it is possible to infringe the ontological requirements because there is a super-theory (as it were) which will pick out from the results those details which are essential and which would scarcely have been discovered without the extra props. So even without the ability to make the intellectual breakthrough of Newton and his successors, something can be done.

The first development, in economics, is in the aggregation problem, and is due to Pokropp.[14] One particular instance will serve to illustrate the aggregation problem. In the Marshallian analysis above we can ignore price, and express today's supply S directly in terms of yesterday's demand D, in suitable units, by

$$S = \frac{K}{D} + \frac{1}{D^2}. \tag{5}$$

But now imagine a two-sector economy, producing two commodities for whom supply and demand are independent, so that

$$S_1 = \frac{K_1}{D_1} + \frac{1}{D_1^2}, \quad S_2 = \frac{K_2}{D_2} + \frac{1}{D_2^2}. \tag{6}$$

The aggregation problem is simply: can one define "aggregate supply" S and "aggregate demand" D so that the macroeconomic variables S, D are related in *some* way as a result of the microeconomic relations. For example, $S_1 + S_2$ will not do as aggregate supply S, since it is obvious that S will depend on D_1, D_2 individually and not simply on some combination D of them.

Such a problem was posed by Klein[15] in 1946, and answered by Nataf 2 years later[16] within a complete calculus context.[17] The criticism of vNM was very apt, for the assumptions required exceeded any economic justification. Twenty years later, Gorman[18] weakened the assumptions to requiring only continuity, but Pokropp argues that even these are otiose, for they imply production processes with the possibility of indefinitely small variations. So Pokropp undertakes the same investigation without assuming even continuity. We need not consider his conclusions in detail; it is sufficient to point to the attempt to meet the ontological criticism *a posteriori.* The labour involved in even this amount of progress is a depressing indication of how much further progress can be expected.

6. NEW DEVELOPMENTS – 2. SIMPLICIAL COMPLEX

The second example, in which the ontological *critique* is taken absolutely seriously, is in the work of Atkin and his fellow-workers[19] in urban structure. To see how it is possible for Atkin to pursue such a rigorous path it is useful to look at some mathematical background. In this century mathematicians have learnt a valuable method of classifying surfaces, known as homology theory. The first step in this classification, whose final stage may picturesquely be described as characterising (for example) the difference between the outside surface of an (American) doughnut and that of an orange,[20] is to draw over the surface a network of triangles, somewhat like the geographers' triangulation but with this difference: the geographers are concerned with local accuracy, the triangles in homology theory need only capture the large-scale behaviour of the surface. (Similar considerations apply to more complex mathematical entities than surfaces.) The continuous surface is replaced by a finite set of triangles, some of which have edges in common, and the characterisation argument then takes place on this finite set.

Atkin reverses this motivation; if we can begin with the finite set, he argues, we need never worry about the continuous surface. Our discussion will be independent of whether it exists or not. The general finite set is a "simplicial complex" and Atkin relates this to social structure by regarding an urban community as "a collection of mathematical relations which naturally exist between physical cells and human activities". For example, he considers the set of retail trades and the set of buildings in the town centre, with the relation which identifies those trades to be found in each building.[21]

Now it is necessary to be careful not to allow the concealed assumption of "underlying surface" to be made; so, Atkin says, "an important aspect of this (and of previous) study is the acknowledgement of what constitutes 'data'. In this we are uncompromising. Data can only be the result of set-membership questions; data is yes/no observations." So his approach is to avoid "isolating" some particular problem – thought *a priori* to be able to be considered alone – because this pre-empts the question. "We do not know what depends on what until we know what depends on what." But this approach suffers certain limitations; while it is very useful in gathering hard data in the particular field of social structures, its use in economics is not at all clear. And its root and branch rejection of theorising in the usual sense of a bold hypothesis, its consequences and an attempt at their refutation implies an austerity which may limit our speculation more than it deserves.

7. NEW DEVELOPMENTS – 3. CATASTROPHES

The third straw in the wind, which takes a compromise path about the ontological criticism, arises in mathematics itself in René Thom's concept of *catastrophe*. It was foreshadowed earlier in our discussion of the Marshallian model in Section 2. Do the defects of this model arise from the specific equations chosen, or are they intrinsic? This can be answered if we know a little more about the general theory of models, and this is what catastrophe theory provides.

To illustrate the mathematics first, consider as a simple example, a quadratic equation

$$x^2 + 2bx + c = 0 ,$$

which has solutions $x = -b \pm \sqrt{(b^2 - c)}$, at least so long as $b^2 \geqslant c$ (otherwise the square root is not a real number). If $b^2 = c$ there is only one solution. Let us plot all this in a b, c plane (Fig. 6). Every point (i.e. every pair of numbers b, c) represents an equation. The points lying on the

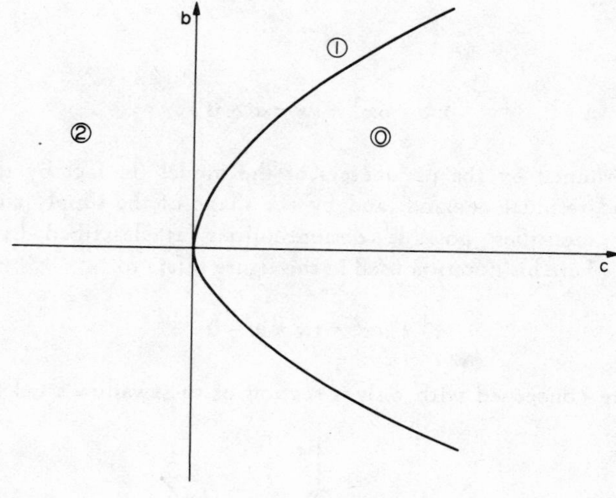

Fig. 6

parabola $b^2 = c$ are those corresponding to equations with one solution, those outside it (that is, to the left) give equations with two solutions and those to the right have no solutions. So much is elementary; Thom's insight here is topological: any point P not on the parabola is a "regular point" with the property that all nearby points (if we restrict "nearby" sufficiently) represent equations of the same general characteristics as P's. But the parabola is the set of catastrophe points, points such that any immediate neighbourhood gives equations of quite different character. Thom's great contribution is to show that, under some very general conditions, all catastrophe sets can be classified, they are all of a finite number of types, and simple examples of each are available.

We can get some idea of Thom's results by looking in more detail at the final Marshall example in Section 2 with both turn-over price P_m and residual demand, an example before which we quailed previously.[22] By suitable choice of variables, the equilibrium price can be shown to be determined by the solutions of the equation

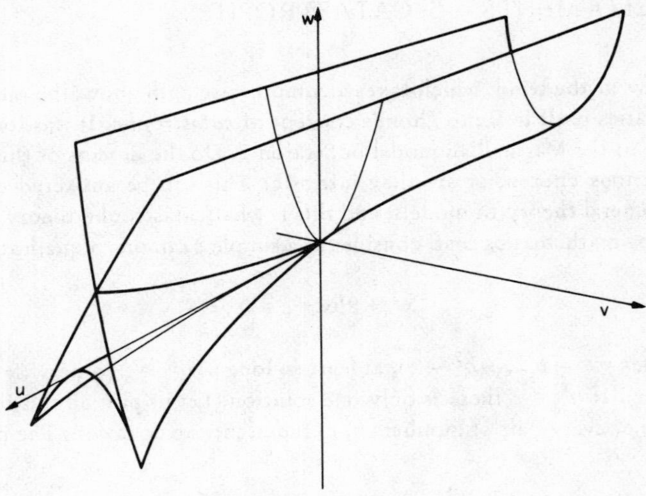

Fig. 7

$$x^4 - 6x^2 + vx + w = 0 ,$$

where v, w are determined by the parameters of the model (in fact by the relative sizes of response demand and residual demand, and by the shape of the supply curve). Reference to Thom's work now identifies possible discontinuities as described by the swallow-tail catastrophe (Fig. 7), where his notation used in the figure refers to

$$x^4 + ux^2 + vx + w = 0 .$$

Since $u = -6$ we are concerned with only a section of the swallow's tail (Fig. 8). Again the

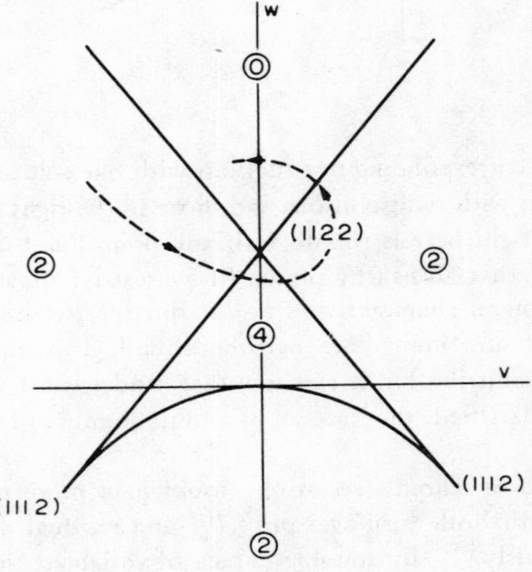

Fig. 8

curve divides the plane up according to the number of solutions (0, 2, 4) of the equation, with repeated solutions at the corners as described (1122 mean two pairs of equal solutions).

There are at least two ways of making use of this analysis. Different numerical models of the same general type correspond to different values of v, w, and so they can be compared in terms of general behaviour. The most dramatic feature is then the top wedge-shaped portion in which no equilibrium prices exist at all. This is much more serious than the former lack of stability, and arises if w is large (actually $w > 9$) and v is small. An approximate analysis shows that the restriction to the wedge corresponds to a lower limit on the response demand compared with the residual demand.(23) But instead of comparing different models we can interpret the analysis as a description of the effect of long-term trends in the market, which will be represented by slow variation of the parameters and so of v, w. Then if, as a result of outside interference (government intervention, taxation, changes in business confidence and the like) the market moves on a path like the dotted curve, unusual effects may arise whenever any edge of the curvilinear triangle is crossed. In particular most uncomfortable market conditions will arise on the later part of the path where it is impossible to fix prices which will match supply and demand.

This is only the more practical aspect of Thom's work. The more important theoretical aspect is the proof that all models of a general class involving three parameters can exhibit only one of seven elementary catastrophes, of which the swallow-tail is one. So our model is simply a particular case of all swallow-tail models; the mathematical complexity of some may be extreme but no qualitatively new phenomena will arise from these complexities. In this way, we establish an uneasy compromise with the ontological criticism; for if it requires us to alter our theories, we can conclude that the alteration will not affect the general qualitative phenomena.

Of the three indications of progress surveyed here, this last one seems to me the most promising. Much remains to be done in tackling the unanswered questions notably that of whether Thom's "general class" of models really includes all those which we want to consider in the social sciences. But it is an approach that gives promise of development, which will certainly attract many in the next decade. In this way the puzzle posed at the beginning will be answered; the reason for lack of success to date will be seen to be the lack of mathematical tools specifically designed for the purpose, and the lack of effect of vNM will be seen as caused by their failure to distinguish method from ontology.

REFERENCES AND NOTES

1. J. von Neumann and O. Morgenstern, *The Theory of Games and Economic Behaviour*, Princeton, 1944.
2. Over both mechanics and the calculus, vNM may be slightly over-generous to Newton. This does not affect the argument at all.
3. Alfred Marshall, 1842-1924, the great English economist and teacher of Keynes.
4. The coefficients a, b, c have ready economic interpretations. Since $c = PD$ it represents demand measured in money terms; so this money-measured demand is assumed constant, which might perhaps arise by a constant money-supply and constant velocity of circulation. Next, a gives the slope of the S, P curve for $P = 0$, a measure of the response of producers to price changes for low P, i.e. marginal supply-price ratio at low prices. (This might well be thought zero, or nearly so, in a plausible model.) Lastly, $2b$ is the rate of change of the marginal supply-price ratio with price, a measure of the "pick-up" in the market as prices rise.
5. It is easily calculated that this is a system in which the economist would recognise a price elasticity of supply e, given by

$$e = 1 + [1 + (a/bP)]^{-1}$$

and a price elasticity of demand of unity.

6. For different K we have different values of e in note 5; a table of solutions of cubic equations enables one easily to calculate that the elasticities at equilibrium are

K	e
0	2
0.16	1.87
1.59	1.44
7.94	1.29
15.87	1.28
Very large	1.28

7. This is the economists' "cobweb" result; the general conclusion being well known to be that, for stability, the slope of the demand curve at x_0 must be greater than that of the supply. Putting $D = dc^{2/3} b^{1/3}$, $S = sc^{2/3} b^{1/3}$, we have

$$d = 1/x, \quad s = Kx + x^2 ,$$

and the cobweb condition becomes $1/x_0^2 > K + 2x_0$, i.e. $1 > Kx_0^2 + 2x_0^3$. Since, however, $1 = Kx_0^2 + x_0^3$, this condition can never be satisfied.

8. More positively, this model shows the impossibility of one price-determining mechanism; such a use of models is commonplace now in economics. A striking example is P. Sraffa, *The Production of Commodities by Means of Commodities*, Cambridge, 1960, where an elegant model of production as a circular process does not pretend a close fit with the real economy, but is used to show that even with favourable assumptions like a uniform rate of profit it is impossible that rate of profit should be determined by marginal productivity of capital.

9. A simple formula for the curved S is not available, but for the dotted curve S' we may surmise

$$S' = bP^2 - dP^3 ,$$

where d is a suitable positive constant, and S, S' will agree so long as the prices involved are significantly less than P_m, where P_m gives $dS'/dP = 0$, i.e. $P_m = 2b/3d$. (This defines d as $2b/3P_m$ where P_m is "turn-over price".) The equation to determine equilibrium prices is now

$$f(P) \equiv dP^4 - bP^3 + c = 0$$

and $f(P)$ has a single minimum value at $P = 3b/4d$. So there are two solutions so long as $f(3b/4d) < 0$, i.e. $cd^3 < 81b^4/256$. But one such solution must be less than P_m, so $f(P_m) < 0$, which strengthens the condition to $cd^3 < 8b^4/81$. Now the cobweb result requires (it is easily seen) $dP^4 > c$ at the equilibrium value, so $f((c/d)^{1/4}) > 0$, which implies that $cd^3 > b^4/16$. In all, then, the new model has stability of prices so long as the one numerical parameter $A = cd^3/b^4$ satisfies $0.0625 < A < 0.0988$.

10. If e is the residual demand, we then have

$$D = \frac{c}{P} + e, \quad S = bP^2 - dP^3$$

so that the equilibrium is determined by

$$f(P) \equiv dP^4 - bP^3 + eP + c = 0 .$$

To what extent will e make any significant difference? Certainly the one numerical parameter A will no longer suffice to determine behaviour. Another arises as follows:

$$f'(P) = 4dP^3 - 3bP^2 + e, f''(P) = 12dP^2 - 6bP$$

and so f' has turning values at $P = 0, b/2d$ and so f' has maximum value e and minimum $b^3/d^2\ (B - 1/4)$ where $B = ed^2/b^3$. Hence f' has two positive zeros if and only if $B < 1/4$ and this in turn may be used to investigate whether f has a unique positive zero less than P_m. Are A and B sufficient? (They should be, of course, since we can always remove the coefficient b by adding a suitable constant to the roots, and dividing by d then gives an equation with only two parameters.)

11. E. W. Bastin and C. W. Kilmister, "The concept of order, I. The space-time structure". *Proc. Camb. Phil. Soc.* **50**, 178 (1954).

12. The idea of a mathematical theory of games of strategy seems to be due to E. Borel in 1923, but von Neumann proved the fundamental theorem (the existence of an optimal mixed strategy for any

two-person zero-sum game) in 1928.

13. In fact, the average pay-off will be

$$2pq + 2(1-p)(1-q) - 3p(1-q) - q(1-p)$$

$$= 8(p - \tfrac{3}{8})(q - \tfrac{5}{8}) + \tfrac{1}{8}.$$

The optimal strategies for P, Q are $(H, T) = (\tfrac{3}{8}, \tfrac{5}{8})$ or $(\tfrac{5}{8}, \tfrac{3}{8})$ respectively.

14. F. Pokropp, *Aggregation von Produktionsfunktionen*, Springer, Berlin, 1972.
15. L. R. Klein, *Econometrica* **14**, 303 (1946).
16. A. Nataf, *Econometrica* **16**, 232 (1948).
17. More explicitly, all the functions assumed in the investigation were differentiable, and had positive partial derivatives.
18. W. M. Gorman, *Rev. Econ. Studies* **35**, 367 (1968).
19. Very full details are given in the series of research reports to the Social Science Research Council on the *Urban Structure Research Project* (University of Essex, 1972, 1973 (two reports), 1974) and the continuation *Methodology of Q-analysis* in 1975. But see also: R. H. Atkin, *Mathematical Structure in Human Affairs*, Heinemann, 1974; R. H. Atkin, *Environment and Planning* B, **I**, 51 (1974) and **II**, 173 (1974).
20. D. B. Scott (a verbal communication).
21. The abstract definition of a simplicial complex, in terms of two finite sets X, Y and a relation λ between them, is that the complex $K_Y(X, \lambda)$ is (i) a collection of simplices σ_p, where (ii) each σ_p is defined by being a subset of $(p + 1)$ elements of X for which there exists a y in the relation λ to all of them and (iii) the σ_0's being identified with the members of X.
22. The equation for equilibrium prices in note 10 above can be rewritten, if $P - b/4d = y/4d$, as

$$y^4 + py^2 + qy + r = 0,$$

where

$$p = -6b^2, q = 8(8ed^2 - b^3),$$

$$r = 64bed^2 + 256cd^3 - 3b^4.$$

Putting $y = bx$ and using the parameters A, B of notes 9, 10 we get

$$x^4 - 6x^2 + vx + w = 0,$$

where $v = 8(8B - 1)$, $w = 64B + 256A - 3$.

23. The corners of the triangle are easily seen to be at $(0, 9)$ corresponding to $x^4 - 6x^2 + 9 \equiv (x^2 - 3)^2$, and at $(\pm 8, -3)$ corresponding to $(x + 3)(x - 1)^3$ or $(x - 3)(x + 1)^3$, respectively. The *fairly* straight sides of the triangle can therefore be approximated to by straight lines $w = \pm \tfrac{3}{2}v + 9$. The wedge-shaped region is then approximately

$$w > \tfrac{3}{2}v + 9, \quad w > -\tfrac{3}{2}v + 9.$$

It is simplest to deal with the two cases separately. If $B > \tfrac{1}{8}$, so that $v > 0$, the restriction is $w > 9 + 3v/2$, i.e. $A > \tfrac{1}{8}B$. If $B < \tfrac{1}{8}$, so that $v < 0$, then $w > 9 - 3v/2$, $A > \tfrac{3}{32} - \tfrac{5}{8}B$. If we take P_m as unit of money, so that $P_m = 1$ and $b/d = \tfrac{3}{2}$, then

$$A = (\tfrac{2}{3})^3 (c/b), \quad B = (\tfrac{2}{3})^2 (e/b)$$

and so A, B are measures of response demand and residual demand in terms of the "pick-up" in supply b.

Heini Halberstam

Professor of Pure Mathematics, University of Nottingham.

Member of the Council of the London Mathematical Society and Vice-President 1962-3, 1974-7. Vice-President of the Council of the Institute of Mathematics and its Applications 1972-4, Member of the U.K. National Committee for Mathematics.

SOME UNSOLVED PROBLEMS IN HIGHER ARITHMETIC

> *. . . a mathematical problem should be difficult in order to entice us, yet not completely inaccessible, lest it mock our efforts. It should be to us a guidepost on the tortuous paths to hidden truths, ultimately rewarding us by the pleasure in the successful solution.*
>
> D. Hilbert

1. Number theory, or the higher arithmetic, is one of the oldest branches of mathematics. The bland appearance of the sequence of integers is profoundly deceptive; it masks an inexhaustible multitude of beautiful patterns and tantalising questions which have fascinated scholars and amateurs alike since earliest times. Usually such problem situations emerge from experimentation, and are easy to describe; but to account for them is apt to be hard and often extremely difficult. It is common nowadays to assess the development of a branch of mathematics by the extent to which its primary matter has been incorporated in a general and comprehensive intellectual edifice; but while many early arithmetical investigations have indeed grown into imposing theories, the basic appeal of number theory is still, and always will be, that most questions do not fit readily into an existing theory. Knowledge and experience are important, of course; but it is the opportunity for *ad hoc* innovation, for the new idea or fresh insight, for novel connections, that draws inquiring minds to the mysteries of number. In this respect (as also in the importance attached to experiment) the higher arithmetic is close to the natural and biological sciences: we know a great deal, but the extent of our ignorance is limitless.

In view of what I have said, it is hardly surprising that there already exists a literature concerned with charting the unknown in the theory of numbers; and the interested reader should consult[(a)] the articles of Erdös[(1,2)] and the books of Shanks,[(3)] Sierpinski[(4)] and Mordell[(5)] for further reading. Indeed, within the space available I can do little more than present a selection from these works which reflects my own interests.

2. The prime numbers are the bricks from which all the positive integers (other than 1) are constructed by multiplication. Their supply is infinite, as was already known to Euclid; for if $p_1 = 2, p_2 = 3, \ldots, p_n$ are the first n primes, the integer $p_1 p_2 \ldots p_n+1$ must be divisible by a prime other than $p_1 p_2, \ldots,$ or p_n. It follows that p_{n+1} exists; moreover, it satisfies

$$p_n < p_{n+1} \leqslant p_1 \ldots p_n + 1. \tag{2.1}$$

Observe that 2+1 = 3, 2.3+1 = 7, 2.3.5+1 = 31, 2.3.5.7+1 = 211, 2.3.5.7.11+1 = 2311 are all primes! But this "pattern" breaks down eventually; indeed, it is not known even whether $p_1 \ldots p_n+1$ is prime infinitely often, nor whether it is infinitely often composite. As every school child knows, and Gauss was the first to prove, every integer $m > 1$ has a unique "canonical" representation[(b)]

$$m = p_1^{\alpha_1} \ldots p_r^{\alpha_r} \qquad (p_1 > \ldots > p_r) \tag{2.2}$$

as a product of primes, where $\alpha_1, \ldots, \alpha_r$ are positive integers. The proof of this basic result (the so-called Fundamental Theorem of Arithmetic) is elementary but delicate; it is significant that the proof depends on both the multiplicative *and* additive structure of the integers. An interesting way of stating the theorem is that if $p_1, \ldots, p_r$ are distinct primes then

$\log p_1, \ldots, \log p_r$ are *linearly independent* over the rationals in the following sense:

$$\text{if } \lambda_1 \log p_1 + \ldots + \lambda_r \log p_r = 0 \quad (\lambda_1, \ldots, \lambda_r \text{ rational})$$

then $\lambda_1 = \ldots = \lambda_r = 0$. It follows at once, for example, that log 3/log 2 is an irrational number. If, on the other hand, $\lambda_1, \ldots, \lambda_r$ are not all zero, then the *linear form*

$$\lambda_1 \log p_1 + \ldots + \lambda_r \log p_r$$

cannot be too small (in absolute value). To obtain good lower bounds for such forms (and indeed for more general linear forms involving algebraic numbers) is the subject of lively current research, spearheaded in recent years by Professor A. Baker of Cambridge, and has many important arithmetical applications.[c] Some of these will be mentioned later, but let me describe a particularly striking instance right away.

In 1894 Catalan enunciated the conjecture that the equation

$$x^m - y^n = 1$$

in integers $x > 1$, $y > 1$, $m > 1$, $n > 1$ has no solutions other than $m = y = 2$, $n = x = 3$. Last year R. Tijdeman succeeded in showing that, at any rate, this equation has at most a finite number of solutions, by applying refinements of Baker's famous method in a surprisingly simple way. A striking aspect of the method is that it gives explicit bounds for possible solutions, so that, in principle, all solutions may be identified by a computer search. Unfortunately, the bounds are so huge that no existing computer is sufficiently powerful. The corresponding question for the equation

$$ax^m - by^n = c \quad (a,b,c \text{ constant})$$

is still open, although Y. V. Čudnovsky in Russia claims to have dealt with the case $a = b = 1$.

The number $\tau(m)$ of all positive integer divisors of m is (cf.(2.2)) given by $\tau(m) = (\alpha_1+1) \ldots (\alpha_r+1)$. Clearly $\tau(m) = 1$ if and only if $m = 1$ and $\tau(m) = 2$ if and only if m is prime. As one might expect, writing down the values of $\tau(m)$ as m runs successively through the positive integers gives no impression of regularity; nevertheless, we meet here the statistical phenomenon that $\tau(m)$ does behave well on *average*; in fact, it behaves like $\log_e m$, in the sense that, as Dirichlet proved long ago,[d]

$$T(x) = \sum_{m \leqslant x} \tau(m) = x \log x + (2\gamma-1)x + O(x^{\delta}) \quad (x \to \infty)$$

with $\delta = \frac{1}{2}$. But this beautiful mean value formula is still the subject of research, for it is known to hold with values of $\delta < \frac{1}{2}$ and even with $\delta < \frac{1}{3}$; on the other hand, we know from a limitation principle of G. H. Hardy, that $\delta < \frac{1}{4}$ is impossible. The gap between $\frac{1}{3}$ and $\frac{1}{4}$ seems small but has proved formidably hard to bridge, despite the beguiling apparent simplicity of the problem; for $T(x)$ is just the number of points with integer coordinates in the shaded hyperbolic region (Fig. 1). Put in this way, we see that this problem is one of a whole class of similar questions, all characterised by the same difficulty. The no less fascinating companion problem involving a circular boundary is at about the same stage of development.

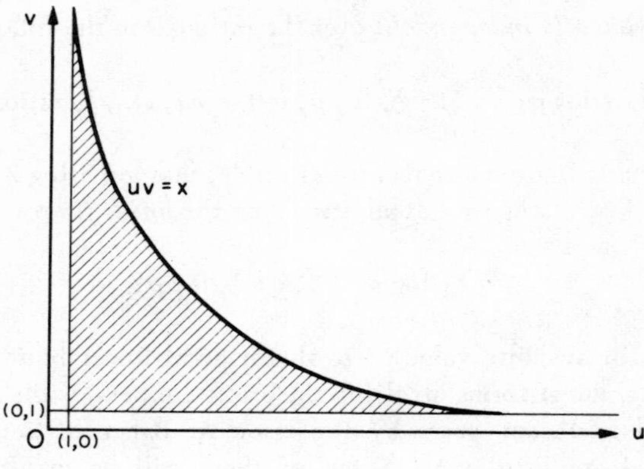

Fig. 1.

Before leaving $\tau(m)$, let us mention an old conjecture of Erdös, that there exist infinitely many values of m for which $\tau(m) = \tau(m+1)$. This would follow at once if we could prove that, for infinitely many primes p, $2p+1$ is a product of two distinct primes (for example, $2.7 + 1 = 15 = 3.5$), since then $\tau(2p) = 4 = \tau(2p+1)$; but this latter question is profoundly difficult (see paragraph 5) and will be discussed later in its proper context.

One of the oldest, and still one of the hardest, problems about primes also goes back to Euclid and derives from another divisibility question. Let $\sigma(m)$ denote the sum of all positive integer divisors of m. Following the Greeks, we say that m is *perfect* if $\tau(m) = 2m$, *deficient* if $\tau(m) < 2m$ and *abundant* if $\tau(m) \geqslant 2m$ – thus 6 is perfect ($1+2+3+6 = 2.6$), as is 28, any prime power (e.g. 8) is deficient, and any number of the form $2^{\alpha-1}(2^\alpha-1)|$ is abundant ($\alpha = 2,3, \ldots$). This classification of the integers was regarded as significant in numerology well into medieval times (one might encounter statements like "the Creation was perfect because God created the world in *six* days" or "the second creation of the world is descended from the *eight* souls in Noah's ark and is therefore inferior to the first", eight being a deficient number, etc.); it remains of interest because the underlying arithmetical questions are extraordinarily difficult. We know from Euler's completion of Euclid's result that an *even* m is perfect if m is of the form

$$2^{\alpha-1}(2^\alpha-1) \textit{ where } 2^\alpha-1 \textit{ is prime.}$$

Hence the set of perfect numbers is infinite if there exist infinitely many primes of the form $2^\alpha - 1$; we conjecture that the latter is true, but have no idea how to set about proving it. It is easy to see that[(e)] $M_\alpha = 2^\alpha - 1$ cannot be a prime unless α is one too; and in fact M_2, M_3, M_5 and M_7 are prime numbers. However, M_{11} is composite and it is known that the only other primes $p \leqslant 257$ for which M_p is prime are 13, 17, 19, 31, 61, 89, 107 and 127. (M_{4423} is a prime; until about 10 years ago it was the largest known prime number!) I mention these numerical facts to indicate how difficult it is to factorize large numbers, even ones having an apparently simple structure. It is plausible that the immediate neighbours of highly composite numbers – that is, numbers like M_p, or the Fermat numbers $2^{2^n} + 1$, or the Cullen numbers $n2^n + 1$, or $n! + 1$ or numbers of the form $2^{2^{\cdot^{\cdot^{\cdot^2}}}} + 1$ – have relatively few prime factors, and one might expect therefore to be able to show at least that, if $P(m)$ denotes the largest prime factor of m, then $P(M_\alpha)$ is large as a function of α. However, even this question is exceptionally difficult; while it is known that $P(M_\alpha) > \alpha$ for every positive integer α, proving that $P(M_\alpha) > 2\alpha$ (for $\alpha >$

12) is already rather complicated. We suspect that

$$\lim_{\alpha \to \infty} P(M_\alpha)/\alpha = \infty,$$

and as a pointer in this direction we have the recent result of Stewart that there exists an integer sequence $\alpha_1 < \alpha_2 < \ldots$ on which $P(M_{\alpha_n})/\alpha_n \to \infty$ as $n \to \infty$; moreover, for all sufficiently large primes p, there is the sharper lower estimate

$$P(M_p) > \frac{1}{2} p (\log p)^{1/4}.$$

Stewart has also proved recently (using lower estimates for linear forms in logarithms mentioned above in all these results) that, for all positive integers n,

$$P(2^{2^n}+1) > cn2^n,$$

with $c > 0$ some absolute constant. Erdös and Stewart have shown also, by elementary means this time, that, for all positive integers n,

$$P(n!+1) > n + (1-\epsilon_n) \log n/\log \log n$$

where $\epsilon_n \to 0$ as $n \to \infty$; and that, for infinitely many positive integers n,

$$P(p_1 \ldots p_n+1) > p_{n+k}, \quad \text{where } k > c \log n/\log \log n$$

with $c > 0$ some absolute constant.

Returning to perfect numbers, it is conjectured that there exist no *odd* numbers of this kind. Experimental evidence shows that even among abundant numbers odd integers are relatively rare (945 is the first odd abundant number), and it has been shown that if an odd perfect number exists it must be exceedingly large (for example, it must be greater than $e10^{36}$ and have at least 2700 distinct prime factors). It appears that this ancient question is still exceptionally intractable.

3. The best known divisibility theorem of elementary number theory is Fermat's congruence: For every integer a, $a^p - a$ is divisible by p; or, as we may write using the language of congruences,

$$a^p \equiv a \bmod p. \tag{3.1}$$

For a long time it was believed that the converse is true also – in other words, that if m is an integer such that

$$a^m \equiv a \bmod m \tag{3.2}$$

for every integer a, then m is prime. This turns out to be false, as we can check in the case of $m = 561 = 3.11.17$, and raises the question whether there are many, perhaps infinitely many, *composite* m for which (3.2) holds for all a. Such numbers m are called *Carmichael numbers* –

5.29.73, 7.31.73, 5.17.29.113.337.673.2689 are other examples – and we do not know the answer.

For a fixed integer $a > 1$, however, we can show without much difficulty that

$$m = \frac{a^{2p}-1}{a^2 \ -1} = \left(\frac{a^p-1}{a \ -1}\right)\left(\frac{a^p+1}{a \ +1}\right) \tag{3.3}$$

satisfies (3.2) for each odd prime p not dividing $a(a^2-1)$, so that there exist infinitely many composite numbers m satisfying (3.2). In the special case of $a = 2$ we refer to all such numbers m as *pseudo-primes*; from (3.3) we see that all integers of the form $\frac{1}{3}(4^p-1), p = 5,7,11, \ldots$, are examples of *odd* pseudo-primes, and the least of these (corresponding to $p = 5$), 341, is actually the first pseudo-prime because it can be checked that if $n \leqslant 340$ and 2^n-2 is divisible by n, then n is prime. Each of the numbers $m = \frac{1}{3}(4^p-1)$ $(p \geqslant 5)$ can itself generate an infinite class of odd pseudo-primes, for one can show that if m is an odd pseudo-prime then so is 2^m-1. It was not until 1950 that D. H. Lehmer discovered an *even* pseudo-prime, 161,038, but in the following year Beeger showed that, in fact, there exist infinitely many such numbers. Nevertheless, it is unlikely that all classes of pseudo-primes have been identified yet, and although we do know that they are relatively rare we have still to determine finer details about the way in which they are distributed.

Returning to (3.1) in the case $a = 2$, we now write it in the form

$$2^{p-1} \equiv 1 \bmod p \qquad \text{for every prime } p > 2,$$

and we ask: do there exist primes p for which

$$2^{p-1} \equiv 1 \bmod p^2 ? \tag{3.4}$$

Two examples are $p = 1093$ and $p = 3511$, and we know there are no others below 100,000. It may be that there are only finitely many solutions of (3.4), but we do not know. The question is of considerable interest in connection with one of the most famous unsolved problems in the whole of mathematics, Fermat's conjecture that if $n \geqslant 3$ the equation

$$x^n + y^n = z^n$$

has no solution in positive integers x,y,z. (In the case $n = 2$ there are infinitely many solutions – the so-called Pythagorean triples, of which 3,4,5 and 5,12,13 are well-known illustrations.) Although formidable tools have been developed to attack Fermat's conjecture[(f)] – much of modern algebra derives from such attempts – we still do not know even that there exist infinitely many n for which the conjecture is true. The case $n = 4$ can be disposed of in elementary fashion, by Fermat's own highly effective method of "infinite descent"; but the case $n = 3$, which also has been settled, requires in addition quite sophisticated algebraic ideas. In view of this, we may clearly suppose, in discussing Fermat's equation, that n is an odd prime $p \geqslant 5$. The so-called "first case" of Fermat's conjecture refers to the equation

$$x^p + y^p = z^p \quad (p \geqslant 3) \tag{3.5}$$

in which x,y,z are positive integers, and none of x,y,z is divisible by p. It was proved by

Wieferich and Miramanoff that if a solution of this kind exists, then (3.4) must hold and also

$$3^{p-1} \equiv 1 \bmod p^2 .$$

All experimental evidence suggests that if such primes exist they must be exceptionally rare. We might mention here also the criterion of Vandiver: if (3.5) has a solution in positive integers, then we may clearly suppose that such a solution triple x,y,z exists with highest common factor 1; and in that case

$$x^p \equiv x \bmod p^3, \quad y^p \equiv y \bmod p^3, \quad z^p \equiv z \bmod p^3 .$$

We end this discussion of Fermat's congruence with one more open question. It is easy to deduce from (3.1) that

$$1^{p-1} + 2^{p-1} + \ldots + (p-1)^p + 1 \equiv 0 \bmod p.$$

It has been conjectured, and checked for all integers up to 10^{1000}, that this divisibility characterizes primes only; but we have no idea how to set about a proof.

4. Consideration of even perfect numbers led us to ask whether there are infinitely many primes in the sequence $2^\alpha - 1$ ($\alpha = 2,3,\ldots$), and we saw that this is a very deep question. In fact, the same question is difficult in relation to any sequence of integers (not excluded by trivial considerations). Let us begin with the simplest of these, the sequence 2,3,4. . . of natural numbers > 1. If $\pi(x)$ denotes the number of primes not exceeding x, we do know, of course (see paragraph 2), that $\pi(x) \to \infty$ as $x \to \infty$, but many difficult questions about the distribution of primes present themselves.

For a start, one might ask how rapidly $\pi(x)$ tends to infinity with x. By systematic use of empirical evidence Gauss and Legendre were led at the end of the eighteenth century (independently of each other) to the conjecture that $\pi(x)$ behaves, for x large, like $x/\log x$; and a hundred years later (in 1896) Hadamard and de la Valleé Poussin (also independently of each other) confirmed this by proving that

$$\lim_{x \to \infty} \frac{\pi(x)}{x/\log x} = 1 . \tag{4.1}$$

Gauss (still in his teens!) conjectured even that

$$\text{li}\, x = \int_2^x \frac{dt}{\log t}$$

is a better approximation to $\pi(x)$ than $x/\log x$, and here too he was right; writing

$$E(x) = \pi(x) - \text{li}\, x,$$

we know that $E(x)$ is at most of the order of magnitude of

$$xe^{-c\sqrt{\log x}} \quad (c \text{ a positive constant}) \tag{4.2}$$

and with much hard work one can do even a little better. However, the true order of magnitude is still unknown; it cannot be much smaller than $\sqrt{x}$ – that we know from the pioneering work of Littlewood – and we expect this to be close to the truth. But the gap between (4.2) and $\sqrt{x}$ is enormous.

The asymptotic formula (4.1) is known as the Prime Number Theorem, and we may regard it as a global statement about the *density* of primes. It tells us little about the finer distribution of primes among the integers – for instance, how often are consecutive primes close to one another, and how far apart at worst can they be? These are questions of tantalising difficulty. We think that prime pairs such as 5 and 7, 11 and 13, 17 and 19, 29 and 31 occur frequently – about $x/(\log x)^2$ times between 1 and x (x large) – but so far we have been unable to prove even that these prime "twins" occur infinitely often. On the other hand, while there are certainly arbitrarily long runs of consecutive composite integers (just take the succession, $m! + 2, m! + 3, \ldots, m! + m$ for any integer $m \geqslant 2$) we believe that for every $m > 1$ there is a prime between m^2 and $m^2 + m$; but we cannot prove it. The best result (due to M. N. Huxley) known at present is that there is a prime between m^2 and $m^2 + m^\theta$ for any number $\theta > 7/6$ provided $m \geqslant m_0(\theta)$ (where $m_0(\theta)$ is a sufficiently large number depending on θ). We believe, on the basis of a statistical argument of Cramer, that if p_n, p_{n+1} are successive primes, there exists an infinite sequence of integers n on which

$$\lim_{n\to\infty} \frac{p_{n+1} - p_n}{(\log n)^2} = 1,$$

and that, given any $\epsilon > 0$,

$$\frac{p_{n+1} - p_n}{(\log n)^2} < 1 + \epsilon$$

for all $n \geqslant n_0(\epsilon)$.

What we know in the other direction (from the work of Rankin) is that for infinitely many n

$$p_{n+1} - p_n > c(\log n)(\ln_6 n)/(\ln_3 n)^2$$

where c is a positive number a little smaller than 1/3 (and $\ln_r n$ is here the r-fold iteration of log, so that $\ln_3 n = \log\log\log n$). On the basis of other work of Cramer, Erdös has conjectured that

$$\lim_{N\to\infty} \frac{1}{N(\log N)^2} \sum_{n=1}^{N} (p_{n+1} - p_n)^2$$

exists and is positive, but this seems hopelessly out of reach at the present time[g]. One could go on conjecturing along these lines indefinitely, but perhaps enough has been said to indicate how far we are from understanding fully the way in which the primes are distributed among the integers.

5. Let us, nevertheless, venture into still deeper water: how are the primes distributed relative to integer sequences other than the sequence of all the natural numbers; recall that we have already raised this question with respect to the sequence $2^n-1 (n = 1,2, \ldots)$? The next simplest case, and from the point of view of applications perhaps the most important, is an arithmetic progression, say

$$\ell, \ell + k, \quad \ell + 2k, \quad \ell + 3k, \ldots;$$

we require that HCF $(\ell,k) = 1$, since otherwise all terms except possibly the first are composite. We know from Dirichlet's seminal memoir of 1837 that such a progression contains infinitely many primes, and we know from later work of de la Valleé Poussin even that each of the progressions with common difference k contains roughly the correct proportion[h] $1/\phi(k)$ of all primes. More precisely, if $\pi(x;k,\ell)$ denotes for a given k the number of primes $p \leqslant x$, $p \equiv \ell \bmod k$, then the difference

$$E(x;k,\ell) = \pi(x;k,\ell) - \frac{1}{\phi(k)} \operatorname{li} x$$

is certainly no larger in order of magnitude than the expression (4.2) which we met in connection with the Prime Number Theorem, and again, the true order of magnitude is expected to be much smaller. However, for application we need to know about the distribution of primes, as measured by $E(x;k,\ell)$, relative to not just one fixed progression but to *all* progressions with common differences $k < x$, and here we run into serious difficulty. All we have at present is information of an *average* kind: for all k almost up to $\sqrt{x}$, $|E(x;k,\ell)|$ is on average no larger than $\sqrt{x}$ in order of magnitude (Bombieri's theorem); and for all k almost up to x, $|E(x;k,\ell)|$ is on average, but only in the "mean square" sense, like $\sqrt{(x/k)}$. It is fortunate that for some very important applications such average results, especially the theorem of Bombieri, have proved sufficient. There are all kinds of other questions one might raise in connection with $\pi(x;k,\ell)$ and $E(x;k,\ell)$. For example, given k, how large does x have to be to ensure that $\pi(x,k,\ell) > 0$; in other words, how large is the *least* prime $P(k,\ell)$ in the arithmetic progression $\ell \bmod k$? We believe that $P(k,\ell)$ cannot be much larger than $k \log k$ and we know this is true in a certain average sense. According to a famous result of Linnik, $P(k,\ell) < k^C$ for some positive constant C, and considerable effort has gone into determining admissible values of C. The present "record" is $C \leqslant 80$; this is due to M. Jutila, but recently S. Graham has shown that $C = 36$ is admissible. Or we may ask for comparisons between $\pi(x;k,\ell_1)$ and $\pi(x;k,\ell_2)$. Littlewood proved, contrary to impressive numerical evidence, that $\pi(x;4,3)$-$\pi(x;4,1)$ changes sign infinitely often, but we have, so far, no unconditional results of a similar kind for general k, despite the impressive studies of Turán and Knapowski in so-called "comparative prime number theory". Absolutely nothing is known about the fluctuations in $|E(x;k,\ell)|$ for a given k as ℓ ranges over values between 1 and k coprime with k, and even numerical evidence about this would be extremely hard to compile.

The linear polynomial $kn + \ell$ assumes prime values; and we may, of course, ask these same questions in relation to any integer valued polynomial (not excluded by trivial considerations). For example, is it true that $n^2 + 1$ is prime for infinitely many values of n? We believe this must happen quite frequently, but the best we can do at present is to show that, infinitely often, $n^2 + 1$ is a number having at most *three* prime factors. There are corresponding results and conjectures for polynomials of higher degree; but for these we refer the reader to Halberstam and Richert, *Sieve Methods* (Academic Press, 1975).

We may ask also whether $kp + \ell$ ($\mathrm{HCF}(k,\ell) = 1$, $2|k\ell$) is prime for infinitely many primes p; here $k = 1$, $\ell = 2$ gives the prime twins problem raised in the previous section, and $k = 2$, $\ell = 1$ the problem mentioned in Section 2. Again, we do not know. In 1974 Chen proved that $kp + \ell$ is infinitely often a number having at most *two* prime factors, so that, apparently, we are quite close. However, the gulf separating "two" and "one" is here gigantic.

A closely related question is the famous Goldbach conjecture, now at least 200 years old: is every even number > 4 the sum of two (odd) primes? From Chen's recent work we know that every sufficiently large even number can be written as $p_1 + p_2$ or as $p_1 + p_2p_3$ in many ways; but as has already been said, removing the second possibility presents formidable difficulties.

The reason that Goldbach's conjecture is so intractable is that it is a "binary" problem – it asks for representation of even n as the sum of *two* numbers of a certain kind. For odd numbers the corresponding problem asks: is every odd integer greater than 2 the sum of at most *three* primes? Here we know from the work of I. M. Vinogradov that this is true for all sufficiently large odd numbers.[i] I have used here, and several times before, this phrase "for all sufficiently large integers". The point is that many analytic methods begin to be effective only for large values, and of course it is the large values that, in principle, give the greatest difficulty because we cannot reach them by direct enumeration. Nevertheless, one does hope eventually to get rid of this limiting phrase; and in this particular instance of representation of numbers as sums of primes, readers may like to know that R. C. Vaughan has recently shown that *all* integers > 1 are sums of at most 27 primes! Thus the primes may be viewed also as an efficient means of building and positive integers > 1 *additively*.

6. There are other integer sequences from which all integers may be constructed by addition of a bounded number of terms. If $\mathcal{B}$ is an integer sequence and every positive integer > 1 is the sum of at most γ elements of $\mathcal{B}$, we say that $\mathcal{B}$ is a *basis of order* γ – of exact order g, if g is the least possible γ for which this is true. Thus the primes form a basis of order 27, and if the full Goldbach conjecture is true then the exact order of the set of primes is 3. In 1770 Lagrange proved, building on the work of Euler, that the natural squares constitute a basis of exact order 4; and in the same year Waring proposed also that, for each $k = 3,4,5, \ldots$, the sequence

$$\mathcal{B}_k : 1^k, 2^k, 3^k, \ldots$$

is a basis. Hilbert was the first to prove, in 1909, the so-called Waring's problem, but the value he obtained for the order $\gamma(k)$ of $\mathcal{B}_k$ was very large. In the years that followed Hardy and Littlewood (building on the work of Hardy and Ramanujan) developed their powerful analytic "circle" method to tackle the problem more efficiently, and later still I. M. Vinogradov sharpened this method in several important respects – his celebrated "three primes" theorem (see Section 5) was one of the triumphs of his improvements. The ramifications of the work on Waring's problem are extensive, and the book of Hardy and Wright – *The Theory of Numbers* (Clarendon, Oxford, 4th ed., 1960) – gives an excellent exposition in chapters XX and XXI. Here we shall just give the gist of the current state of play. Because of computational difficulties, the values of $g(k)$, the exact order of $\mathcal{B}_k$, are still largely unknown – the only precise result we have still is that $g(2) = 4$. The main thrust of endeavour has been towards determining $G(k)$, the least γ such that $\mathcal{B}_k$ is a basis of order γ for all sufficiently large integers. But here too, the only precise result we have, due to Davenport, is that $G(4) = 16$. We know that $G(3) \leqslant 7$ but probably $G(3) = 4$ or 5 so that there is still a large gap to close. Hardy and Littlewood conjectured that

$$G(k)\begin{cases}\leqslant 2k+1 & (k\neq 2^m \text{ with } m>1)\\ = 4k & (k=2^m \text{ with } m>1)\end{cases}$$

but so far this has been confirmed only for $k = 3$ and 4. The best we have, for large k, is Vinogradov's result that $G(k)$ is no larger than about $2k \log k$. To improve substantially on Vinogradov's estimate – to prove, for example, that there exists a positive constant C such that $G(k) \leqslant Ck$ – awaits some new ideas, which would need to be of such power as certainly to lead to progress in various other directions too. After half a century of intensive struggle, work in this fascinating area is all but at a standstill.

7. The questions we discussed about the occurrence of primes among the integers, in arithmetic progressions and polynomial sequences generally, may be formulated also for many number sequences other than the primes. An exhaustive account is impossible, but we single out two cases of special interest by way of illustration: (i) the sequence of s-free numbers, i.e. integers not divisible by the sth power of a prime (the so-called "square-free" integers correspond to $s = 2$); (ii) the sequence of numbers that are sums of two squares. The k-free numbers are in plentiful supply – a positive proportion of all integers are s-free (for example, about $6/\pi^2$ of all integers are square-free) – and we should therefore expect all the questions to be much easier here. They are not. We know that the least s-free number in an arithmetic progression mod k occurs before about $k^{1+(1/s)}$, but this falls well short of the expected truth, a constant multiple of k. We know that there is always a k-free number between n and $n+n^{1/(2k)}$ (if n is large enough), but again, this falls far short of what is likely to be true. We know that an integer-valued quadratic polynomial takes infinitely many square-free values; but we should expect the same to be true for any integer-valued polynomial (not excluded by obvious considerations), and we are not able to prove this yet. We know only (from a recent result of M. Nair) that if g is the degree of such a polynomial, then infinitely often it is $[\lambda g]$-free, where $\lambda = \sqrt{2}-\frac{1}{2}$, provided g is large enough (for example, it is $(g\text{-}2)$-free if $g \geqslant 18$). Turning to numbers that are sums of two squares, these are only a little more common than the primes (there are about $cx/(\log x)^{\frac{1}{2}}$ of them between 2 and x), and very little indeed is known about their finer distribution. For example, it is almost trivial to see that there is always such a number between n and $n+n^{\frac{1}{4}}$, yet literally nothing better is known; and the state of ignorance with respect to all the other kinds of questions is no less profound.

8. In paragraph 2 we mentioned inequalities of linear forms in logarithms of primes and several important applications of these. This deep and powerful method relates to a much more general situation and was conceived originally in connection with the problem of identifying classes of *transcendental numbers*, i.e. numbers that are *not* solutions of algebraic equations with integer coefficients.[j] We know that, in a certain precise sense, most numbers are transcendental, yet the problems of deciding whether a given number is transcendental is profoundly difficult.

For example, it is known that e and π are transcendental and that so is e^{π}; but we do not know how to show the same for $\pi + e$ or πe or π^e. Euler's constant γ, mentioned in paragraph 2, is given by

$$\gamma = \lim_{n\to\infty}\left(1+\frac{1}{2}+\ldots+\frac{1}{n}-\log n\right),$$

and arises in many mathematical contexts. There is an old story that G. H. Hardy declared himself prepared to resign his Chair in Cambridge in favour of anyone who proved that γ was irrational, let alone transcendental! We know that $2^{\sqrt{2}}$ is transcendental and, more generally, that α^β is transcendental if $\alpha(\neq 0,1)$ is algebraic and β is algebraic and irrational (proved by Gelford, Schneider independently of each other in 1934); and Baker has shown with his method that the same is true of $\alpha_1^{\beta_1} \ldots \alpha^{\beta_r}$ (with the α's, β's similarly constrained provided that there is no linear relation

$$n_0 + n_1\beta_1 + \ldots + n_r\beta_r = 0 \quad (n_i\text{'s integers not all } 0)$$

connecting the β's). However, if β is transcendental, classification of α^β is much harder; for example,

$$\beta = \frac{\log 3}{\log 2}$$

is transcendental yet $2^\beta(=3)$ is clearly an integer. In this direction we can say that if β is transcendental, at least one of

$$2^\beta, 3^\beta \text{ and } 5^\beta$$

is transcendental – so that, in particular, one of

$$3^{\log 3/\log 2}, \quad 5^{\log 3/\log 2} \tag{8.1}$$

is transcendental – and more general results along these lines can be formulated. Of course, we suspect that each of the numbers (8.1) is transcendental, but this we cannot prove at present.

For this fundamental area of work, the classification of numbers, where the achievements to date have required some of the most difficult and complicated methods in the whole of mathematics, we see clearly the abiding fascination of numbers; nowhere is their opacity more manifest – or more challenging.

Perhaps the reader will understand now what Kronecker had in mind when he compared those mathematicians who concern themselves with number theory to Lotus-eaters who "once having consumed this food can never give it up".

REFERENCES

1. P. Erdös, "Some unsolved problems", *Publ. Inst. Hung. Acad. Sci.* 6, 221-59 (1961).
2. P. Erdös, "Some recent advances and current problems in Number Theory", *Lectures on Modern Mathematics*, Vol. III, edited by T. L. Saaty, Wiley, 1965.
3. D. Shanks, *Solved and Unsolved Problems in Number Theory*, Spartan Books, Washington D.C. 1962.
4. W. Sierpinski, *A Selection of Problems in the Theory of Numbers*, Pergamon Press, 1964.
5. L. J. Mordell, *Diophantine Equations*, Academic Press, New York, 1969.

NOTES

(a) See also L. E. Dickson's *History of the Theory of Numbers* (3 vols.) and the six volumes of *Reviews in the Theory of Numbers* (ed. W. J. leVeque), published by the American Mathematical Society.

(b) p, with or without suffices, always denotes a prime; but $p_1, p_2, \ldots$ denote all the primes written in ascending order only where explicitly stated, as above in (2.1).

(c) See, for example, A. Baker, *Transcendental Number Theory*, Cambridge, 1975.

(d) γ denotes Euler's constant (see paragraph 8). $F(x) = O(x^\delta)$ means that $F(x) \leqslant Ax^\delta$ for some positive constant A.

(e) Primes of the form M_α are known as Mersenne primes.

(f) Perhaps it is of interest to report that quite recently K. Inkeri and A. J. van der Poorten, also M. Nair, have shown that, given a natural number k, all solutions of (3.5) in positive integers x, z and odd primes p, subject to H.C.F. $(x,k) = 1$ and $y - x = k$, are bounded by effectively computable constants depending only on k. The method of proof is, essentially, that used in the proof of Catalan's conjecture (see section 2, p. 193).

(g) A. Selberg proved in 1943, that, if the Riemann Hypothesis is true, then

$$\sum_{n=1}^{N} (p_{n+1} - p_n)^2 = O(N \log^2 N).$$

Recently D. R. Heath-Brown has proved, unconditionally, that the sum on the left is $O(N^{4/3} \log 10^4 N)$. The Riemann Hypothesis, one of the best known conjectures in the whole of mathematics, asserts that the function $\zeta(s)$, defined for $\mathrm{Re}\, s > 1$ as $\sum_{n=1}^{\infty} n^{-s}$ and defined in the rest of the complex plane by analytic continuation, has all its zeros (in the half-plane $\mathrm{Re}\, s > 0$ located on the line $\mathrm{Re}\, s = \frac{1}{2}$.

Apropos of the conjectured existence of primes between m^2 and $m^2 + m$, we know from recent work of Chen that there exist between these numbers many integers each of which is a prime or a product of two primes, provided only that m is large enough.

(h) Euler's function $\phi(k)$ is the number of integers from 1 to k coprime with k, so that there are exactly $\phi(k)$ progressions ℓ mod k with $1 \leqslant \ell < k$ and H C F $(\ell,k) = 1$. If $k = p_1^{\alpha_1} p_2^{\alpha_2} \ldots$, then $\phi(k) = k(1-1/p_1)(1-1/p_2) \ldots$.

(i) This can be shown to imply that integers from some point onwards are sums of one, two, three or four primes.

(j) Thus all rationals, and all surds, and all numbers derived from these by the elementary operations of addition, subtraction, multiplication and division, are algebraic, i.e. solutions of algebraic equations.

Sir John Kendrew, C.B.E., F.R.S.

Director-General of the European Molecular Biology Laboratory, Heidelberg and President of the International Council of Scientific Unions.

Has contributed extensively to the scientific literature and held many distinguished positions.

In 1962 was awarded the Nobel Prize for Chemistry.

INTRODUCTION

In spite of the most sophisticated modern techniques we are still totally ignorant of the mechanism, and sometimes even of the purpose, of many familiar and universal biological phenomena. Two examples discussed in this book are sleep – what is its purpose and what is the underlying mechanism – and memory, about which Buchtel and Berlucchi write: " 'Where or how does the brain store its memories? That is the great mystery.' This statement, taken from Boring's (1950) classical work on the history of experimental psychology, is still valid today, despite a quarter of a century of intensive work."

The purpose of this part is to explore areas of biological and medical research in which our ignorance remains profound, just as its companion part examined areas of ignorance in the physical sciences.

The causes of our ignorance are of several different kinds, some applying to one field and some to another. Many of them are encountered both in the physical and in the biological sciences. For example, the system under scrutiny may be inaccessible in space or time; thus it is difficult to study the origins and evolution of the Universe itself because we cannot perform experiments on it nor can we actually visit any but a tiny part of it (but note that the Moon and the nearer planets, only 25 years ago considered totally inaccessible, can now be visited by men and instruments); the origin of life on earth is inaccessible in time and we can only get faint clues of the most indirect kind by studying contemporary living organisms which certainly differ profoundly from their early progenitors. Again, our ignorance may result from the lack of an adequate theoretical framework in the light of which to order and interpret the relevant facts, a problem that has in more or less degree retarded progress in fields as diverse as particle physics, pure mathematics, the study of cancer, and the functioning of the central nervous system. Or the experimental techniques may simply be extremely difficult (and often very expensive) as in nuclear physics, in the exploration of the microcosmos of biological tissue, and in developmental biology where Crick in his essay emphasizes the sheer difficulty of carrying out the experiments that would help to solve the outstanding problems, and Wigglesworth comments that "in the past, the physiologists . . . have come near to saying that the physiology of embryonic development is unknowable".

These and other reasons for our ignorance have all been illustrated in the first part, and we meet them again in the present one. But in biology there are some special kinds of difficulty that assume a particular importance.

One of these is the enormous complexity of even a single living cell, and still more of a complete organism consisting of a million million cells all co-operating to form a single whole with the most intricate structure and behaviour. The techniques of physics and chemistry, which antedated those of biology, were intended to extricate simple principles from complex phenomena; in biology the complexity is itself an intrinsic condition and characteristic of the phenomenon, and today we simply do not have at our command techniques that will enable us to gain an intellectual comprehension of it. The enormous complexity of a human brain has already been referred to and forms the subject of several essays. At the other end of the spectrum, Holdgate and Beament in discussing ecology, a subject of the greatest practical importance for the future of the human race, stress the conceptual limitations of the field and think that ecological systems, studied in their totality, may be simply too complex for full comprehension. As Maynard Smith remarks in his essay, "if biology is difficult, it is because of the bewildering number and variety of things one must hold in one's head".

Another major difficulty is that there are some experiments which just cannot be performed. Tracey gives excellent examples in his article on human nutrition where he points out that some long-term effects, harmful or otherwise, of substances ingested in the food may be quite different in humans and in other animals. Direct experiments on humans are not ethically tolerated; to acquire adequate statistical information on mice may require an impossibly large number of animals and may in any case give the wrong answer as far as man is concerned. Here we appear to have a total block to progress in an area of the most direct practical importance to man.

Perhaps the most fundamental of all the difficulties encountered in biological research is that the investigator cannot detach himself from the system under study because he himself forms part of that system or partakes of its nature in such a way that objectivity is impossible. A similar difficulty arises in particle physics where any conceivable experiment itself affects the object investigated; it has been enshrined in the Uncertainty Principle as a fundamental law of nature. In biology the problem appears in many different forms in different areas. In the social sciences and in psychology the investigator is himself one of the subjects under study. At a more fundamental level the problem of the relationship between mind and matter and of the nature of consciousness seems impossible of solution because the investigator himself is conscious mind and there is a complete conceptual gap between that mind and the physical objects in which it resides.

Do these difficulties, some of a more practical and some of a more fundamental nature, mean that there are some aspects of biological organisms of which we shall for ever remain ignorant, just as the finite speed of light means that there is a horizon in the physical universe beyond which we can never penetrate? If there are such horizons in biology too, they are far away, so far that we cannot be sure that they even exist. In the meantime the difficulties present to the biologist challenges that make his field one of the most exciting intellectual endeavours of the human race as well as one of the most important for its survival on earth.

Heidelberg John Kendrew

Roy J. Britten

Senior Research Associate, California Institute of Technology and Staff Member, Carnegie Institution of Washington.

Member of the U.S. National Academy of Sciences.

THE SOURCES OF VARIATION IN EVOLUTION

QUESTIONS ABOUT THE MECHANISM OF CHANGE

A brief catalogue of ignorance or list of questions about the mechanism of evolution makes a good starting-point. Many of these questions may seem foolish or unanswerable when we understand the process of evolution. However, I feel they are appropriate from a molecular biologist's point of view. The nature of living systems is so intimately intertwined with their evolution that most areas of biological ignorance should be listed but are not. As a matter of style and out of ignorance of the voluminous evolutionary literature no references are included. I use incautious and non-professional language in order to raise questions which cannot yet be properly formulated.

Population genetics explores the role of populations in evolution and the way in which natural selection operates on the individuals. It is a field with a large mathematical structure, many specific problems and questions but no overall solutions. The quantitative importance of large and small populations and the ways in which populations are genetically isolated from each other are not known. Are major populations the principal sources of change or are they potential fossils with all of the action occurring in small, almost invisible, offshoot populations? No quantitative explanation is yet available for the ubiquitous large protein polymorphism that is observed in most modern species.

For myself the question of questions is what I term the *immediacy* of natural selection. Does it operate only through the score of reproduction of adapted offspring that in turn reproduce? Are there ways other than population count for integrating the effect of selection over many generations? Can there be integration over long lineages passing through many successive daughter species? How is it possible for future evolutionary flexibility to be preserved when the exigencies of survival apply strong immediate selection pressure? Is preservation of flexibility due to variability in the environment and the currently useful variation in the genetic constitution of the populations of a species? Is it simply chance that some species preserve evolutionary flexibility while others do not? Is the stability of populations due to well-balanced ecological and competition limits? If there are other mechanisms that lend stability such as behavioral and hormonal responses to crowding, how many generations are required for the integration of an appropriate reproductive score? All of these questions suggest that natural selection is a subtle process and that a significant part of the genetic information may not be subject to short-term selection. How could such information be stored and over what period of time is it effectively selected? There are aspects of the fossil record which suggest parallel evolution of species lines that have long been separate. Such convergent or parallel evolution does not have an easy explanation and also suggests long-term storage of genetic information.

On a molecular level there are also suggestions of freedom from selection pressure, or longer periods of integration. For example, mammals contain enough DNA per cell to code for an excessive number of potential genes (though most of this DNA is surely something other than structural genes). The amount of mammalian DNA per haploid cell is 3000 million nucleotide pairs which would code for 10 million short protein molecules. This is a lot, but consider amphiuma. It is a salamander, assuredly no more complex than a man, which has 20 times as much DNA per cell as you or I. These large quantities of DNA are not simply multiple copies of structural genes since most genes are coded by single copy DNA regions. There are also a number of cases in which closely related animals have widely different amounts of DNA per cell. There is obviously a lot of DNA in the genome of higher organisms that we can not account for. This has been termed the C-value paradox. To add to the mystery most of the single copy DNA in primates changes so rapidly in evolution that it is probably under little or no selection pressure. We do not know what unexpressed potentialities exist in all of this "extra" DNA.

We are nearly totally ignorant of the way "external" events such as environment or population affect evolution. Obviously we are not much better off in regard to the "internal" events. I divide internal and external domains arbitrarily at the skin of the animal. Although behaviour and many other things cross this barrier, thinking of these domains separately has value. Most internal events are decided molecularly and the basis is DNA. There follows a brief review of the current dogma. The DNA of each cell of an individual is the same and carries the genome of the species. The sub-set of these genes that are expressed determines each cell type and state. For simplicity I assume that the process by which genes are expressed is understood. DNA codes for a protein sequence. The sequence is transcribed into messenger RNA and the message is translated on ribosomes into the protein molecule. The amino acid sequence determines the three-dimensional structure of the protein with appropriate consideration of the influence of other molecules and structures in the cell. As a result, all of the activities, rate constants and equilibrium constants fall out correctly. If all of the appropriate gene products are produced without error and in the right quantities the result is a cell, differentiated for its role. If all of the appropriate cells are present in the right places we get a living organism with all of its complex structure and behavior. For the time being I am willing to by-pass the question of the validity of this model and ask if we can use it with appropriate subtlety to go after the mechanism of evolution.

Underlying all is the control of gene expression, appropriate to the time in development for each cell of the embryo and in later life. While the regulation of gene expression is basic, it is only one of many levels at which regulatory processes occur. Moreover, the systems regulating gene expression are often themselves under control from the cytoplasm of the cell and from other tissues. The events of regulation of gene expression are imbedded in a network of feedback loops on several levels of control, including transcription, translation, biochemical function, intercellular communication and embryonic induction. However, there must be a system for the integration of the control of gene expression which resides in the organization of the DNA and the nucleus. I call this basic source of control and integration the "gene regulation system". Understanding this underlying part of the control system will probably allow us to see the crucial aspects of the feedback network. A knowledge of the genetic aspects of the gene regulation system will permit us to see the ways in which it can be modified in evolution. When we understand the significant aspects of all of these control processes we will understand the nature of life. Or more cautiously, I cannot from present knowledge ask more subtle and over-riding questions, though the answers would not satisfy a poet.

Consider what a mutation or change in the regulatory system must face to be successfully included in the genome of the species. If it has no effect on the form or chemistry of the organism, the change is termed neutral and chance rules in the following way. Each surviving offspring will either carry the mutation or not and after a number of generations equal to the number of individuals in the population its fate is rather likely to have been decided. Either the change will have disappeared or will have become fixed and be a part of the genome. The time required for this decision may be long and a lot of genetic variation is normally present. If a change or mutation actually affects the organism then chance is modified by a gantlet of requirements involved in "selection". The change must not be expressed inappropriately in the development of the embryo and should be under control of a system of regulation which will establish the appropriate timing, location and rate of its expression. Otherwise the effect would be almost certainly deleterious, regardless of advantages conferred. In fact any change must be a consistent part of a full solution to the problem of embryonic development. If the change is significant and leads toward a new organ, structure or biochemical system it must contain within itself a set of regulatory relationships which form a consistent part of the pattern of development. All of these control relationships are absolutely required and still further the mutant molecules must be able to work properly and be consistent with the behaviour pattern.

We know that evolution works! However, we know almost nothing of how it works and so I must make the obvious statement: many different parts of biological knowledge independently demonstrate the reality of evolution. A few of these: the fossil record and the relationships of form and structure to modern species; the detailed comparison of amino acid sequences of related proteins among related species; the coherence of the times of primate divergence from the fossil record and the sequence divergence of their DNA. Chimp DNA differs from human DNA in only about 1 per cent of the bases of the single copy DNA. This is only a small difference. It must ultimately be important to our examination of the mechanism of change in evolution.

So we can believe that evolution works, but is natural selection a part of the process? Yes, inescapably, but how does it operate? Is it immediate (effectively scored generation by generation) or longer term? In what way are the changes inherited? Of course, through mating and all of the richness of stored variability permitted to a population of diploid individuals, with chromosomes that exchange parts. In addition it is now permissible to consider infection. Viruses transfer DNA sequences between distantly related species. It is not known whether genes or control system parts can be transferred among higher organisms but transfer among bacteria is well known. How important was transfer by processes other than inheritance in the long course of evolution?

What are the sources of variation? They are almost certainly heritable changes in the DNA. Are they ordinary mutations (base substitutions in the genes coding for proteins) or are they more complex changes in the DNA? Are visible chromosomal rearrangements more important? Are there many rearrangements of short DNA sequences? What about changes in the "system of gene regulation"? Changes in the regulation of genetic activity would be very important to form and development. Rearrangements could add and subtract given genes from a coordinately expressed set of genes. Changes on the level of the regulatory system itself could have immense potential power. Large blocks of genes could be affected. Whole branches of the regulatory system could be turned on or off that contain extensive information capable of specifying an organ or organ part. These "turned-off" sets of regulatory system information could be preserved for extensive periods. In this way fossil organs, organ systems or system parts could be preserved in unexpressed or occasionally expressed form as preserved regulatory patterns. The reexpression of combinations of such patterns could produce novel structures with unexpected potentiality. Interesting suggestions, but the problem before us is the development of some direct evidence.

DNA SEQUENCE ARRANGEMENT AND GENE EXPRESSION

In this section the knowledge of DNA sequence organization is reviewed since it has deep implications both for the system of gene regulation and for the evolution of the genome. The unexpected discovery that higher organism DNA has a lot of sequences which are repeated many times per cell has opened a way to examine DNA sequence organization. Many DNA measurements have been made, induced by the concept that the repetitive sequences have something to do with the "gene regulation system". We have found that a typical genome contains about three-quarters single copy DNA and about one-quarter sequences present in 100 to 10,000 copies in the DNA of a single cell. The individual repeats are more or less imperfect and copies differ by as much as 10 to 20 per cent of their bases.

The majority of the typical repeated sequences are interspersed among the single copy sequences. In other words, the repeated and single copy sequences are linked together in an alternating pattern throughout the long DNA molecules. The repeated sequences average about 300 bases long while the single copy sequences range from less than a thousand bases to several thousand in length. The majority of the single copy DNA is found in this pattern of interspersion. Since 1000 bases could code for a typical protein of about 300 amino acids or 30,000 molecular weight, these single copy sequences are about gene-size.

This pattern of sequence organization is nearly universal in higher organisms. DNA has been studied from twenty species chosen to represent many branches of the phylogenetic tree and almost all show the interspersion of short repeats and single copy DNA sequences. Only one dipteran (fruitfly) and hymenopteran (honey bee) show few if any short repeats. The interpretation of these exceptions is uncertain since other insects and even another dipteran (house fly) show that typical pattern of interspersion of short repeated sequences with single copy sequences.

Most structural genes are present in only a single copy. The number of genes expressed and the type of sequences adjacent to these genes has been examined in sea urchin embryonic development at the gastrula stage (which is just beginning to form its gut and contains about 600 cells). About 14,000 genes are expressed in messenger RNA. *The DNA sequences from which this RNA is transcribed are almost all adjacent to interspersed repeated DNA sequences.* These repeats are apparently a special set and differ in sequence from the majority of interspersed repeated sequences. This arrangement could not occur by chance and the existence of a special set of repeats adjacent to expressed genes implies that they have a direct role in gene expression. It is not known yet if this is a specific role. In other words, the adjacent repeats could at this stage of knowledge be general transcription start or end signals or specific controlling parts of the gene regulation system.

It is not possible to leave the subject of gene expression without mentioning the paradox of nuclear RNA. In all higher organisms that have been studied a very large part of the RNA synthesized at any one moment is nuclear RNA of fairly large and heterogenous size. This RNA is rapidly degraded and apparently does not leave the nucleus. In sea urchin embryos about 10 times as much length of single copy DNA is transcribed into nuclear RNA than is transcribed from structural genes. The role of the nuclear RNA remains a mystery while its universal occurrence shows that it is important. Perhaps it contains regulatory RNA molecules or messenger RNA for regulatory proteins, and will one day be known as an obvious expression of the action of the gene regulation system.

RATE OF CHANGE OF SINGLE COPY DNA AND RAPIDLY EVOLVING PROTEINS

What is the evolutionary significance of different kinds of changes in the DNA? The best measurements that we have show changes in DNA sequence which are dominated by DNA base substitution. Measurements have not yet been made of other possible changes such as DNA sequence rearrangement. The best measured sequence relationships are for single copy DNA of primate species. In such measurements the fraction of the bases that differ between two related DNA sequences (the sequence divergence) is calculated from the measured stability of a reassociated pair of complementary DNA strands from the two species. Labeled DNA from one species is paired with unlabeled DNA from another and the melting temperature compared with that of well-matched DNA strands, with the sort of results that are shown in Fig. 1. During much of the period of primate evolution differences have accumulated at about 10 per cent per 50 million years since separation of the species lines. This corresponds to a rate of substitution in each species of about 0.1 per cent per million years or about four base substitutions incorporated per year. Thus chimp and man differ by about 1 per cent or 30 million bases out of a total of 3000 million. This suggests a more recent separation of human and primate lines than that indicated on Fig. 1, or effects of generation time.

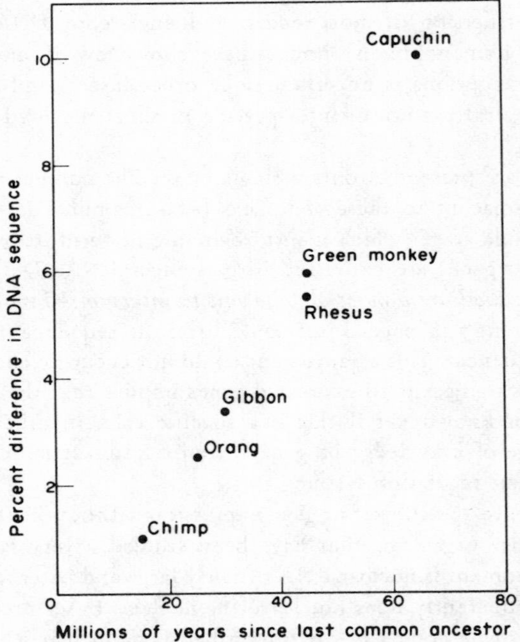

Fig. 1. The difference between human DNA and the DNA of several primates. The per cent of the bases that differ between any two DNAs is calculated from the average of the single copy DNA melting temperature measurements published by Kohne, Chiscon and Hoyer in the *Carnegie Institution of Washington Yearbook* 69 (1970), and Hoyer, van de Velde, Goodman, and Roberts, *ibid.* 71 (1972).

The comparison of the rate of DNA substitution with the rate of amino acid substitution in proteins is quite informative. Many proteins show strong effects of selection and the rate of amino acid substitution is diminished, presumably because most changes in important locations are deleterious. In fact histones appear to evolve at a rate which is less than one ten-thousandth of the rate of the most rapidly changing proteins. However, to get at the significance of the underlying DNA changes I have chosen the most rapidly changing protein amino acids for comparison with DNA. The fibrinopeptides are fragments of blood-clotting proteins which are cut away from the functioning proteins and degraded. It is unlikely that they reflect much selection pressure and many amino acids could probably replace those present in most positions. Measurement of fibrinopeptide sequences among related animals supports this concept as shown in the second figure. For example, green monkey fibrinopeptides differ in about 17 per cent of their amino acids from those of man. How much has the coding DNA changed? The minimum change in the DNA that would cause these amino acids to be replaced is 6 per cent since most changes in the three codon positions would cause an amino acid change. The estimated rate of change in the coding sequence for fibrinopeptides shown by the circles is very similar to the average rate of change in the single copy DNA. A similar comparison is made for a carefully selected set of amino acids of hemoglobin. Hemoglobin, of course, has a critical function and as a whole its amino acid sequence evolves relatively slowly. However, about one-third of the amino acid positions evolve rapidly. This set was selected by counting the number of changes that had occurred in all of the amino acid sequences know for vertebrate hemoglobins. The estimated change in the coding sequences for these amino acids is shown by squares. The probable rate of evolutionary change of these coding DNA sequences is also very similar to the rate of change of the average single copy DNA sequences.

THE "UNSELECTED" RATE OF CHANGE

We have seen that the rate of base substitution in the average primate single copy DNA sequences is about the same as that for the coding sequences of the fibrinopeptides. The rate of

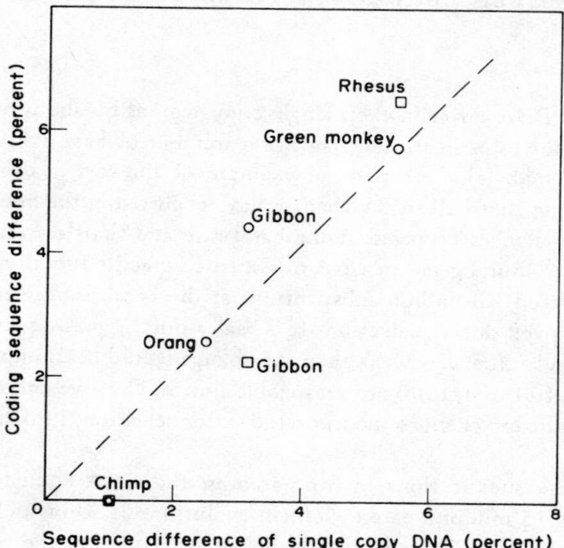

Fig. 2. Comparison of the evolution of the coding sequences of rapidly changing amino acids with the average single copy DNA of primates. Adapted from Britten and Davidson (F.A.S.E.B., 1975, in press). The per cent sequence difference in the single copy is from Fig. 1. The per cent difference in the coding sequences for fibrinopeptides (0) is calculated as the per cent difference in amino acid sequences (Dayhoff, in *Atlas of Protein Sequence and Structure*, Vol. 5, 1972) divided by three. The difference in the coding sequences for a selected set of hemoglobin residues (□) was calculated in the same way. For this comparison the forty-five residues of alpha and beta hemoglobin were chosen which showed more than two different amino acids among the hemoglobin sequences of eleven vertebrates.

substitution in the coding sequences of the selected set of hemoglobin residues is also very similar. If there were any selection pressure on these three different classes of DNA sequence it is unlikely that they would be the same. The most obvious explanation for the similarity of these rates is that there is no selection pressure on any of them. In other words, most substitutions in these sequences would have relatively little effect on the form or chemistry of the animals. We have as yet few cases and it remains possible that these three rates are similar by coincidence. Therefore the conclusion is not certain and is termed the "Basal rate hypothesis". It includes the following propositions: (1) the structural genes form a small part of the total DNA and changes in other single copy sequences have little effect on the animal; (2) most base substitutions occurring in the single copy DNA of the germ cells are subject only to chance loss during inheritance; (3) the resultant evolutionary rate is the "basal" rate of substitution and occurs in all of the DNA; (4) the rate of incorporation of changes in the small fraction which affects the animal is lower as a result of selection; (5) a few actual genes will also show the rapid basal rate since many amino acids can be substituted without affecting the progeny.

If indeed this approach identifies the unselected rate of base substitution we now have the basis for estimating whether base substitution has had an important effect on evolution. Since this is still uncertain the next parts of the argument are principally intended to raise another parcel of questions.

THE NUMBER OF EVENTS IN LATER PRIMATE EVOLUTION TO MAN

Human and chimp DNA differ in their single copy regions by about 30 million bases. From this number we attempt to estimate the significant number of base substitutions. The "hairy" assumptions required make this estimate an example of the sort of calculation that will one day be feasible. Assume that half of the changes have occurred in the human line. Then assume that these 15 million changes occurred at the basal rate and that less than 10 per cent of the DNA is made up of structural genes or DNA of sequence specific function. The structural genes would then have suffered 1.5 million substitutions at the basal or unselected rate. Very few of these would have survived due to selection. It is really only a guess as to what selection factor should be used. If the selection were weak then the change would be unimportant. For important changes I feel that 1/100 or 1/1000 are reasonable limits. Thus we end up with an estimate of 1500 to 15,000 significant changes incorporated, after selection, into the human DNA in 15 million years.

Are these few base substitutions incorporated in the DNA enough to be the source of variation for the last 15 million years of human evolution? It seems unlikely unless they had just the right kinds of effect. We can think in terms of changes in the gene regulatory system that would affect the form or function of an organ. But how many base substitutions can have such effects? Amino acid substitutions in typical proteins – no way. I feel that it would take billions of small biochemical lesions to add up to the multiple changes that have occurred in form. Even billions might not be enough, owing to a low probability of the proper combinations of events. Once we know the organization of the gene regulatory system we will be able to discuss the types of modifications that are likely to have occurred and begin to answer the question of whether these changes would form a sufficient source of variation.

SPECULATIONS ABOUT REGULATORY EVOLUTION

I have juxtaposed three concepts: an apparently insufficient rate of base substitution during evolution: the highly organized state of the DNA: and the existence of a gene regulatory system that operates in an unknown way. The purpose of this juxtaposition was to indicate that we are ignorant of the sources of variation in evolution and to establish a basis for speculating on the effects of DNA sequence rearrangement.

If the gene regulation system is established by the sequence organization of the DNA, as now seems likely, appropriate events of rearrangement of DNA sequences would change the sets of genes which are activated in particular conditions. A good model for a significant change is an unequal crossover event that moves a control DNA sequence. A receptor sequence involved in transcription initiation might be moved into a position adjacent to a structural gene or away from it. At a higher level in the gene regulation system a DNA sequence which is transcribed to yield a regulatory molecule might be moved in such a way as to change the control of its own transcription. In this way a set or battery of structural genes could be simultaneously affected. Are events of unequal crossover sufficiently frequent in evolution? We do not know but it is clear that such changes would usually be deleterious and be subjected to selection pressure.

The large excess of DNA of unknown function may be involved in the gene regulation system. Most of this DNA shows the typical sequence organization of short repetitive sequences alternating with single copy sequences. This interspersion of DNA sequences is probably the result of many past events of change in sequence arrangement. If the interspersed sequences are a part of the gene regulatory system their arrangement must reflect this history, primarily as a selected functioning pattern. It is very unlikely that such a highly organized pattern would result from chance and I assume that the function and arrangement of this DNA are closely related. In fact the observed arrangement reflects an appropriate balance of evolutionary flexibility and stability of function. In other words, the sequence arrangement has evolved so that it successfully evolves. I assume that living organisms as a whole have this quality and the gene regulatory system shares this exquisite adjustment to the realities of both surviving and evolving. As a part of the mechanism I visualize large amounts of inactive "spacer" DNA sequences which would permit events of rearrangement without excessive damage to active DNA sequences. In addition there could exist repetitive recognition sequences which would control the location of unequal crossover events so that certain parts of the regulation system could be more flexible than others. It would be an advantage if the ancient specifications for early embryonic development were protected from excessive experimentation while other regulatory systems could show a more radical behavior.

In conclusion I wish to emphasize that while ingenious speculations can be made about the gene regulation system and its relationship to the source of variation in evolution our ignorance of the actual sources of variation is just as abysmal as our ignorance of the formal properties of the regulatory system that is evolving.

Many of the concepts and measurements described here are the result of collaboration with my partner Eric H. Davidson, though he has no responsibility for the form and some of the wilder speculations. This chapter is intended to lie outside the pattern of formal publication and ignores the issue of scientific priority.

James C. Lacey, Jr.

Arthur L. Weber

Kenneth M. Pruitt

James C. Lacey, Jr.

Associate Professor of Biochemistry at the University of Alabama in Birmingham.

Major research interests include protein synthesis, cancer and study of the genetic code.

Arthur L. Weber

Postdoctoral Fellow at the Salk Institute, La Salla, California.

Kenneth M. Pruitt

Associate Professor of Biochemistry and Biomathematics at the University of Alabama, Birmingham.

THE EDGE OF EVOLUTION— A MOLECULAR HISTORICAL PERSPECTIVE

To be ignorant of many things is expected.

To know you are ignorant of many things is the beginning of wisdom.

To know a category of things of which you are ignorant is the beginning of learning.

To know the details of that category of things of which you were ignorant is to no longer be ignorant.

Phenella in *The Unwritten Comedy*

While we frequently speak in general terms about the evolutionary process, it is possible to speak of evolution in many other ways, for example, with respect to a specified individual person or thing as simply the sequential events in time related to that person or thing. As pointed out by Phenella, the learning process involves steps in which the person's brain passes from a completely naïve state through a sequence of events until the person's brain has generated a mental construct of the details of some feature of existence and arrived at a condition of comprehension. Conceptually, the person's brain has *evolved* in the process and has in a literal sense accommodated itself to the environment by reacting to it and mentally stimulating it.

As a result of this brief evolutionary process, a learning experience, the individual undergoing the experience has an advantage in the process of natural selection. By accommodating itself to the reality of the environment, the individual can better persist in that environment.

Although Phenella was speaking of a person, she has captured the whole sense of the evolutionary process of all living things. She might have written instructions to the first life forms, "Know your environment, it is your master, but to persist you must also make it your slave." It would seem a rather ridiculous charge to a beginning life form, but that's obviously the way it was and is. The environment is simultaneously the dictator and the provider. Evolution has involved passage from a time when the environment was complete dictator to the present when man is, in considerable measure, in control of his environment, man made pollution notwithstanding. This transition is the conceptual trademark of the process of evolution, i.e. a living thing survives by coming to know its environment and in knowing it, withstanding it and using it.

The main thrust of this article will be to treat first *conceptually* how a non-living thing could come to know its environment and then *specifically* discuss a molecular mechanism of how this could take place. Needless to say, we are appropriately ignorant of whether our speculations, in fact, had reality in the evolutionary process.

The molecular processes whereby a present-day person learns something, or knows something, are still not known, and these processes may be at a different level than the process a living thing goes through to persist. More precisely, we should ask, what does it mean for a living cell to "know" its environment? We believe this means for the cell to have information within itself which concerns the environment. For example, if a system were exposed to a food supply of glucose and the cellular system had never seen glucose before, it could be said to be naïve with respect to that environment. However, if the cellular system is to survive, it must come to be able to use the glucose energy supply. This requires that the cell be able to produce a series of enzymes, not necessarily all at once, but eventually, to degrade the glucose. In terms of present-day cells, this would mean that the system contained information in its DNA regarding the degradation of glucose. Now that this information is included in the DNA, what is the significance?

The presence of this genetic information tells us that: (1) glucose has been experienced by this particular system at some time in the past, (2) it was able to somehow generate the necessary information in order to survive and (3) since the time of that original generation of genetic information, the utilization of information to degrade glucose has been frequently required. This idea of considering DNA as a molecular history of the creature is not new to us, but has been discussed by Zuckerkandl and Pauling.[1] It is important to point out that the information a cellular system carries at any one time does not necessarily represent its entire history. Rather, the information most likely relates to the recent history of required information for survival. That is, some of the genetic information acquired in the past may not have continued to be necessary and thus would be lost because it gave no survival value.

Having discussed the concepts of information acquisition and how the information a system retains is a reflection of the environment, we will now consider a specific mechanism whereby a naïve cellular system can acquire new DNA information *de novo*. What we will consider is a molecular mechanism whereby the information may be generated in a non-specific way, but which may then give the system that contains it some survival value.

Here is the basic problem. It is one thing to generate nucleic acid information, and it is quite another to have that information selected. Natural selection processes select systems on the basis of what they can do. Now, the nucleic acid can do virtually nothing except carry information. Therefore, the information in nucleic acid cannot be directly selected, rather it must first be converted into something that is functional – protein. It is the protein that is selected.

In contemporary systems, which carry elaborate machinery for the conversion of DNA information into protein, there is no problem with selecting DNA. However, in the most primitive systems there was no mechanism for converting nucleic acid information to protein. Therefore, we need to consider how such a system could ever get started. It is another chicken-and-egg-type question. In order to discuss this problem, we need to deal, at least briefly, with the problem of specificity between amino acids and nucleotides. We cannot yet deal with this problem precisely, but information from several sources is beginning to show:

1. At the very least there is a direct correlation between the properties of amino acids and their anticodonic nucleotides. (These are the nucleotides that are in the anticodonic loop in all RNAs which carry the amino acids into protein synthesis.) Data show, for example, correlations between the hydrophobicity of and hydrophilicity of amino acids and their anticodonic nucleotides.[2–4] We have suggested that the genetic code originated because of these specifities.
2. There may, in fact, be affinities between amino acids and their anticodonic nucleotides. These affinities have been displayed in interactions between thermal proteinoids and polynucleotides and concern primarily our work with Professor S. W. Fox [5, 6] as well as in certain other studies. Taking hydrophobic amino acids as an example, peptides containing the hydrophobic amino acids, tryotophane or tyrosine, interact with the most hydrophobic homo-polynucleotide, polyadenylic acid (poly A).[7] We have found that phenylalanine increases the solubility of adenosine.[8] Razka and Mandel[9] found that phenylalanine interacts best with poly A. Hydrophobic amino acids do seem to interact with adenine.

If we accept that there are specific interactions between amino acids and nucleotides, then we might well imagine that an activated form of the amino acid could be formed by recognizing and reacting with the triphosphate form of its anticodonic mononucleotide triphosphate. For example, the amino acids listed in Table 1 would be activated by the high-energy triphosphates shown.

TABLE 1. Specificities in Amino Acid Activation

	Amino acid	Activating nucleotide	Activated form
Decreasing hydrophobicity ↓	Phe	ATP	Phe-ADP
	Pro	GTP	Pro-GDP
	Gly	CTP	Gly-CDP
	Lys	UTP	Lys-UDP

Actually, in a very primitive form, all hydrophobic amino acids might be activated by ATP, all charged amino acids by UTP,† polar uncharged ones by CTP and intermediate polarity by GTP. The proposed activated form of the amino acid is the amino acid anhydride diphosphate. There are several reasons for proposing this form.

1. It is well established that amino acid nucleotide anhydrides contain sufficient energy for the formation of peptides.[10]
2. Polynucleotides can be formed from nucleotide diphosphates using the enzyme, polynucleotide phosphorylase.[11] The process is non-template and the composition depends on the composition of the input mononucleotides.
3. The proposed compound would then have sufficient energy to form both peptides and polynucleotides.
4. Beljanski[12] has described a bacterial enzyme which forms peptides, requires all four nucleotide triphosphates and converts them to the diphosphate.

Rich[13] suggested the simultaneous polymerization of amino acids and nucleotides some time ago, and Fox *et al.*[14] have actually demonstrated the formation of both peptides and oligonucleotides from ATP and phenylalanine. The mechanism of these simultaneous syntheses is not known. We would like to consider in some detail the possible mechanisms and evolutionary consequences of such a simultaneous polymerization.

The suggestion only has significance if there are specificities between the amino acids and nucleotides. An example of the polymerization we suggest is shown in Fig. 1. The initial composition of the polymers generated by such a system would depend on availability of monomers and the sequences would in part be dictated by the composition. Nevertheless, the appearance of a particular peptide would be a matter of probability. The important part of the proposal, however, is that for every peptide generated, a nucleic acid of the same length would also be generated and the composition and sequence of the polynucleotide would be anticodonically consistent with the peptide. Here is a system which is "randomly" synthesizing peptides and equivalent polynucleotides (on a one-to-one basis, not yet three to one as in the genetic code).

Now, let us recall that natural selection process selects on the basis of function, i.e. proteins. What could happen to isolated microsystems which are "randomly" synthesizing peptides and equivalent polynucleotides? Let us again take the example of glucose. Suppose the

† UTP was shown to interact with charged species.[3]

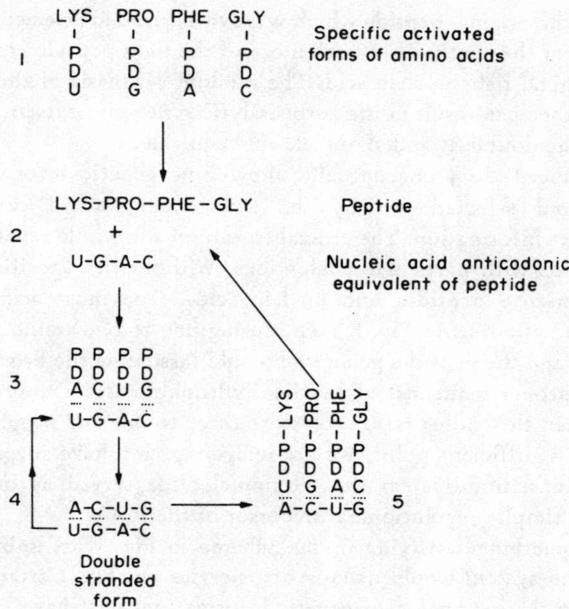

Fig. 1. Schematic of processes whereby a system could generate information *de novo*.
1. Amino acids are activated by reaction with specific nucleotides, e.g. ATP activates all hydrophobic amino acids, GTP activates amino acids of intermediate hydrophobicity, CTP activates all hydrophilic amino acids and UTP activates charged hydrophilic amino acids. The proposed activated form is the aminoacyl nucleotidyl diphosphate anhydride in which the carboxyl group of the amino acid is attached to the terminal phosphate of the nucleotide.
2. The amino acids polymerize with each other yielding a peptide and the nucleotides polymerize with each other yielding a nucleic acid which is anticodonically equivalent to the peptide in a one to one relationship.
3. The nucleic acid portion can serve as a template for the synthesis of a complementary strand of nucleic acid which bears the same relationship to the peptide that a contemporary messenger RNA bears to its protein.
4. A double-stranded nucleic acid results from no. 3. Disruption of the double strand gives the two strands, either of which can serve as a template to synthesize more double strand.
5. Or the upper strand can now serve in a messenger capacity associating with specifically activated amino acids to generate the coded synthesis of more of the original peptide as in no. 2.
Whereas the synthesis of the original peptide in step 2 produced a peptide by simple probability, natural selection process allowed the system containing that peptide and its nucleic acid complement to persist. This persistence then allowed the other steps to take place, yielding more of all components of the system.

system has glucose as its only energy source. If in the random generation of peptides and polynucleotides, a peptide is generated which decomposes glucose, yielding some energy-rich compound that can be further used, the system containing that peptide would be able to persist, while similar systems not containing the proper peptide would not persist. The selection of the particular system with the right peptide would simultaneously be selecting the polynucleic acid information for that peptide.

The further consequences of the above described selection process would be as follows:

1. The polynucleotide could serve as a template for the formation of a complementary polynucleotide which would then result in a double-stranded or "protected" form of the polynucleotide (Fig. 1, step 3).
2. The newly synthesized complementary strand would be the equivalent of a messenger

RNA for the original peptide which was synthesized and selected and could serve as a template for the synthesis of both more of the right peptide (Fig. 1, step 5) and more of the original right nucleic acid. The random synthesis of the right peptide for the situation then can result in the autocatalytic synthesis of itself, the equivalent nucleic acid and the double-stranded nucleic acid equivalent.

The scheme in Fig. 1 does conceptually show how genetic information generated on a probability basis and selected for by the environment could cause the autocatalytic multiplication of that information. The crucial feature of the whole scheme is the specificity of activating amino acids with particular nucleotides. Without this specificity the entire scheme would generate nonsense peptides and nucleic acids. One must acknowledge that such a primitive translation system as in Fig. 1, step 5 using one to one amino acid/nucleotide coding would be imprecise, and the peptides generated would (assuming the broad selectivities in Fig. 1 caption) only have proper sequences of varying hydrophobicity. The precise amino acid could only be selected when the coding ratio evolved to three to one and proper enzymes, ribosomes, etc., had evolved to a sufficient point. We are imagining, and have suggested earlier,[2, 3] that in this primitive translation system the mononucleotide served as both the activator and adaptor and was the simplest evolutionary precursor of the RNAs.

While there is something satisfying in the scheme in Fig. 1, its unbridled operation could not be tolerated, the system would exhaust its energies quickly. Certainly pretty early, some means of controlling the expression of genetic information must have arisen. The information needed to control the expression of other information is in a real sense super information. This new kind of information could only be selected after the primary information relating to external functions had been generated and selected. The primary DNA information, although inside the cell, now represents part of the environment for selecting the super information. The super information could also be generated on a probability basis but selected on the basis of functional control of the primary information.

The suggestions that we make here, we believe, agree with the theoretical suggestions of Eigen[15] and provide specific molecular mechanisms for initiation of the hypercycles he discusses.

REFERENCES

1. E. Zuckerland and L. Pauling, "Molecules as documents of evolutionary history", *J. Theoret. Biol.* 8, 357-66 (1965).
2. J. C. Lacey, Jr. and A. L. Weber, "The origin of the genetic code: an amino acid-anticodon relationship", in *Protein Structure and Evolution* (L. Fox, Z. Deyl and A. Blazej, eds) Marcel-Dekker, Inc. N.Y. (1976) p. 213-222.
3. A. L. Weber and J. C. Lacey, Jr., "Genetic code correlations between amino acids and anticodon nucleotides", Submitted.
4. A. L. Weber and J. C. Lacey, Jr., "Genetic code correlations between amino acids and their anticodon dinucleotides", Submitted.
5. S. W. Fox, J. C. Lacey, Jr. and T. Nakashima, in *Protein-Nucleic Acid Interactions*, edited by Dr. Ribbons and F. Woessner, North-Holland Publ. Co., Amsterdam, 1971, p. 113.
6. J. C. Lacey, Jr., D. P. Stephens and S. W. Fox, "Thermal proteinoids and polynucleotides: some specificity observed". Submitted.
7. J. Dimicoli and C. Helene, "Interaction of aromatic residues of protein with nucleic acid". *Biochem.* 13, 714-23 (1974).
8. J. C. Lacey, Jr. and M. N. Downing, "Amino acid anticodon interactions: L-phenylalanine increases the solubility of adenosine", American Chemical Society Southeastern Regional Meeting, Gatlinburg, Tenn. Oct. 1976.

9. M. Razka and M. Mandel, "Is there a physical chemical basis for the present genetic code?" *J. Molec. Evol.* 2, 38-43 (1972).
10. G. Krampitz and S. W. Fox, "The condensation of the adenylate of amino acids common to protein," *Proc. Nat'l. Acad. Sci. USA* 62, 399-406 (1969).
11. M. Grunberg-Manago, "Polynucleotide phosphorylase", in *Prog. Nucl. Acid. Res.* 1, 93-133 (1963).
12. M. Beljanski, "Participation of an RNA fraction in peptide synthesis in the presence of a purified enzyme system from *A. faecalis*," *Biochem. Biophys. Res. Commun.* 8, 15-19 (1962).
13. A. Rich, in *Horizons of Biochemistry*, edited by M. Kasha and B. Pullman, Academic Press, N.Y. (1962), p. 115.
14. S. W. Fox, J. R. Jungck and T. Nakashima, "From proteinoid microsphere to contemporary cell," *Origs. of Life* 5, 227-37 (1974).
15. M. Eigen, "Molecular self-organization and the early stages of evolution," *Quart. Rev. Biophys.*, 4, 2 & 3, 149-212 (1971).
16. M. Paecht-Horowitz and A. Katchalsky, "Synthesis of aminoacyl adenylates under prebiotic conditions," *J. Molec. Evol.* 2, 91-98 (1973).

NOTE

Data consistent with the model in Fig. 1 have been reported by Paecht-Horowitz and Katchalsky (16). These authors found that, in the presence of a synthetic zeolite, ATP reacts with alanine to form aminoacyl adenylates (probably the ADP mixed anhydride). Furthermore, when montmorillonite was added, peptides were formed. However, they did not have evidence of oligonucleotides being formed in this system.

E. W. F. Tomlin, C.B.E.

Fellow of the Royal Society of Literature. A full-time writer and editor who has combined for many years a career in the public service with authorship of books on philosophical and literary subjects.

Former British Council Representative and Cultural Attaché in Turkey, Japan and France and Professor of Philosophy and Literature at the University of Nice.

Has a strong interest in the philosophical implications of the Life Sciences.

FALLACIES OF EVOLUTIONARY THEORY

Although evolutionary theory is supposed to have brought about great changes in outlook, there is not much evidence that the ordinary man has modified his behaviour as a result of learning that all life sprang from a common ancestor or a single primordial form, such as a flagellated cell or something more elementary still. Hypotheses which embrace the whole of nature do not as a rule make an immediate impact upon individual conduct, unless the challenge to prevailing orthodoxy be an offence punishable at law. In the case of evolution, the shock to human pride is sometimes thought to have been considerable; the psychiatrist, Karl Stern, places it second of the three great threats to human complacency of which the first was represented by Marxist reductionism and the third by Freudism.(1) Yet the common man does not seem to have been unduly humiliated or cowed by the revelation that he had descended from apes rather than from angels. Admittedly, Disraeli declared that he was "on the side of the angels", but Disraeli can hardly be described as a common man. The general view was perhaps summed up by the wag who said that it mattered little if we had sprung from apes, so long as we had sprung far enough.

Historians are now disposed to agree that the theory of evolution caused less of a hullabaloo than has been commonly assumed. In *The Education of Henry Adams* (1906), the author makes clear that evolution was regarded less as "an idea that changed the world" than as an attitude favoured by the sophisticated as a vindication of their belief that existence lacked purpose. In fact, evolution as an *idea* was by no means new. Millennia before Darwin (who never used the word evolution; it was popularized by Herbert Spencer), the notion of *cosmic* evolution had been entertained, and it was certainly familiar to the Babylonians. The discovery, announced 5 years before that of evolution, of post-Euclidian geometry by Bernard Riemann was in many respects a more epoch-making event, since it paved the way for Einstein and modern physics and the release of atomic energy.

Despite Copernicus and Galileo, the Sun still "rises" and "sets". Despite Einstein, clocks measure time as they always did. And despite Darwin, we feel no closer to the animals than our fathers did – perhaps less so. With evolutionary theory, however, the absence of general impact may be due partly to the vagueness and elusiveness of the central concept; for if this theory disposed of the idea of special creation, it introduced an idea almost as difficult to grasp, namely, the "interiorization" of the evolutionary process at the stage of man, whereby conscious intelligence succeeded an apparently unconscious and "meaningless" process. In short, cultural evolution succeeded biological evolution, but how it did so remained obscure.

The truth is that *evolution was an hypothesis which hardened into dogma before it had been thoroughly analysed.* Hence it mothered a number of fallacies. It was easy to say that the idea of change or transformation in nature had been substituted for that of immutability; but what sort of change was involved? If species were no longer regarded as immutable, the fact remained that they exhibited a measure of stability, or they would not have deserved the name of species. Evolution was conservation as well as transformation. And if the human species possessed a unique character, wherein the evolutionary process acquired inwardness and conscious direction, this still did not prove that evolution had "led" to man. Until man's fortuitous and unaccountable advent, it had appeared to lead nowhere; and this was not merely perplexing, but it placed man outside rather than inside the evolutionary process.

This contradiction remained at the heart of the orthodox evolutionary position, and it was not to be removed by the dogmatic assertion that evolution simply "happened like that": an assertion which was made still more suspect by the further claim that the processes anterior to man exhibited only "apparent purpose" (Julian Huxley)[2] or "purpose-like processes" (Sommerhoff):[3] for this was to derive reality from its appearance.

Even sophisticated Darwinians such as Konrad Lorenz assume without question that the origin and formation of species can be explained as a succession of fortuitous variations and mutations passing through the mesh of selection. The oddity of this theory is partially concealed by its mode of presentation. The order and coherence of organisms are so obvious as to demand an explanation which is both ordered and coherent; and this is supposedly met by the contention that, in the struggle for survival, the unfavourable mutations are severally eliminated. Such a view, as we shall see, depends upon certain well-concealed assumptions. And if it is advanced to account for the development of species, or phylogenesis, it must apply also to the development of the individuals of which the species is composed, namely ontogenesis. It is impossible to assume that there are two separate forms of morphogenesis, one for the species and the other for the individual. Evolution is a single, unified process; and in this is supposed to reside its uniqueness.

If we were to imagine a species arriving at its final form by a series of chance but "favourable" mutations, each individual being stamped with the specific form, we should have an explanation of morphogenesis which was at least concise. But it would not be an explanation of the facts. The facts, at least concerning forms of life higher than the virus (which may possibly be "struck off" in this manner), are precisely the facts of embryogenesis. No one supposes that in every example of embryogenesis the adult form develops by chance mutation. Each form is constituted or reconstituted by the kind of orderly process which may be called thematic; for an organism resembles a temporal melody. Now if the higher forms of life are declared to be the product of chance mutation, how is it that their individual representatives develop one by one in an orderly fashion from an embryonic state *below* the level thus fortuitously reached? How could life, having achieved a certain level in the "struggle for existence", go through that struggle again for each individual? Why is this individual "struggle" so obviously not a chance process, while the struggle to arrive at a *species* is deemed to be so? In what sense can individual morphogenesis depend upon the morphogenesis of the species if the one is totally different in character from the other? Have we not here the very thing we declared to be impossible, namely two forms of morphogenesis, neither of which serves to explain the other?

This dilemma, though only gradually perceived, served to temper the initial wave of Darwinian enthusiasm. In due course, it gave rise to the movement known as Vitalism, which culminated in the ideas of Bergson and the emergent evolutionists. And this movement in turn gave place to another, which appeared to restore Darwinism to favour. The new ally of Darwinism was genetics, a science which arose from the discoveries of Mendel, first published in 1900. Genetics was hailed as providing the key not only to the evolution of species but also to individual morphogenesis.

If we abandon the notion that the morphogenesis of species is the direct *cause* of individual development, and thereby escape from the dilemma which makes the one the product of genetic mutation and the other an ordered, thematic process, the way is open to a more intelligible explanation. In order that there shall be a genuine evolution or origin of species, there must be a passing over of one species to another. Now this passing over must be undertaken by a vast number of individual formations undergoing approximately simultaneous and united change. Since the modifications which give rise to a change in species are suffered or undergone by these individuals themselves, or at least by a sufficient number of them, *it is ontogenesis which creates phylogenesis and not the other way round.*[4]

We now perceive the insufficiency of the popular idea of evolution as moving forward in a trial-and-error fashion and throwing up new species from time to time as a result of the operation of fortuitous variations – strokes of luck becoming strokes of genius – and exhibiting purpose-like behaviour which, at the super-stroke of genius, man turns into purpose-*ful* behaviour. Fluctuations cannot negotiate in the void for the production of organic wholes or unities, or for the generation of order, any more than business transactions can be negotiated in the absence of economic norms or a common commercial code. We have to presuppose some element, some primal activity or principle, to which accidents "happen". To evolve, in any rational sense, is precisely to *dominate* accident: so that even if a favourable mutation should arrive fortuitously, its maintenance, its status as favoured, cannot be fortuitous. How, if it is not recognized as the favourite, can it be favoured? This means that mutation and selection, or the selection of mutations useful in promoting the life of the species, cannot satisfactorily be accounted for by the "official" view of evolution; and although disinterested scientists such as Huxley and Lorenz claim that this view is now even more convincing than in Darwin's time, the theory as it stands simply will not bear the weight they put upon it. On the orthodox view, selection is purely negative, the elimination of the unfavourable. Thus it cannot be true to say with Lorenz that "selection (is) one of the great constructors of evolution. The other constructor is mutation."[5] For the negative cannot be made to assume such a positive role.

The capacity of orthodox evolutionists to entertain a kind of double-think is further observable in the idea of "apparent purpose" being reflected at levels inferior to the human. Throughout his voluminous writings on evolution, Julian Huxley never satisfactorily defines what he means by "apparent purpose", nor does Sommerhoff's attempt to translate it into mathematical terms or relations between variables clarify it any better. What is remarkable in particular about Sommerhoff's argument is that he is so evidently impressed by purpose-like behaviour in nature that, in the course of his book, he frequently forgets to use the word "apparent" (would not "pseudo" have been better in any case?) and speaks of purpose and finality outright. He seems to labour under the impression that if it is made initially clear that purpose and finality are only *apparent*, and that he does not really believe in it, he may proceed to assume its operation with impunity. An even more striking example is that provided by C. H. Waddington. Waddington lays stress upon what he calls the "creodic" character of organic process, that is to say, the manner in which distinct processes converge and meet together, and he usefully defines an organism as a "trajectory", acting as "an attractor for other trajectories"; but, in fear of incurring the label of teleologist, he enjoins us to consider the "whole time trajectory" – the *trajet*, so to speak – rather than "the final stage". At best, he maintains, following Huxley, we can speak of quasi-finality, and he goes so far as to admit the word "teleonomic".[6] To flirt in so dangerous a manner with the vocabulary of finality, while denying a place for teleology in biological thinking, is to engage in a kind of intellectual prurience. In fact, many theoretical biologists tend to dwell on finality with an enthusiasm equal to that displayed by the most ardent providentialists. What they clearly recognize, along with men such as R. B. Braithwaite,[7] is that the processes of nature cannot intelligibly be discussed except in terms *borrowed* from the vocabulary of finality. This was the view of Immanuel Kant; but Kant, unlike the modern apostles of mechanism, did not leave it at that. He based his world of phenomena on a "kingdom of ends", where, by definition, finality held undisputed sway. The modern empiricist has no such noumenal world to call in to redress the balance of the world of phenomena.

If, then, we recognize that the origin and evolution of species depend upon active thematism (which is another name for finality), and that such thematism is incarnate in the morphogenesis of billions of individual beings, then we are driven to accept the view of Polanyi that "the process must have been directed by an *orderly innovating principle*, the action of which could only have been *released* by the random effects of molecular agitations and photons coming from outside, and the operation of which could only have been *sustained* by a favourable environment".[8]

It is not the purpose of this essay to inquire into the nature of this "orderly innovating principle", but rather to examine the arguments of those who claim to advance a mechanistic interpretation of biological phenomena. We are concerned here with analysis, not metaphysics.

It will now be clearer why the succession of cultural evolution to organic evolution, upon which Huxley and also Waddington[9] lay such stress, is unintelligible without presupposing continuity between the two. If evolution becomes "interiorized" at the level of man, *this is because it was always an interior process.* Human culture was preceded by biological culture. It will not do to say with Dobzhansky that "cultural evolution was added to biological evolution", and then to declare that "evolution has no purpose; man must supply this for himself".[10] Man was not "added" to the ape; he developed or evolved from that creature. This is another example of professed evolutionists refusing to adopt a truly evolutionary view. Granted, only man *behaves* in the explicit sense; but since such behaviour or conscious activity, guided by thematism, entails the use of organic tools, these tools themselves must have been developed according to some form of thematism. They cannot have come into being by a series of mutations due to mechanical faults of copying: and the same applies to the brain and the nervous system. These organs are the means whereby higher evolution is directed: to describe their development to the play of blind forces is to suspend rational judgment and to betray the cause of science. It is legitimate to go further and to call it, with Karl Stern, "crazy". Stern adds: "I do not mean crazy in the sense of slangy invective but rather in the technical meaning of psychotic. Indeed, such a view of the history of the world has much in common with certain aspects of schizophrenic thinking".[11]

In addition to the problem of evolution as a unified process, there is the even more disputed problem of how this process began. Up to the nineteenth century, theories of the origin of life varied from pure creationism to notions such as that minute organisms were carried to the Earth by meteorites, or that life arose spontaneously (abiogenesis). None of these theories is wholly extinct. Despite the conclusive demonstrations of the Florentine Francesco Redi (1626-97), many persons continued to believe that the lower forms of life were generated by non-vital elements: maggots from filth, fleas from plaster, etc.[12] And on a higher intellectual level it is assumed that at some point there must have been a jump from the multiple to the unified, from the particle to the organism.

In approaching a problem of this complexity, it is perhaps best to lay down a general principle, especially as it is implicit in what has already been said. No evolution is possible from particle to organism, from the multiple to the unified, unless we take leave of science and have recourse instead to scientific myth. As Frey-Wyssling has said, there is evolution only from the organic to the organic: *structura omnis e structura.*[13] A generation or so ago, and even in some textbooks in use today, it was assumed that the egg contained a *homunculus* whose transformation into a homomegalus constituted the growth of the organism. The view to succeed this notion differed only in being less anthropomorphic; it envisaged a mosaic pattern in the egg from which the various characteristics were derived point-by-point. Then followed the idea of a "code script" and the transmission of genetic "information". Despite the substitution of impersonal pattern for mythical entities – a change similar to that in physics from occult powers to physical forces – the new theory was still a brand of *preformation*: that is to say, the idea that "little forms" are present in the egg or the germ cell. And preformation of whatever sort is a form of mechanism incompatible with true biological creativity or epigenesis. As we now know, the egg is a complicated unity of which the constituents are themselves organic, for the cell is composed of "organules" which in turn depend upon macro-molecules. These latter are capable of a special form of reproduction and replacement of parts, implying organic activity even at this level.

The properties of the molecule are now much better understood than even a few years ago. and from what we now know, it is apparent that the molecule is truly formative rather than functional, though many molecules, such as that of adenosine phosphate (ATP), exist for only a fraction of a second.

Granted, to describe molecular activity as organic and formative may seem to be flying in the face of the most recent conclusions of molecular biology, and therefore to be challenging, in an arbitrary manner, the most sophisticated of modern physical disciplines. It is true that the hypotheses of molecular biology have disposed of any semblance of Vitalism; but Vitalism is a view of organic development no less misleading than mechanism itself, if it is not a kind of mechanism in disguise. The point at which the theory of molecular biology is open to criticism (and we must stress that we are not calling in question the remarkable and revolutionary discoveries of modern molecular biologists) is precisely that at which it refuses to be sufficiently biological; for the assumption that chemical *functioning* can at a certain point moment become biological *formation* is as arbitrary as any to be found in Vitalism, and amounts to the assumption that biological phenomena are simply a higher complexity of physico-chemical function. What the molecular biologist must finally admit is that there is no jump from chemical activities to biological activities – from formlessness or multiplicity to form; there is only continuity from form to form. Thus the orthodox evolutionist is to be criticized not for interpreting the whole of nature in the light of the concept of evolution, but *for limiting that concept to the relatively restricted macro-biological sphere, leaving out of account the immense world of microscopic and submicroscopic nature*. Evolution is to be criticised for not being evolutionary enough. Once the evolutionary process is comprehended in its fullest reach and range, the orthodox view is seen to fall short of a satisfactory interpretation. It is rather as if we were to generalize from history while excluding the entire palaeolithic period. As Bernal observed: "Those who wished to see in the fossil record more facts from which it was warranted or impious to draw conclusions as to any process of development, find parallels to-day in those who refuse to draw conclusions from the patterns of typical biochemical molecules or reactions."(14) For in the light of the new ideas of chemical morphology raised by quantic chemistry, chemical relations are not static and geometrical but akin to structural states, so that an atom of a particular element is not a *thing* but an *activity*, depending in turn upon the dynamic interaction of electrons none of which is strictly localizable. (Heisenberg was perhaps the first to point out that ordinary spatial conceptions do not apply to the electron.)(15) Such structurizing activity or comportment (a word used to distinguish such activity from behaviour proper), is difficult to describe, first, because the concept is a new one, and secondly because it is impossible satisfactorily to illustrate by means of models. All we can affirm is that the concept enables us to dispense with "missing links" or unaccountable jumps such as the preachers of indeterminism descry in nature.(16) *It also enables us to dispense altogether with the notion of an inorganic world.*

If the concept of evolution needs to be extended to cover the vast world of the submicroscopic, the concept of life necessarily requires a comparable extension. And if there appears to be an anomaly in describing the world of the atom and the molecule as *alive*, this is due to the habit, long ingrained, of regarding life as a local burgeoning in a vast inorganic crucible. We need to recall the chief characteristics of living things: the existence of invisible but *cognizable* relations, and the capacity for self-maintenance, self-regulation and reproduction, etc. In the light of the discoveries of modern chemistry and microphysics, these characteristics may be fittingly ascribed to the atom and the molecule, revealing their comportment to be formative in the manner we have described.

Thus the world consists of individualities or beings: entities which have what the Existentialists call an *en soi*. and there is a "line of continuity" whereby such beings take visible form at a certain point as historical spatio-temporal structures of increasing complexity.

Only in the light of such an extended definition of organism is the evolutionary process lent that "interiority" of which Huxley and Waddington speak; and because the interiority is primal and not an unaccountable intrusion at the stage of man, it implies finality – a finality which, along the greater part of the evolutionary thrust, these thinkers regard as only apparent.[17] Naturally, the ascription of finality to nature at the pre-human state does not carry the ethical implications of its later presence. The refusal to ascribe consciousness or rather psychism to the stage preceding conscious intelligence is the correlative of the refusal to extend life beyond the "organic", though the word psychism must be disinfected of its spiritualistic connotations. Both refusals are the result of seeking to put new scientific wine into old philosophical bottles. The fallacies engendered by orthodox evolutionary theory were initially due to the early capture of that theory by a science dominated by the ethos of Classical Physics; and it is a fact patent to the historian of thought that this ethos, though long superannuated, continues to exercise authority. "It is an amusing paradox", wrote the philosopher A. D. Ritchie in 1948, "that the fertility of classical theory is greatest at present in certain branches of biology, so that its most fervent advocates are actually people who are by way of being biologists. *Yet it is in biology that the insufficiency of the theory is most glaring.*"[18] And an American scientist, writing 12 years later, remarked that "among all scientists, biologists . . . are strangely . . . the most inclined to accept without question a causal, mechanistic interpretation of nature."[19] A study of contemporary biological literature shows that, with a few exceptions, the situation is largely unchanged.

SUMMARY

The theory of evolution, as first propounded, may not have produced such a convulsion as has been supposed. From the start it was a somewhat vague concept, and the most puzzling innovation of the Darwinians was the idea of the "interiorization" of the evolutionary process at the stage of man. Secondly, there was an ambiguity concerning the classical idea of Natural Selection, and no satisfactory explanation was given as to how the "favourable" mutations were preserved. Thirdly, if mutationism applied to species, it applied also to individuals. Thus, instead of a unified evolutionary process, the evolutionists appeared to posit two forms of morphogenesis; for as embryogenesis was not a process of chance mutation, there was a contradiction between individual ontogenesis and the "fortuitous" morphogenesis of species. In fact, species are generated by the gradual variation of multitudes of individuals, so that ontogenesis creates phylogenesis and not the other way round. Evolutionary selection implies the *dominance* of accident. The concept of "apparent purpose" (Huxley, Sommerhoff, Waddington) is likewise characterized by double-think. A positive evolutionary principle at work even at the sub-human level is an inescapable conclusion. All evolution is an "interior" process, characterized by thematism.

Evolution is from the organic to the organic. The entire microphysical world is now seen to come within the organic sphere, revealing a "line of continuity" between all individuals or beings. This is to widen the sphere of evolution. The present *impasse* in evolutionary thinking, productive of so many fallacies, is due chiefly to the interpretation of biological fact in terms of out-of-date physical theory.

NOTE

1. See *The Third Revolution: a Study of Psychiatry and Religion* (1955).
2. See his *Evolution, the Modern Synthesis* (1942).
3. See his *Analytical Biology* (1951).
4. Cf. Ruyer, *La Genèse des Formes Vivantes* (1958), p. 46.
5. *On Aggression* (1966), p. 9.
6. "The basic ideas of biology", in *Towards a Theoretical Biology* (1967), pp. 2,10,13,15.
7. For example, *Scientific Explanation* (1953), especially p. 327.
8. *Personal Knowledge* (1953), p. 386 (my italics).
9. See *The Ethical Animal* (1960), pp. 104-7.
10. *The Biological Basis of Human Freedom* (1960), p. 31.
11. *The Flight from Woman* (1966), p. 290.
12. Cf. *The Living World* (1966), by C. L. Duddington, p. 14.
13. *Submicroscopic Morphology* (1953), p. 373.
14. "The origin of life", *New Biology*, April 1954, p. 28.
15. See also Bertrand Russell, *The Analysis of Matter* (1954 ed.), pp. 321, 326.
16. For example, C. G. Simpson in his *Meaning of Evolution* (1950), p. 235.
17. The authors of "Organisms as physico-chemical machines" (*New Biology*, April 1964), H. G. Bray and K. White, also hold this view.
18. "The biological approach to philosophy", in *Essays in Philosophy* (1948), p. 104 (my italics).
19. Bentley Glass, reviewing Niels Bohr's *Atomic Physics and Human Knowledge* in the *Quarterly Review of Biology*, American Institute of Biological Sciences, Johns Hopkins University (Vol. 35, No. 3, September 1960).

John Maynard Smith

Professor of Biology at the University of Sussex since its foundation in 1965.

Graduated first in Engineering, then in Zoology, after which he studied the genetics, physiology and behaviour of the fruitfly.

Present interests are in the evolution of sex and of altruistic behaviour, and the relationship between evolution and ecology.

THE LIMITATIONS OF EVOLUTION THEORY

There are a lot of things we do not know about evolution, but they are not the things that non-biologists think we do not know. If I admit to a non-biological colleague that evolution theory is inadequate, he is likely to assume at once that Darwinism is about to be replaced by Lamarckism and natural selection by the inheritance of acquired characters. In fact, nothing seems to me less likely. In common with almost everyone working in the field, I am an unrepentant neo-Darwinist. That is, I think that the origin of evolutionary novelty is a process of gene mutation which is non-adaptive, and that the direction of evolution is largely determined by natural selection. I am enough of a Popperian to know that this is a hypothesis, not a fact, and that observations may one day oblige me to abandon it, but I do not expect to have to. Indeed, everything that has happened during my working life as a biologist, and in particular the development of molecular biology, has strengthened rather than weakened the neo-Darwinist position.

The difficulties of evolution theory seem to me to arise from another direction. The essential components of the theory are mutation (a change in a gene), selection (differential survival or fertility of different types) and migration (movement of individuals from place to place). The theory tells us that each of these processes, at a level far too low to be measurable in most situations, can profoundly affect evolution. For example, consider selection. Suppose that there are two types of individual in a population, say red and blue, which differ by 0.1 per cent, or 1 part in 1000, in their chances of surviving to breed. If the population is reasonably large (in fact, greater than 1000), this difference in chances of survival will determine the direction of evolution, towards red or blue as the case may be. But if we wished to demonstrate a difference in the probability of survival during one generation – that is, we wished to demonstrate natural selection – we would have to follow the fate of over one million individuals, usually an impossible task.

Similar difficulties of measurement arise with mutation and migration. When a gene replicates, there is a chance of the order of 1 in 100 million that a particular base will be miscopied. It is possible to measure these astonishingly low rates of error in very special circumstances in some microorganisms. It is also clear from the theory that rates of this order are sufficient to provide the raw material of evolution. But in most natural situations, mutation rates cannot be measured. Finally, consider migration. Suppose that a species is subdivided into a number of populations, and we wish to know how far the evolution of any one population is influenced by immigration from the others. Theory shows that if a population receives on the average one migrant from outside in each generation, this can have a decisive effect. Yet in practice we could not hope to measure such a low rate of migration.

Thus we have three processes which we believe to determine the course of evolution, and we have a mathematical theory which tells us that these processes can produce their effects at levels we cannot usually hope to measure directly. It is as if we had a theory of electromagnetism but no means of measuring electric current or magnetic force. Now things are not quite as bad as I have painted them. Surprisingly often, it is possible to measure natural selection directly, because the selective differences are not 1 in 1000, but 1 in 10 or thereabouts. But the measurement difficulty is serious. It means that we can think up a number of possible evolutionary mechanisms, but find it difficult to decide on the relative importance of the different mechanisms we have conceived.[(1)] I hope that this will become clearer if I give some examples.

1. WHAT DETERMINES THE RATE OF EVOLUTION?

Suppose that at a time 200 million years ago, during the age of reptiles, some event had occurred which doubled the rate of gene mutation in all existing organisms; we must also suppose that for some reason the rates did not fall back to their original levels. What would have been the consequences? Would the extinction of the dinosaurs, the origin of mammals, of monkeys and of man have taken place sooner, so that roughly the present state was reached in only 100 million years? Or would the rate of evolution have stayed much the same? Might it even have been slower? The short answer is that we do not know.

The difficulty is that we can picture evolution in two ways. According to the first picture, a species spends most of its time more or less "at rest" in an evolutionary sense; the species is well adapted to the contemporary environment, and selection acts to maintain its characteristics rather than to change them. Occasionally some change in the environment – an ice age or the appearance of a new predator – imposes new selective pressures, and the species responds by a burst of evolution. To do so, it does not have to wait for new mutations, because the species will already have a large amount of genetic variation which was generated by mutation while it was at rest. If this picture were true, an increase in the mutation rate would have little effect on the rate of evolution, which depends on environmental change.

I think this picture arises from our knowledge of the effects of artificial selection on domestic or laboratory animals. If you take a population of, say, mice and select for increased size, the average size will increase rapidly, far more rapidly, by a factor of perhaps 100,000, than metric characters of this kind normally change in evolution. It can be shown that the change does not depend on new mutations occurring since the start of selection, but on genetic differences already present in the population from the outset. What is true of size in mice is true of almost any character you like to name in any outbreeding species. To base a picture of evolution on such experiments, however, overlooks two facts. First, in evolution a number of different characteristics usually change simultaneously, and second, changes in response to artificial selection do not continue indefinitely, but slow down and stop when the initial supply of genetic variability has been used up.

An alternative picture of evolution has been suggested by Van Valen.[(2)] He points out that the main feature of the "environment" of most species consists of other species, which it eats, or which eat it or compete with it for food. If so, when any one species in an ecosystem makes an evolutionary advance, this will be experienced by one or many other species as a deterioration of their environment; their food will have got more difficult to find or their predators more difficult to escape. If so, the other species will evolve in their turn, and cause environmental deterioration for still other species. The result will be that every species will be evolving as fast as it can, simply to keep up with the others. Van Valen calls this the Red Queen hypothesis; the Red Queen, you will remember, told Alice: "Now here, you see, it takes all the running you can do to keep in the same place."

If this picture is true, then our original question reduces to this; if the mutation rate is doubled, does this double the maximum rate at which a species can evolve? I wish I could tell you that this question has a simple and agreed answer, but I cannot. My own opinion,[(3)] however, based on some rather messy algebra, is that it does, provided that the population is not very large (roughly, not greater than the reciprocal of the mutation rate; if it were greater than this, then every possible mutation would occur in every generation anyway). But my reason for raising the question in the first place was not to answer it. It was to make the point that a theory of evolution which cannot predict the effect of doubling one of the major parameters of the process leaves something to be desired.

This discussion raises two other questions. What determines the mutation rate itself? Do species go extinct because they cannot evolve fast enough? As I will show, we do not know the answer to either of them.

2. WHAT DETERMINES THE MUTATION RATE

The mutation rate – that is, the probability of error in gene replication – is not a fundamental constant like Planck's constant or the velocity of light. It depends on the enzymes which replicate DNA and which correct errors as they go. Since the enzymes are produced by genes, the mutation rate is under genetic control.

Two views can be taken about the mutation rate. According to the first, the mutation rate is as low as natural selection can make it; after all, 1 in 100 million per base replication is fairly low. The logic behind this is that the vast majority of mutations are harmful, so an individual with a low mutation rate will leave more offspring.

The alternative view is that the actual mutation rate we observe is an optimum. If it were higher then the harmful effects just mentioned would predominate, but if it were lower there would be some other disadvantage; for example, too low a mutation rate would inhibit evolutionary adaptation.

In some microorganisms (some bacteria and yeasts) there is now a strong reason for preferring the second view. Genetic strains can be isolated which have a lower mutation rate than typical "wild" strains.[4] Since a lower mutation rate is possible, the actual rate must be an optimum rather than a minimum. Of course, this argument would fail if the strains with a lower mutation rate suffered some other disadvantage; for example, they might replicate DNA very slowly, just as a neurotically careful proofreader would slow down publication of this article. As far as we can see, there is no corresponding disadvantage in the low mutation-rate strains, but we cannot be sure. More serious, there is no way at present of saying whether what is true of bacteria and yeasts is true of all organisms.

3. WHY DO SPECIES GO EXTINCT?

A corollary of the Red Queen hypothesis is that those species which are unable to evolve rapidly to meet changed circumstances will go extinct. But is this an important reason for extinction? I recently wrote a mathematical analysis of the Red Queen hypothesis, and sent a copy of it to Dr. G. C. Williams. He replied as follows. Suppose, he said, there had been a flea peculiar to the passenger pigeon. As the pigeons became rarer, the fleas would have been under stronger and stronger selection pressure to exploit an ever-dwindling resource. But no matter how rapidly they improved their adaptation to their way of life, nothing could prevent their extinction.

When a species goes extinct, could it have survived if it had been able to evolve more rapidly? Or does extinction happen because the way of life of a species disappears and there is no other sufficiently similar way of life to which it can adapt? Clearly, both kinds of extinction must sometimes happen. But which is more frequent? We have no idea. As so often in evolution theory, we can imagine two mechanisms but cannot estimate their relative importance. There is, however, a reason why we would like to know more about the causes of extinction; as we shall see in the next section, Williams had a reason for inventing the flea on the passenger pigeon.

4. SEX, ALTRUISM AND THE ORIGINS OF SOCIETY

In most discussions of evolution, the "unit of selection" is the individual organism. We explain the evolution of horses' legs by arguing that those individual horses which had legs of a particular form were more likely to survive and to produce children like themselves. But in some discussions of evolution a different argument is used, or at least implied. For example, the origin and maintenance of sexual reproduction has been explained by saying that species (such as the dandelion, or the lizard *Lacerta saxicola*) which consist entirely of parthenogenetic females cannot evolve so rapidly to meet changed conditions, and so in the long run will go extinct.(5) In this argument, the "unit of selection" is the species; it is not the individual which evolves or goes extinct, but the species.

Selection depending on the differential survival of species or other populations has been called "group selection". Whenever a characteristic is explained as conferring an advantage on a species, rather than on the individuals which compose it, group selection is being assumed. The snag with such explanations is that the process is a very weak one: if there is a conflict between individual and group selection, individual selection is likely to win. This has led people to seek some short-term individual advantage for sexual reproduction. G. C. Williams(6) has been one of the strongest opponents of group selection, which may be why he is reluctant to think that species go extinct because they cannot evolve fast enough.

For the present, it is fair to say that we do not have an agreed explanation of how sex originated or how it is maintained. The same difficulty arises for all aspects of the "genetic mechanism" – for example, the amount of recombination between genes, the number and arrangement of chromosomes, the pattern of self- and cross-fertilisation, and the mutation rate. In discussing these features, biologists have fallen into the bad habit of assuming that if a particular property can be shown to be advantageous to the species, then this will account for its evolution, without asking themselves whether a species advantage could overcome an individual disadvantage. Discerning readers will have noticed that I used, deliberately, an argument of this kind myself when discussing whether the mutation rate is optimal. Loose thinking of this kind is found not only in discussions of genetic mechanisms, but also of ecology and of animal behaviour.

A related difficulty arises because it is not individuals which reproduce but genes. The individual is simply a device constructed by the genes to ensure the production of more genes like themselves. Normally, a gene will only increase in frequency if it produces an individual of greater than average fitness; to this extent, the argument above about the legs of horses is sound. But sometimes the multiplication of genes is not tied to the multiplication of the individuals carrying those genes. This is particularly true in bacteria; consider, for example, the "resistance transfer factors" which can transfer the genes for drug resistance from one species of bacterium to another. The evolutionary processes of bacteria are still a largely unexplored field, but it is one I do not feel competent to discuss.

The fact that it is genes and not individuals which replicate has some intriguing consequences for the evolution of higher organisms. I can best introduce the idea by quoting a remark of J. B. S. Haldane's; he announced that he was prepared to lay down his life for two brothers or eight cousins. The point behind the remark is as follows. Suppose there existed a mutant gene which caused any individual carrying it to be prepared to sacrifice its life if, by so doing, it could ensure the survival of more than two brothers which otherwise would have died. Such a gene would increase in frequence in the population. This is because there is a resemblance, an "identity by descent", between the genes of relatives; in particular, half my brother's genes are identical by descent to my own.

Thus a gene which reduces the probability of survival of an individual carrying it but produces a corresponding increase in the fertility or probability of survival of relatives can increase in frequency. The process has been called "kin selection". It has been used to explain apparently "altruistic" behaviour (strictly, "nepotistic" would be a better word), such as the giving of alarm notes by birds or the self-sacrificing behaviour of worker bees. That kin selection will affect evolution would be admitted by anyone who understands the laws of inheritance and who is capable of logical thought. But it is much harder to say whether the process has been an important one. Opinions vary from those who regard it as, at the very most, relevant to understanding a few peculiar special cases, to those who see it as a major key to an understanding of animal societies, including our own.[7]

5. DRIFT OR SELECTION?

I find that I have so far failed to mention the most protracted debate in evolution theory, which concerns the relative importance of chance and selection. This is because I find the argument less interesting than the ones I have discussed. It does, however, illustrate the point that our difficulty in evolution theory is not so much to think of processes which would explain what we see, as to evaluate the relative importance of different possible processes.

The basic problem is as follows. In an infinite population, if one can imagine such a thing, natural selection would always determine which of two types would be established and which eliminated. But real populations are finite, and in a finite population it is quite possible for the fitter of two types to be eliminated and the less fit established, provided that the fitness differences are small. Such random changes, not produced by selection and sometimes contrary to selection, are referred to as "genetic drift". There is no question that drift will occur, but deep disagreement about its importance. In recent years the argument has taken a new form, under the titles "neutral mutation theory" or "non-Darwinian evolution". It has been argued, particularly by Kimura, that a large part of the variation which we observe, both within and between species, in the structure of proteins has no effect on survival, but has arisen by drift.[8]

The question is difficult to answer because selective forces too small to measure would still be more important than genetic drift in determining genetic change. It is sometimes difficult to understand the heat which has been generated, since both sides, neutralists and selectionists, agree that the evolution of adaptations is the result of natural selection, and it is adaptation which is the most striking feature of the living world.

6. EVOLUTION AND DEVELOPMENT

So far I have discussed difficulties which are in a sense internal to evolution theory. There are other difficulties which arise because of ignorance in other fields of biology, and whose solution will depend on research in those fields. This is most obviously true of the relation between evolution and development.

During the last century ideas about the development of individuals and the evolution of populations were inextricably mixed up in people's minds. When Weismann formulated his theory of the independence of germ line and soma – that is, that the processes whereby a fertilized egg gives rise to germ cells which are the starting point of the next generation is independent of the process whereby a fertilized egg gives rise to an adult organism – he was in effect saying that it is possible to understand genetics without understanding development. He thus set the stage for the growth of the science of genetics during this century; sadly, we still do not understand development. It follows that our understanding of evolution is necessarily partial, because genes are selected through their effects on development.

This difficulty arises in many ways. The most topical concerns the evolution of "structural" and "regulator" genes. The term "structural gene" can be misleading; it does *not* mean a gene concerned with an aspect of adult morphological structure. A structural gene is a gene which codes for a protein, which may be an enzyme or which, like collagen or the muscle proteins, may form the substance of the body. We now know a good deal about the evolution of particular kinds of proteins – for example, the haemoglobins which carry oxygen in our blood, or the hormone insulin – and can therefore infer a good deal about the evolution of structural genes. "Regulator" genes are a much vaguer concept, particularly in higher organisms. We know that there is a lot more DNA than is needed to code for proteins, and we guess that some of the rest is concerned with regulating the activities of other genes, and so controlling cellular differentiation and growth. In higher organisms we know rather little about how such regulation works, and even less about how it evolves. But it seems clear that structural and regulator genes evolve at very different rates. One rather striking feature of the evolution of structural genes is that the rate of evolution is surprisingly uniform in different groups of organisms, at least for a given class of structural genes – for example, those coding for haemoglobin. The reason for this uniformity of rate is a matter of controversy, but the uniformity itself seems well established. For example, frogs are, as a group, about twice as ancient as placental mammals, and the range of differences between their proteins is correspondingly about twice as great. Yet in morphological structure the differences between mammals are far greater. This presumably reflects a more rapid change in their "regulator" genes, although we have no direct evidence. Incidentally, it also seems to be true that the numbers and shapes of chromosomes change much more rapidly in mammals than in frogs, but the relation between this and morphological change is still obscure.[(9)]

These difficulties arise because, although we have a very clear idea about how genes specify proteins, and can therefore speak with confidence about the evolution of proteins, we have no corresponding idea about how genes specify adult morphological structures.

7. HOW WILL OUR IGNORANCE BE DIMINISHED?

If it was easy to answer the questions I have raised, they would have been answered long ago. It is rarely possible in evolution theory to think of a single decisive experiment or observation which will settle a controversy. Our understanding of evolution depends on a combination of clearly formulated theories and wide comparative knowledge. For example, to understand the evolution of genetic mechanisms, we need on the one hand clear theories which predict that particular types of mechanism should evolve in particular situations, and on the other we need comparative knowledge of the associations between the ecology and taxonomy of particular species and their genetic systems. Unfortunately, the kind of scientist who is good at developing clear theories often finds it difficult to remember facts, whereas those who know the facts tend to jib at the algebra. It seems to me that there is no single idea in biology which is hard to understand, in the way that ideas in physics can be hard. If biology is difficult, it is because of the bewildering number and variety of things one must hold in one's head.

REFERENCES

1. I first became fully aware of this difficulty during discussions with Dr. R. C. Lewontin; the matter is discussed further in his book, *The Genetic Basis of Evolutionary Change*, Columbia University Press, 1974.
2. L. Van Valen, "A new evolutionary law." *Evolution Theory* 1, 1-30 (1973).
3. J. Maynard Smith, "What determines the rate of evolution?" *Am. Natur.* (in the press, 1976).
4. J. W. Drake, "The role of mutation in microbial evolution." *Symp. Soc. Gen. Microbiol.* 24 (1974).
5. The idea that sex is an adaptation for evolutionary change was proposed in the last century by Weismann. R. A. Fisher (*The Genetical Theory of Natural Selection*, Oxford University Press, 1930), regarded sexual reproduction as the only characteristic of organisms which had evolved by what we would now call "group selection". The idea that genetic mechanisms in general have evolved by group selection was strongly reinforced by C. D. Darlington, *The Evolution of Genetic Systems*, Cambridge University Press, 1939.
6. G. C. Williams, *Adaptation and Natural Selection*, Princeton University Press, 1966; *Sex and Evolution*, Princeton University Press, 1975.
7. The concept of "kin selection" was stated by both J. B. S. Haldane and R. A. Fisher. Recent interest in the process, and in particular in its role in the evolution of animal societies, is largely due to W. D. Hamilton, "The genetical evolution of social behaviour", *J. Theor. Biol.* 7, 17-52 (1964), and more recently to E. O. Wilson, *Sociobiology*, Harvard University Press, 1975.
8. The "neutral mutation theory" was first proposed by M. Kimura, "Evolutionary rate at the molecular level", *Nature*, **217** 624-6, (1968) and by J. L. King and T. H. Jukes, "Non-Darwinian evolution", *Science*, **164**, 788-98 (1969). For a recent account, see Lewontin, *ibid.*, or, less technically, J. Maynard Smith, *The Theory of Evolution*, 3rd edition, Penguin Books, 1975.
9. A. C. Wilson, *et al.*, *Proc. Nat. Acad. Sci. U.S.A.* 71, 2843-7 and 3028-30 (1974).

D. C. Johanson

Curator of Physical Anthropology and Coordinator of Scientific Research, Cleveland Museum of Natural History, Ohio. Associate Professor of Anthropology, Case Western Reserve University, Cleveland.

Has carried out extensive research in Ethiopa on the origins of Man. Interests include Paleoanthropology, Dental Anthropology and Primatology.

RETHINKING THE ORIGINS OF THE GENUS HOMO

Investigations related to unraveling the intricacies of mankind's earliest stages of evolution have proliferated during approximately the last 15 years. It has become increasingly clear that although the storehouse of human paleontology is considerably fuller now than in the past, we still must await additional evidence before final decisions can be made concerning human evolution and taxonomy. However, it is useful to review, in a general framework, the evidence which is currently influencing the development of new thinking on the occurrence of early hominids, presumably belonging to the genus *Homo*, which are most likely to be direct ancestors of modern man. In this essay I shall not attempt to present the details of complicated taxonomic arguments, but rather will review the evidence which now favors consideration of an early appearance of the genus *Homo* perhaps approaching four million years ago, and the possibility that even older specimens may some day be recognized.

The family of man and his ancestors is known as the Hominidae. Within this zoological family, different scientists recognize various hominid forms, some within the direct ancestry of mankind, and some as side branches. The presence of *Ramapithecus* fossils in Kenya, Pakistan, India, and Hungary from time horizons dating from about 9 to close to 14 million years are recognized by some as the earliest occurrence of recognizable members of the hominid family (Simons, 1972; Pilbeam, 1972). *Ramapithecus* is clearly distinctive from that of the Pongidae, the family of apes and their ancestors, others question the validity of such a conclusion because of the limited nature of the evidence (Greenfield, 1974). Clearly, if *Ramapithecus* is considered a hominid (or definite hominid ancestor), it is probably from a time period not far distant from when the hominid and pongid families split from one another. In a paleontological sense this poses numerous difficulties. Le Gros Clark (1962) has remarked that until sufficient evolutionary time has elapsed, both lineages will share numerous resemblences. For example, the earliest representatives of the hominid family probably more closely resembled their pongid-type ancestors than later hominid forms. It is a difficult task for the anthropologist to ascertain relationships between such fossils. While *Ramapithecus* retains certain primitive (pongid) characters in its dentition such as a sectorial lower third premolar, characteristics of dental wear, enamel thickness, jaw morphology, etc., demonstrate resemblances with later fossils of the Hominidae. Final resolution, I believe, will not be possible until we have additional, more complete material including pelvic and lower limb bones. In this regard, our discovery of a partial skeleton named "Lucy" from Ethiopia may be significant (Johanson and Taieb, 1976). Although she is only about 3.0 million years old, much younger than the known *Ramapithecus* material, her lower jaw and dentition exhibit provocative resemblances with the earlier specimens' jaws. Detailed anatomical studies of "Lucy" will be highly instructive and exceedingly important.

At this time we feel her closest affinities are with *Australopithecus africanus*, the gracile early hominid know from Sterkfontein in South Africa. However, other aspects of her anatomy, apart from the primitive mandible, such as longer forelimbs relative to the hindlimbs, and certain primitive aspects of her pelvis suggest a certain degree of divergence from typical *A. africanus* conditions.

The paleontological record is rather empty during the long span between *Ramapithecus* and the occurrence of hominids in the late Pliocene. With the exception of two isolated molars from the region of Lake Baringo in northern Kenya, dating to about 10 and 6-7 million years, the record is bare. A hominid jaw fragment from Lothagam (about 5.5 m.y.) (Patterson, *et al.* 1970) and an arm bone fragment from Kanapoi (about 4.0 m.y.) (both in northern Kenya) do not give us much insight into the problems of human origins because these specimens are so fragmentary that accurate identification as *Australopithecus* or *Homo* is difficult. It is not until the period between 3 and 4 million years ago that more complete fossil remains are found which provide a clearer understanding of human evolution.

SOUTH AFRICAN FOSSIL HOMINIDS

Before presenting the fossil evidence from eastern Africa it is necessary to briefly mention the South African early hominid sites. It was the well-known discovery of the Taung baby in 1924 by Raymond Dart (1925) which gave us the first definite evidence of "primitive" forms of human ancestors in Africa of considerable antiquity. His specimen was named *Australopithecus africanus*, the "southern ape" of Africa. Although not immediately accepted by his peers, this discovery was of great importance. Continued investigations in different cave sites in the Transvaal of South Africa such as Sterkfontein, Swartkrans, Kromdraai, and Makapansgat produced abundant hominid specimens for study.

It was through the efforts of Robert Broom, originally a specialist of mid-wifery, his student John Robinson, Phillip Tobias, Alan Hughes and others that lime deposits from the caves were examined for fossil remains before they were incorporated into the manufacture of cement.

From the sites of Sterkfontein and Makapansgat came a collection of hominid fossils considered by most experts to be closely related to Dart's Taung baby. Thus, with the mounting evidence, former critics were soon convinced of the hominid status of *Australopithecus*. These creatures were apparently bipedal walkers, possessed large teeth, small brains and may have been omnivorous.

Fossil specimens from Kromdraai and Swartkrans, while obviously belonging within the Hominidae, differed in certain distinct ways from *A. africanus*. These specimens had exceptionally large cheek teeth, relative to the anterior teeth, large massive jaws, heavily buttressed faces, sagittal crests on their crania and other evidence of powerful muscles of mastication. Although these forms were also bipedal some believe they possessed a peculiar, less advanced form of walking (Robinson, 1972). It was thought that such differences could only reflect a different status for the Swartkrans and Kromdraai hominids and hence they were called *Paranthropus robustus* or *Australopithecus robustus*, depending on whether the taxonomic distinction was considered to be at the generic or specific level.

John Robinson, while completing his dissertation on the dentition of the South African forms of *Australopithecus*, formulated a hypothesis to explain the pronounced differences between the two species (Robinson, 1963). He suggested that the larger, more robust type was a vegetarian, while the smaller, more gracile type, was omnivorous. This hypothesis has been questioned by a number of workers (Tobias, 1967), but it would appear that dietary differences influencing the architecture of the skull and size of the teeth might explain the distinctive morphologies of these two forms.

The precise relationship of these South African hominids to those of East Africa is at present difficult to assess accurately. This is due to the unknown influence of geographic separation and also the uncertainty of the age of the South African sites. While many attempts have been made to determine absolute ages for the sites, all have failed. It is generally agreed, however, that the sites of Taung, Sterkfontein and Makapansgat are probably of greater age than Swartdrans and Kromdraai. If this is the case, I would entertain the possibility that the earlier *A. africanus* forms, perhaps more similar to ancestral hominid types, could have been a precursor of *A. robustus.* This suggestion has not been widely considered, but the relative ages of the sites support such a scheme. Also, there may have been particular ecological pressures operating which favored ultimate dietary specialization of the robust type. As we shall see below, indisputable evidence for robust types in eastern Africa is not found until around 2.0 million years ago.

Other interpretations of the South Africa fossil hominids have been proffered. Some, in the minority, believe the differences between robust and gracile forms are simply a reflection of sexual dimorphism (Brace, 1973): males are robusts and females are graciles. It would be curious that members of the opposite sex lived at different sites (this would be improbable due to the different ages of the sites, also). In addition, other problems such as females (gracile forms) having larger canines than males would have to be resolved. At Olduvai Gorge (in eastern Africa), robust hominids like "Zinjanthropus" are found in an equivalent time horizon with small hominids like *Homo habilis*. This does not lend support to a single-species hypothesis as proposed by Wolpoff (1970). Another problem is that fossils attributed to *Homo*, not resembling *A. africanus*, were recovered from Swartkrans and are apparently contemporaneous with robust australopiths (Broom and Robinson, 1950; Robinson, 1953). Taken as a whole, the present evidence does not substantiate placing all specimens into a single species – the range of variation is too pronounced and the morphological adaptations appear to be quite distinctive.

Other investigators have considered the robust forms as a specialized side branch which eventually became extinct, while later, more advanced hominids (*Homo*) are thought to have evolved out of *A. africanus* (Robinson, 1972). We shall see below that recently discovered finds in eastern Africa have cast doubt on this interpretation.

HOMINIDS OF EASTERN AFRICA

I would like to leave the South African evidence and consider in somewhat more detail the fossil sites in eastern Africa. Closely related to the east African Rift Valley, hominid sites occur in depositional contexts which yield important information on dating, paleoecology, paleogeography, paleobehavior, etc. Such data are of great importance for achieving a proper understanding of the environment in which early hominids lived and interacted.

The discovery by Mary Leakey in 1959 of "Zinjanthropus", a robust *Australopithecus*, at Olduvai was an important and long anticipated event. When accurately dated, the deposits gave an age of about 1.8 m.y. for the specimen. L. S. B. Leakey (1959) was astonished with its age and also with the associated stone tools. He suggested that "Zinj" made the artifacts and was a "true" ancestor of mankind.

However, 2 years later a small mandible and an associated partial cranium from virtually the same horizon caused Leakey to alter his hypothesis. The new specimen, Olduvai Hominid (O.H.) 7, appeared to be more advanced and possessed a larger cranial capacity (about 680 cm^3 compared to around 500 cm^3 or less for *Australopithecus*). Leakey reversed his ideas as to the Olduvai tool-maker, and the new form was christened *Homo habilis* meaning "handy man" (Leakey, *et al.*, 1964). This event marked the beginning of a new emphasis towards the idea that the genus *Homo* might be older than the *Homo erectus* stage of human evolution.

Continued field research at Olduvai Gorge brought to light additional specimens which reinforced Leakey's assertion that *Homo* was living alongside *Australopithecus*. From deposits higher in the sequence a cranium was recovered which was clearly similar to *Homo erectus*. It became known as "Chellean Man" and was dated at about 1.1 m.y., thus post-dating both

"Zinj" and O.H. 7. Other *H. habilis* material was found: O.H. 13 ("Cinderella"), O.H. 16 ("George"), and O.H. 24, badly crushed and referred to as "Twiggy". With these additional finds, *H. habilis* became better known with regard to dentition and skull morphology. The teeth were small, set in a U-shaped arch and the cranium appeared to be more advanced in shape and in cranial capacity than *Australopithecus* (Leakey, *et al.*, 1964).

The suggestion that *Homo* occurred at such an early date was not accepted by everyone. Some critics maintained that *H. habilis* could not be distinguished from *A. africanus* in South Africa while others suggested that some of the material was an advanced form of the gracile hominid and the remainder representative of early *H. erectus* (Robinson, 1965, 1966, 1967; Brace *et al.*, 1973). The controversies were often quite intense and still there are some who doubt Leakey's contention. On the whole, however, the morphology of *H. habilis* reveals strong resemblances to later fossils which have been more widely recognized as members of the genus *Homo*.

Conclusions drawn from the Olduvai material required some additional substantiation and support. This came from discoveries made by L. S. B. Leakey's son Richard and his team who have conducted research along the eastern shores of Lake Turkana (formerly Lake Rudolf in Kenya). Richard Leakey has assembled an impressive sample of hominids from east Turkana, some of which have been identified as *Homo*. Perhaps the most notable specimen is the well-known KNM-ER 1470 cranium. It is clearly distinct from all australopiths not only in its relatively large cranial capacity (over 750 cm^3), but also in its shape (R. Leakey, 1973). Originally this specimen was thought to derive from sediments dated to nearly 3.0 m.y. but new age determinations suggest a date of only around 2.0 m.y. (Curtis *et al.*, 1975). Even with a readjustment of its absolute age, the 1470 cranium has contributed much to our understanding of early representatives of the *Homo* lineage. A number of limb bones found in equivalent stratigraphic levels are suggestive of an advanced, fully bipedal, large-bodied hominid.

Recently, R. Leakey (1976) announced the recovery of a relatively complete cranium (KNM-ER 3733) from deposits securely dated to about 1.5 m.y. This specimen has been assigned to *Homo erectus* and in time-equivalent sediments remains of robust australopiths such as KNM-ER 406 occur. Hence, the single species hypothesis appears even less tenable.

Other discoveries at Lake Turkana, including a number of additional skulls, postcranial bones and jaws support conclusions drawn from 1470. A pelvic bone announced by R. Leakey (1976) is from an horizon older than the *H. erectus* cranium and is of substantial importance due to its definite affinities with *Homo*.

Recent explorations in Hadar, Ethiopia, by myself and M. Taieb have resulted in the recovery of a number of jaws and, more recently, postcranial specimens which resemble very closely the Lake Turkana *Homo* fossils (Taieb *et al.*, 1976; Johanson and Taieb, 1976). Particularly interesting are a large and a small palate which were found very close to one another. These provide some idea of variability in tooth and facial size of early *Homo*. Further jaw material and a unique assemblage of several associated partial skeletons from Hadar may fill in details of our picture of a very ancient *Homo*.

The Hadar site is of considerable interest because it samples a period of time poorly understood in other areas of Africa. The deposits have been dated by various methods including radiometric potassium-argon dating. The *Homo* fossils are apparently from strata ranging in age from about 2.6 to 3.3 m.y. This clearly extends the occurrence of *Homo* even further back in the fossil record.

Renewed research by M. Leakey (M. Leakey *et al.*, 1976) at Laetolil in Tanzania has provided even older evidence for *Homo*. A number of jaws and teeth from this site have been dated to about 3.6 m.y., and are therefore the earliest evidence in the world for the *Homo* lineage.

Hence the fossil record is yielding more and more specimens which are elucidating the evolution of a lineage which some of us believe led directly to modern man. It is not unforseeable that even older fossils of the *Homo* lineage will be unearthed.

SOME COMMENTS

Although I have portrayed the recognition of fossils of early *Homo* as unequivocal, this is not the case. We are faced with numerous problems and questions in paleoanthropology. It is appropriate to make note of controversies and difficulties which we are encountering in identifying early *Homo* ancestors.

If *Homo* is a distinct taxon in eastern Africa and overlaps in time with both forms of *Australopithecus* each form must have exploited distinct ecological settings. Some understanding of this problem has come from research by Behrensmeyer (1976, in press) on fossil distributions at Lake Turkana. She has demonstrated that robust australopiths are three times more abundant than *Homo* remains in fluvial deposits. The frequency of both groups are, however, about equal in lake margin deposits. It is, therefore, plausible that robust hominids inhabitated fluvial habitats characterized more by gallery forests and bush. *Homo*, on the other hand, had preferences for more open, drier areas near lake margins. The fluvial deposits adjacent to the Omo River in southern Ethiopia yield predominantly robust hominid remains and thus add some support to the speculations about hominid distributions at Lake Turkana. What the gracile australopiths were doing is not well understood with regard to their habitat preference. "Lucy" from Hadar is similar to gracile australopiths from South Africa and presumably was coeval with *Homo* (Johanson and Taieb, 1976). However, at the moment we do not have sufficient information to delineate habitat differences between "Lucy" and her contemporaneous, but more advanced relatives.

What were the particular selective forces which brought about the development of *Homo* as a distinct lineage? Did this occur only in eastern Africa? For the moment this appears to be the case. What sorts of daily interactions did the different forms of hominids have with one another? Was *Homo* the sole hunter and tool-maker? Was *Homo* responsible, directly or indirectly, for the extinction of *Australopithecus*?

Was there great competition for food sources? If robust forms were vegetarians and *Homo* more of a hunter, or at least a scavenger, perhaps distinct dietary specializations were sufficient isolating mechanisms. Could occupation and exploitation of ecological areas have been seasonal? Was there severe competition for food sources with other primates, or even other animals?

Taxonomically speaking, what is *Homo*? How is *Homo* distinct from *Australopithecus*? There is certainly no confusion distinguishing *Homo* and *A. robustus*, but in body size, tooth size, bone morphology, etc., *Homo* and gracile australopiths evidently did overlap. It may be difficult to establish whether teeth, and even jaws with teeth in some instances, are *A. africanus* or *Homo*. Maybe some material previously referred to *A. africanus* should be reclassified as *Homo* as Robinson (1967, 1972) has suggested.

All of these are questions which display our ignorance of early hominid diversification, adaptation and phylogeny. We can say that there were at least two distinct forms of hominids during the Plio/Pleistocene time range but exactly how they are related to one another, or to a third or even fourth type, is unknown at this time.

For me the evidence is still far from being completely clear, but I do support the idea of *Homo* in eastern Africa at a fairly early date. This perhaps extends back to Laetolil times and might have an even greater antiquity. The cranial and postcranial fossils are suggestive of this, but there is not universal agreement among paleoanthropologists. Perhaps what is now required is a critical and extensive reappraisal of our definitions of *Homo* and *Australopithecus*. Fossils should be evaluated in new frameworks and ultimately assigned to a particular taxon. Where this is impossible it may be necessary to place such specimens in an indeterminate category. New models of taxonomy and phylogeny should include considerations of morphology above all, but also aspects of ecology, time, geographic distribution, etc.

Although this review is necessarily cursory in its scope, it should serve to enlighten the reader that the study of paleoanthropology is now undergoing considerable changes in approach, both theoretical and practical. In my way of thinking it is highly likely that a *Homo* lineage was distinct some 4.0 million years ago from the other hominid lineages which ultimately became extinct. We shall continue to strive to understand exactly how this came about, where and for what precise evolutionary reasons. Mankind's origins are of great interest to scientists and laymen alike and I have every belief that with continued laboratory and field research the details of our fuzzy picture of the earliest stages of human evolution will come more clearly into focus.

REFERENCES

Behrensmeyer, A. K. (1976) "Taphonomy and paleoecology in the hominid fossil record." *Yb. Phys. Anthrop.* **19**, 36-50 (1975).

Behrensmeyer, A. K. (in press) "The habitat of Plio-Pleistocene hominids in East Africa: taphonomic and stratigraphic evidence." In: *African Hominidae of the Plio-Pleistocene: Evidence, Problems and Strategies,* edited by C. J. Jolly, Duckworth, Cambridge.

Brace, C. L. (1973) "Sexual dimorphism in human evolution." *Yb. Phys. Anthrop.* **16**, 13-49.

Brace, C. L., Mahler, P. E. and Rosen, R. B. "Tooth measurements and the rejection of the taxon *Homo habilis.*" *Yb. Phys. Anthrop.* **16**, 50-68 (1972).

Broom, R. and Robinson, J. T. (1950) "Man contemporaneous with the Swartkrans ape-man." *Am. J. Phys. Anthrop.* **8**, 151-6.

Curtis, G. H., Drake, R. E., Cerling, T. E., Cerling, B. L. and Hampel, J. H. (1975) "Age of KBS tuff in Koobi Fora Formation, East Rudolf, Kenya." *Nature*, **258**, 395-8.

Dart, R. A. (1925) "*Australopithecus africanus*: the man-ape of South Africa." *Nature*, **115**, 195-9.

Greenfield, L. O. (1974) "Taxonomic reassessment of two *Ramapithecus* specimens." *Folia primatologica*, **22**, 97-115.

Johanson, D. C. and Taieb, M. (1976) "Plio-Pleistocene hominid discoveries in Hadar, Ethiopia." *Nature*, **260**, 293-7.

Kretzoi, M. (1975) "New ramapithecines and *Pliopithecus* from the Lower Pliocene of Rudabánya in north-eastern Hungary." *Nature*, **257**, 578-81.

Leakey, L. S. B. (1959) "A new fossil skull from Olduvai." *Nature*, **184**, 491-493.

Leakey, L. S. B., Tobias, P. V. and Napier, J. R. (1964) "A new species of the genus *Homo* from Olduvai Gorge." *Nature*, **202**, 7-9.

Leakey, M. D., Hay, R. L., Curtis, G. H., Drake, R. E., Jackes, M. K. and White, T. D. (1976) "Fossil hominids from the Laetolil Beds." *Nature*, **262**, 460-6.

Leakey, R. E. F. (1973) "Evidence for an advanced Plio-Pleistocene hominid from east Rudolf, Kenya." *Nature*, **242**, 447-50.

Leakey, R. E. F. (1976) "An overview of the Hominidae from East Rudolf, Kenya." In: Coppens *et al.* (organizers).

LeGros Clark, W. E. (1962) *The Antecedents of Man*, 2nd ed., rev. Edinburgh: Edinburgh Univ. Press.

Patterson, B., Behrensmeyer, A. K. and Sill, W. D. (1970) "Geology and fauna of a new Pliocene locality in north-western Kenya." *Nature*, **226** 918-921.

Pilbeam, D. (1972) *The Ascent of Man.* The Macmillan Series in Physical Anthropology. New York: The Macmillan Company.

Robinson, J. T. (1953) "*Telanthropus* and its phylogenetic significance." *Am. J. Phys. Anthrop.* **11**, 445-501.

Robinson, J. T. (1963) "Adaptive radiation in the australopithecines and the origin of man." In: F. C. Howell and F. Bourliere (eds.), *African Ecology and Human Evolution*, Aldine, Chicago, pp. 385-416.

Robinson, J. T. (1965) "*Homo 'habilis'* and the australopithecines." *Nature*, **205**, 121-4.

Robinson, J. T. (1966) Comment on: The distinctiveness of *Homo habilis* (by P. V. Tobias). *Nature*, **209**, 957-60.

Robinson, J. T. (1967) "Variation and taxonomy of the early hominids." In: *Evolutionary Biology*, edited by T. Dobzhansky, M. K. Hecht and W. Steere, New York: Appleton-Century-Crofts, pp. 69-99.

Robinson, J. T. (1972) *Early Hominid Posture and Locomotion*, Chicago: Univ. of Chicago Press.

Simons, E. L. (1972) *Primate Evolution*, New York: Macmillan Company.

Taieb, M., Johanson, D. C., Coppens, Y. and Aronson, J. L. (1976) "Geological and paleontological background of Hadar hominid site, Afar, Ethiopia." *Nature*, **260**, 289-293.

Tobias, P. V. (1967) *Olduvai Gorge*, Vol. 2: *The Cranium and Maxillary Dentition of* Australopithecus (Zinjanthropus) boisei, Cambridge Univ. Press, Cambridge.

Wolpoff, M. H. (1970) "The evidence for multiple hominid taxa at Swartkrans." *Am. Anthrop.* 72, 576-607.

Sir Vincent B. Wigglesworth, C.B.E., F.R.S.

Fellow of Gonville and Caius College at the University of Cambridge. Formerly Quick Professor of Biology at the University of Cambridge and Director of the Agricultural Research Council Unit of Insect Physiology.

THE CONTROL OF FORM IN THE LIVING BODY

New discoveries in science are not made by plunges into the unknown. They are made on the misty fringes of "the known" by observers whose eyes can pierce the fog more deeply than others. I therefore make no apology for devoting much of this essay to what is known. But what *is* "known" in science? The philosophical answer is: "nothing". What a scientist means when he says that something is known, is merely that he has recognized certain consistencies in the sequence of events which lend plausibility to the idea that certain causes are at work, and the observed consistencies suggest that the phenomena follow certain laws. This is well recognized as a religious approach; it rests upon an unquestioning faith that natural phenomena conform to "laws of nature" whose origins are pressed back into the innermost recesses of our consciousness.

If you look at your two hands you will find that they are remarkably similar: mirror images of one another; and yet they are different from the hands of any other person. How is this amazing precision in form achieved? The short answer is simple enough: by the genes. Forty or fifty years ago the genes appeared only to be concerned in adding the details, often more or less trivial details, to a body whose general form was already decided. But "molecular biology", in its impact upon biochemistry and genetics, has changed all that.

The essential substance of the living body is still protein. The building blocks of proteins are still the amino acids. The pattern of this building determines the properties of each protein. The pattern in question is controlled by a special family of ribonucleic acids (RNA). These in turn are synthesized under the immediate direction of specific deoxyribonucleic acids (DNA) located at constant sites in given chromosomes of the cell nucleus. It is these chromosomal sites of DNA which are the genes.

But the genes as a whole constitute an heirarchical system, with genes in groups co-operating to give wide-ranging effects; genes controlling the timing of the activity of other genes; genes concerned in the formation of hormones which can bring about dramatic changes throughout the body; in short, a system of interacting regulators of baffling complexity. To say that the body form is controlled by the genes is hardly more illuminating scientifically than to say that it is controlled by God.

In the past, the physiologists, that is mammalian physiologists, such as that most philosophical of physiologists J. S. Haldane, have come near to saying that the physiology of embryonic development is unknowable. It was the biologists who began to establish some of the principles at work by studying experimentally more simple organisms: slime moulds, algae such as *Acetabularia*, protozoa, sponges, planarians, echinoderms, insects and, of course, amphibians. For the purpose of this essay I propose to limit my remarks to the insects, which have two outstanding advantages for experimental study.

(i) In the later stages of development, that is, after hatching from the egg, the entire visible form of the insect consists of an inert and stable cuticle, the structure and pattern of which persists unchanged until it is shed at the next moult when the cuticle of the new form, or instar, appears. Thus during the period immediately preceding a moult, the epidermal cells below the cuticle pass through a phase of growth and cell division which leads to changes in form and pattern that provide admirable opportunities for experiment.

(ii) Insects also have an unusual mode of development in the egg. The egg is relatively large and rich in yolk. The fertilized egg nucleus, the zygote, which occupies only a very small amount of space within the egg, divides repeatedly, and the daughter nuclei, the cleavage nuclei, proceed to migrate towards the surface. A few remain behind as yolk cells, but most of them settle in the cortical layer of the cytoplasm immediately below the egg shell and here they form a regular epithelium. In the higher insects the cortical plasma seems to carry an invisible mosaic of some unknown nature which has the effect of deciding the fate of the cleavage nuclei which settle in the different zones. In 1909 Hegner had shown that in the potato beetle *Leptinotarsa* those cleavage nuclei which arrived at the posterior pole of the egg become the germ cells that are incorporated into the future gonads. It was later observed that a cleavage cell might divide shortly before arrival at the surface and that one daughter cell entering the zone of "germ cell determinants" became a germ cell, the other daughter cell falling outside this zone did not. If the zone of germ cell determinants is tied off, pricked with a needle, or injured with ultraviolet irradiation, no pole cells are formed and the resulting insects have no germ cells in the gonads.

The mosaic state has been regarded as particularly characteristic of the higher Diptera, such as the house-fly *Musca* or the fruit fly *Drosophila.* In these insects it seemed that even immediately after the egg was laid the cortical plasma was already an invisible mosaic; not only did the posterior pole enforce germ cell formation upon the nuclei entering this zone, but other regions compelled the arriving nuclei to form the head, the thorax or the abdomen. It seemed that even at this stage the precursor of the future organism existed already, laid out in invisible form in the cortex and enforcing the cells to follow determined lines of development and so to contribute to particular parts of the body. The cortical cytoplasm of the egg is formed in the ovary of the mother and is thus exposed to the maternal gene system. It was formerly argued that in the higher Diptera the blueprint of the future embryo was already mapped out at this early stage. But recent work has shown that even in *Drosophila* there is a stage in the newly laid egg at which the mosaic map of determination does not yet exist and in many insects this plastic state of affairs may persist even after the cellular blastoderm has formed.

If the eggs of certain midges (*Smittia*, *Chironomus*) are centrifuged, or exposed to ultraviolet irradiation, many queer disturbances in development can occur: there may be a head at each end of the embryo, connected together, with no abdomen; or there may be larval abdomen at each end, joined together, with no head. And if cortical plasma is removed from the anterior region of the newly laid egg of *Drosophila* and implanted at the abdominal end, it can lead to the development of anterior structures at the site of implantation.

Experiments of this kind have led to the conclusion that there is a graded developmental effect deriving from some point in the system. At the high point of the gradient, head structures are formed; at the low point, abdominal structures. "Determination" at the various regions in the egg represents the establishment of this gradient – and that may happen (in higher Diptera) at the cortical plasma phase of development or (in most insects) after the cellular blastoderm has formed.

What is the nature of the supposed gradient? It has been suggested that levels in the gradient are characterized by the appearance of different determinants, each evoking a different type of development in the gene system. Alternatively it can be supposed that a single chemical determinant exists; and that we are concerned with a concentration gradient of this substance, the different gene activities being evoked by different levels in the concentration of the one substance.

This idea that quantitative differences in the concentration of a chemical could produce such striking qualitative differences in morphology may sound improbable; but we do have the example of the juvenile hormone of insects. This is a relatively simple chemical, a derivative of farnesenic acid, which in high concentrations results in the development and maintenance of larval characters, as in the caterpillar of Lepidoptera. At a low concentration (in the presence, of course, of the moulting hormone which is needed for growth and moulting) it leads to development of the pupa. And when completely absent the insect undergoes metamorphosis to the adult stage with the development of wings and genitalia, and a complete transformation of the mouth parts and the form of the body.

At the present time it is not even established that the gradient is a chemical gradient. If it is so, it is to be hoped that in due time the active agent will be isolated and analysed. There are already the usual candidates: RNA of some kind, acetylcholine, cyclic AMP or cyclic GMP, prostaglandins, etc., waiting in the wings. But it is well to point out that much can be learned before that is done. It has been of great interest to learn the chemical nature of the juvenile hormone – but almost everything we now know about the functions of this group of hormones had already been discovered before that was achieved.

If this general picture of differentiation in the early embryo is correct, in other words, if that process does conform to the ground rules described above, it is highly probable that the same rules will operate in the later stages of development; and in studying later development we may be able to find examples which lend support to these ideas.

Many years ago Carlson observed that at a certain stage in development of the nervous system in a grasshopper, the "neuroblasts", or "nerve mother cells", divide and give rise to two daughter cells, one of which is again a "neuroblast" and the other is a "ganglion cell". He found that if microsurgery was performed on a neuroblast at the time when it was undergoing mitosis, it was possible, when the nucleus was in metaphase, to rotate the mitotic spindle through 180° with a dissecting needle, so that the chromosomes which would have gone to form the future ganglion cell now go into the opposing daughter cell. But this makes no difference to the morphogenetic result: the ganglion cell and neuroblast appear as expected, although the chromosomes have been transposed. It is clearly the cytoplasm which has been changed or "differentiated" and which determines what the nature of the cell will be. This experiment is reminiscent of the fate of the cleavage cells in the higher Diptera on entering the cortical plasma.

The most instructive medium on which to study the control of form during later development of the insect is the epidermis, the single cell layer which lays down the cuticle. Throughout much of larval and pupal development an occasional epidermal cell will give rise to a tactile sensillum, a diminutive sense organ produced by four cells (Fig. 1). The epidermal cell in question divides into four daughter cells: a trichogen cell which lays down the tactile seta, a tormogen cell which forms its socket, a primary sense cell which gives off a dendrite running to the base of the seta and an axon which grows inwards to connect up with the central nervous system, and a neurilemma cell which produces a sheath for the axon and dendrite. This striking process in which a single cell gives rise to four very different daughter cells is probably of the same nature as that of the nerve mother cell and the ganglion cell. That is, the fate of the chromosomes is determined by differences in the cytoplasm in which they find themselves, and it is again the cytoplasm which evokes particular activities in the gene system; but that has not been proved experimentally.

A particularly instructive way of studying the control of form at a late stage in growth is by observing the process of regeneration that follows injury. The epidermis will commonly lay down tactile setae at regular intervals. If the cells are killed over a wide area by holding a heated rod in contact with the cuticle, the surrounding epidermal cells divide and multiply and spread inwards to repair the injury; and at the next moult they lay down new cuticle over the burned area. The cells and cuticle appear quite normal, but the tactile setae are absent. However, when the insect moults again tactile setae appear with the usual intervals between.

Here again we seem to be concerned with a gradient phenomenon. A cell or group of cells determined to contribute to the building of the sensillum seem to inhibit the neighbouring cells from doing the same thing. The nature of this effect is unknown; but it appears to diminish

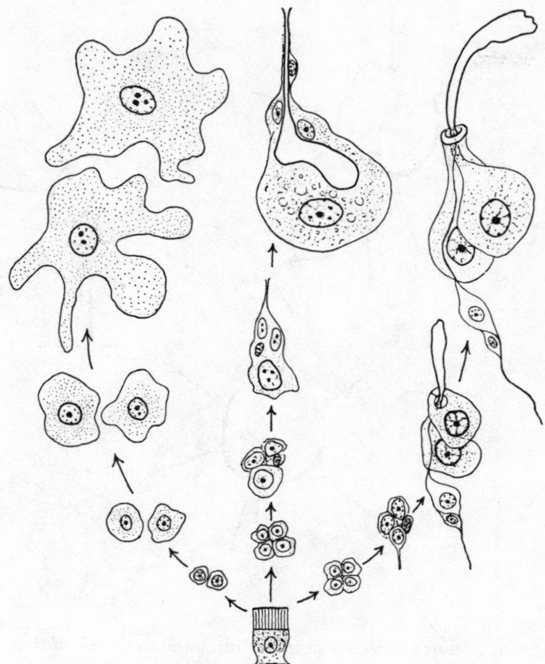

Fig. 1. The types of differentiation of which an epidermal cell in an insect (the blood-sucking bug *Rhodnius*) is capable. To the right is the differentiation into four cells which become respectively a trichogen, a tormogen, a ganglion cell and a neurilemma cell.

with distance; and at a certain distance from one developing sensillum a second may make its appearance (Fig. 2); and the final result is an evenly distributed pattern of setae, to which new setae can be added at the next moult where the existing gaps are sufficiently wide.

The whole of the integument shows a consistent pattern; a pattern of pigmentation, of types of setae, of density of packing of the setae, and so forth. Cells contributing some characteristic element in the pattern, when they migrate and multiply during the repair of a burn, carry with them their special characteristics, be these the capacity to form pigmented cuticle, or to form cuticle with densely packed sensilla (Fig. 3). Clearly the local characters of the cells are transmitted to their daughter cells. Whether this transmission is by way of the nuclear genes or by the cytoplasm is not known.

In many insects, for example in the cockroach, if a leg is removed by cutting through the basal segment, or coxa, in one of the young stages, a new leg of normal form and proportions is regenerated at the ensuing moult. If the leg is cut off through the femur or through the tibia only the more distal parts are renewed. There seems again to be a gradient of regenerative potential. The base of the leg can regenerate an entire limb, but not a new cockroach. Parts lower down the limb can regenerate only those parts that are lower still.

We saw earlier that isolated parts of the embryo may either regenerate the missing parts; or they may form a mirror image of themselves; if the future head region of the blastoderm is divided sufficiently early, two complete embryos may be formed. The same phenomenon is seen during the regeneration of limbs. An injured limb may give rise to a triplicated structure; and when that happens, two of the new distal fragments are always mirror images of one

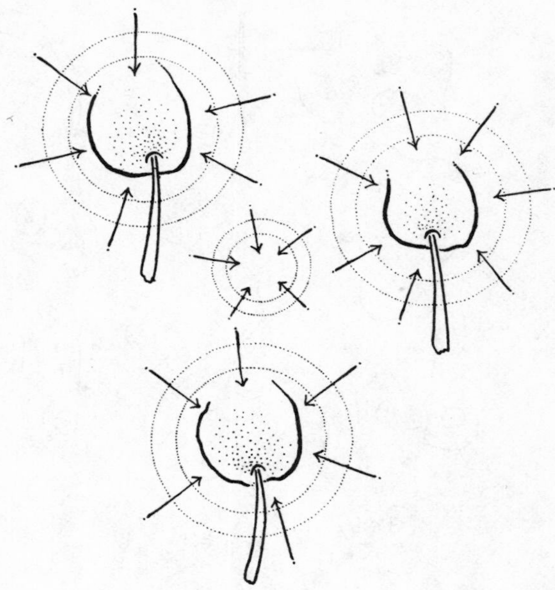

Fig. 2. Theory of determination of tactile setae in the integument of *Rhodnius*. Explanation in the text.

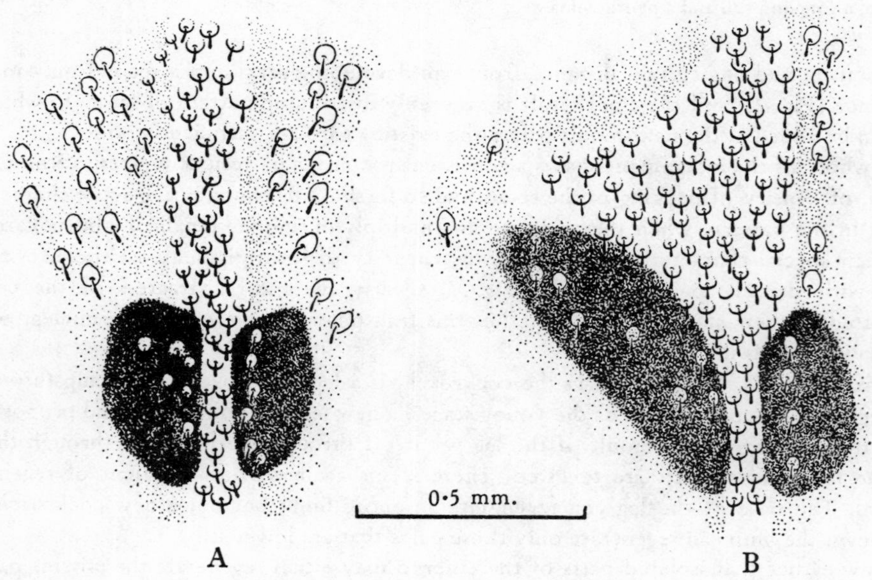

Fig. 3. (A) Surface view of the margin of the abdomen of a normal *Rhodnius* larva showing two pigment spots, and setae which form a closely packed band running between them. (B) The same area in a larva in which the epidermis had been killed by a small burn on the left side. Elements in the pattern, both the pigment spot and the densely packed setae, have extended into the burned area.

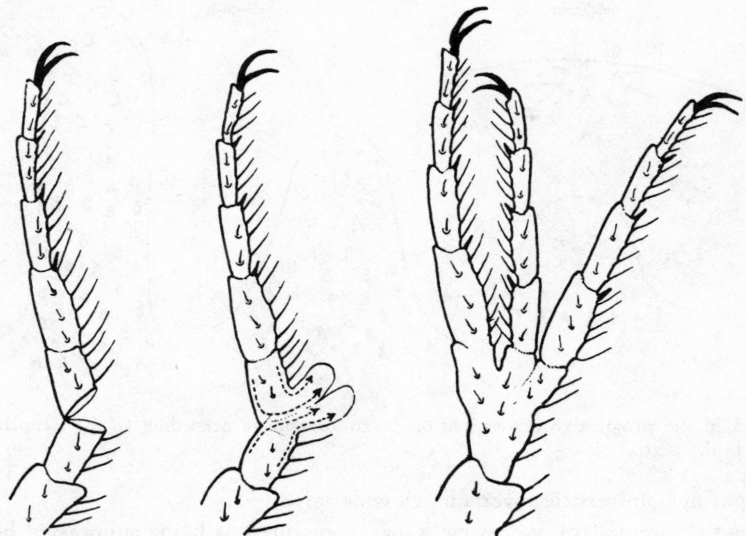

Fig. 4. Drawings of regeneration in a partially severed limb to illustrate Przibram's explanation of Bateson's law. Description in the text.

another (Bateson's law). The explanation put forward by Przibram in 1921 is illustrated in Fig. 4. If the regeneration of the partially detached fragment is considered it can be seen that cells which grow out from the various faces of the limb will carry with them the characters for which these faces have been determined (just as the black pigment spots or other types of cuticle spread during the repair of a burn, as shown in Fig. 3). It follows therefore that the regenerate arising from the distal fragment (see the middle branch on the right of Fig. 4) will be a mirror image of itself, whereas the regenerate from the basal fragment (the right hand branch in Fig. 4) will have the normal orientation of the limb.

The essential feature of these processes of regeneration is the ability of tissue to regenerate only those structures which are at a lower point in the gradient of differentiation in the limb. Some years ago I put forward a conception for the graded differentiation of such a limb, which was derived by analogy from the differentiation of a tactile sensillum in the epidermis (Fig. 2). A certain zone in the epidermis has lost its capacity to form other parts of the body but is fully competent to form a limb; but the cells around this zone can no longer form a limb (just like the cells around a sensillum). The nature of this inhibition is not known. For the sake of argument I suggested that it might result from the uptake by these cells from all the zone around, of materials necessary for limb formation. These cells form the field "A" in Fig. 5. They multiply and start to form the leg bud; and then the same process is repeated leading to a new and more specialized field of determination "B". This in turn, by the same process, leads to a new field "C". And so the process continues, with the uptake by an active centre of the materials necessary for a particular determination and the consequent inhibition in the surrounding zones of centres with the same potential activities.

As I have often pointed out, this process is analogous with the differentiation of institutions in human societies: the rural landscape is marked at intervals with churches, each of which meets the needs of christians in the surrounding area and thus inhibits the emergence of new churches in the immediate neighbourhood; universities, by the uptake of scholars, suppress the

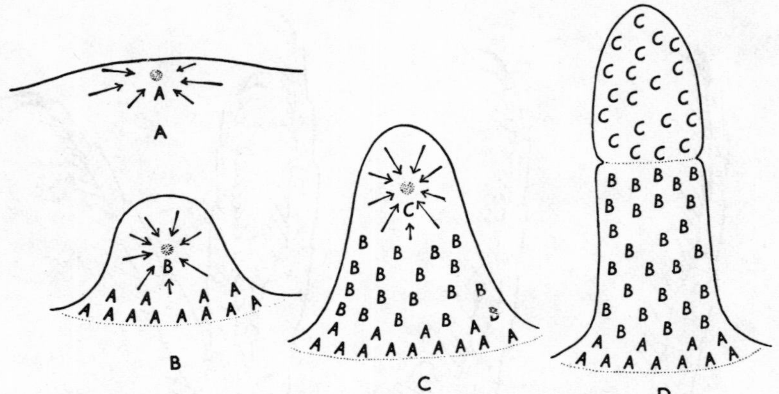

Fig. 5. Stages in the progressive determination in an appendage according to the hypothesis put forward. Explanation in the text.

appearance of new universities over a much wider area.

The growth potential of successive zones is pictured as being suppressed because they are deprived of essential raw material; but, of course, other mechanisms are possible; and, as pointed out earlier, we may be dealing not with a succession of active substances, but with diminished concentrations of a single active substance. There is thus not only a limitation of potencies in each activated centre, there is also a positive evocation of faculties hitherto dormant. The process continues until terminal structures are produced like the trichogen, tormogen, or sense cell of the sensillum, in which as the result of chemical determination followed by differentiative cell division the capacity of the cell is specialized for the performance of a single function.

This autonomous process of activation and suppression of chemical potentiality leading to the production of a mosaic of "fields" each with multiple potentialities and capable of differentiating further by the same procedure, is what I picture as the basic mechanism of the control of body form in all organisms.

It may seem a far cry from the legs of the cockroach to the hands that write these words. But the process of regeneration in the legs of our lowly cousins the amphibia are very similar to those in insects. As in the cockroach it includes the frequent formation of triplicated limbs, two branches of which are mirror images.

At each step in this building process, protein synthesis is under the control of the gene system. Our constellation of genes is absolutely personal to ourselves; no other person alive or dead has had exactly the same assemblage of genes. Very many of these genes will have direct or indirect effects on the growth process. The genes are the same in all parts of the body. It is therefore not surprising that in the graded stages of differentiation which culminate in the production of our body form, the ultimate result proves to be a pair of hands which are exact mirror images of one another.

FURTHER READING

Lawrence, P. (Ed.) (1976) *Insect Development* (Symposia of the Royal Entomological Society of London, No. 8), Blackwell, Oxford.

Wigglesworth, V. B. (1959) *The Control of Growth and Form*, Cornell University Press, Ithaca.

Horace B. Barlow, F.R.S.

Royal Society Research Fellow in Physiology at the University of Cambridge.

Formerly Professor of Physiological Optics at the University of California, Berkeley.

Research interests include the mechanisms of perception and the neurophysiology and neurology of language.

THE LANGUAGES OF THE BRAIN

The object of this essay is to point to a fascinating question whose answer seems almost within reach. With what language does one part of the brain communicate with another? How does the sight of a snake arouse fear, or the sound of a familiar voice cause pleasure? The nervous activity first roused by the image of a snake occurs in regions of the brain far removed from those thought to be associated with emotional states, and the same is true for the auditory centres detecting the familiar voice. How then do the visual or auditory centres tell the rest of the brain that there is a fearsome object or pleasant person in the vicinity?

At first one may think that the word "snake", or the person's name, is communicated, for in our conscious thought we make free use of the symbols of the language by which we communicate verbally with each other. Some information about the neural organisation of these language faculties has been obtained from the clinical study of individuals where they have been disrupted by injury to restricted parts of the brain. We shall return to this, but when we want to move a limb the messages that go down the nerves are very different from the word "move", and when a loud noise occurs what passes up the nerves from the ear is not the word "bang". We want to understand the language of communication between the parts of the brain in the same way that we understand the language by which muscles are controlled and the language by which sense organs tell us what is going on in the world around us.

Now a language is a system of conventions that enables effective communication to occur, and one of the most striking developments of recent years has been the creation of languages that enable man to control computers efficiently. More recently, these developments have been carried to the point where, in special cases, ordinary English can be used both for input and output, and this allows one to understand at a new level what is involved in organising a language system.

The neurophysiological study of sensation, the clinical study of injuries to the language mechanisms of the brain, and the attempts to make computers use human language, are topics in which rapid advances may converge and greatly improve an understanding of the languages of the brain. But there is a quite different reason for being specially interested in the subject, which is that human consciousness resides at the interface between internal and external language. Consider for a moment what it is that most urgently demands your attention; certainly the spoken word would be high on the list, for it enters our consciousness with great vividness and force. It is remarkable that, if words are pitted against direct experience, the word will often win and many will believe it and not what their own senses directly tell them.

Another experience with a conscious vividness comparable to that of the received verbal message is the speech in our own mind clamouring to be spoken, the prepared thought that we wish to make public. External language becoming internal and internal language becoming external are certainly important components of consciousness, and one must also be reminded that one is *not* conscious of many of the internal communications of the brain that are not at the interface of the two languages, because they have not been recently received and are not about to be spoken. This applies to the Freudian subconscious, and also to sensory messages before they have been analysed to the level of the elements of our conscious perception, and to the detailed commands involved in executing a conscious motor decision. If I decide to light a cigarette, I am hardly conscious of my hand going to my pocket for the package and totally unaware of the detailed and intricate sequence of commands to individual muscles; a large proportion of the brain's internal communications must be of this sort. At a more important level, one would not wish to deny conscious status to the thoughts, dreams, or wishes that we experience but have no intention of uttering, nor to the more forceful mental states induced by emotion-laden events and by aesthetic and religious experience, but the major part of consciousness surely resides close to the interface between internal and external language, and this gives our problem special interest.

NEUROPHYSIOLOGY OF THE INNER LANGUAGE

One part of the brain is connected to other parts by means of nerve fibres which conduct all-or-nothing impulses at rates up to nearly 1000/sec. These undoubtedly form the main medium of communication within the brain, and I can imagine someone being puzzled or even irritated by mention of "language". But consider two people having a discussion or argument: neglecting facial expressions or other visible gestures, the main medium of communication between them is the air, which transmits the pressure fluctuation generated in one person's throat to the other's ears. It is interesting and important to know that these sounds are the communication medium, but this fact by itself tells one extraordinarily little about spoken language. Similarly there must be much more to be found out about the language of the brain than the fact that nerve fibres are the communication medium. One must of course take note of the fact that nerve fibres pick up stimuli from a restricted group of cells and transmit their activity to another restricted group of cells. In this respect nerve messages are quite unlike sound waves that propagate indiscriminately in all directions, but nevertheless they are used in particular ways, and it is the system of conventions and restrictions surrounding their use that is referred to as "language". To get a feeling for this let us see first how nerves control muscular contraction.

Control of Muscular Contraction

As you watch a cat jump on to the mantelpiece and wend its way delicately past the ornaments it is hard to believe that the elementary components of the movements are spasmodic synchronised jerks of dozens of muscle fibres. Yet that is the case, for mammalian muscles are composed of groups of about 10 to 500 individual fibres each about 1/30 mm in diameter and several centimetres long. Such a group is called a "motor unit" and is controlled by a single nerve fibre. When a single impulse travels down this fibre it causes synchronous activation of all these fibres, resulting in a brief twitch lasting about 1/20 second. How are these twitches formed into smooth movements?

The answer to this question would require much more than can possibly be considered here, but one aspect will be isolated to illustrate how the language for muscle control depends on particular ways of using nerve impulses. The aspect to be considered is how a smoothly sustained tension of any required value can be produced from these twitches. One variable is the frequency of impulses travelling down a nerve fibre, but there is a difficulty in using this, for below about 10/sec. the muscle fibres will simply jerk in time with the impulses. This might be useful for a pianist executing a trill, but it is not what is required for producing smooth sustained tension. From frequencies of 20 to about 50/sec, the twitches are no longer separated and a fairly smooth contraction is obtained which becomes stronger as the frequency is increased, but it is only possible to reduce the tension from its maximum by a factor of 3 or 4 before the contraction becomes very tremulous.

Another method of adjusting tension would be to vary the number of nerve fibres, and hence the number of motor units partaking in the contraction. Now the muscle is innervated by a fairly small number of nerve fibres which varies from muscle to muscle, but let us say 100 for the sake of discussion. Each of these produces a particular tension when activated at such a rate that a reasonably smooth contraction is produced. Which nerve fibres are to be activated to produce a particular tension? The situation is rather like putting weights on a balance. One could have 100 equal weights and use whatever proportion was required to produce the desired force. Alternatively one could have a binary series, each weight being half the value of the next larger and twice the value of the next smaller. This would enable a much finer gradation of tension to be obtained; in fact with these unequal weights, equivalent to unequal numbers of muscle fibres in the motor units of different nerve fibres, one could obtain the same accuracy of gradation with only 7, instead of 100 equal motor fibres, but there would be a penalty in the requirement of a rather complex system for producing the combination to be activated for any particular required tension. In fact the method actually employed for muscle control is to have variable-sized motor units, but they do not form a binary series, nor are they recruited in a complex sequence. As increasing tension is required they are recruited always in the same order; the first recruited has fewest muscle fibres, and the latest recruited is always larger than any previous one. They are graded so that each newly recruited unit adds a roughly constant percentage to the tension developed by those already active and thus each newly recruited fibre increases total tension by a roughly constant factor. As with 100 equal-sized motor units, only 100 gradations are possible, but the gradations are fine for weak tensions, coarser for strong tensions. The fixed order of recruitment obviously simplifies the control required compared with the binary, combinatorial system.

Much more could be said about the control of muscular movement, but this detail does show that there is more to the problem than the answer "It is done by nerve impulses". The system of using nerve impulses is also interesting and important.

Sensory Messages

To find out how conversation is conducted within the brain itself one needs to intercept messages passing from one part to another within it, and there is no great difficulty in doing this. However, interpreting the records obtained is a serious problem because one does not usually know what the intercepted messages were supposed to achieve. To understand a new language it is not enough just to have some messages in it; one also needs their meanings, translations into a language one already knows. For this reason it is helpful to see how ordinary sensations are conveyed to the brain by nerve impulses, for one does know something of the meaning of these messages both from one's own subjective experience of them, and because the physical nature of sensory stimuli is usually well understood. The visual system has special advantages, but let us first consider other sensory messages.

As the cat executes its graceful movements it is kept informed of its progress not only by its eyes, but also by messages from its pads and elsewhere in the skin, its vibrissae, the organs of balance in its inner ears, and the sense organs of joints and muscles mentioned above. The first point to note about these messages is that, although the nerve impulses that carry them all look similar when recorded by a probing electrode, their meaning varies greatly, and depends upon the particular nerve fibre conducting the impulses. In one fibre they may signify that a whisker has brushed against an obstruction, in another that the surface upon which the cat has placed its pad is cold, in a third that a gust of air has ruffled the fur on its flank. The meaning of these nerve impulses depends upon the anatomical connections of the fibres conveying them, for the stimulus that activates a fibre depends upon what part of the body it comes from, and what particular sensory organ it innervates in that part, while the impulses are all alike.

It is not easy to believe that all our experiences of the external world, even the most vivid and interesting of them, are conveyed by these small and brief electrical disturbances in nerve fibres. But these anonymous impulses are the units of activity in a system of dazzling complexity, and it is the system that gives the language of sensation its meaning.

Vision

When we come to the visual system we can record not only from the receptors and their immediate connections, but also from nerve cells that are several synaptic stages removed from them, though still almost exclusively concerned with the visual input. Clearly we must be extremely interested in what goes on in these synaptic stages, for we can feel reasonably confident that the transformations brought about serve the purpose of presenting the visual image to the rest of the brain in a suitable form. In other words the physical image is being transformed to the psychic image, and this is represented in the language we are interested in. We thus have an opportunity to gain real insight into the nature of this language.

The image formed by the cornea and lens is focused upon a thin layer containing well over 100 million cells at the back of the eye. Most of these are receptor cells that absorb light and generate a response which is picked up by a class of neurons called "bipolar cells", which in turn pass on excitation to "ganglion cells". These have "axons", long fibre-like processes passing up to the brain which together constitute the optic nerve. All visual sensations are carried along these one million nerve fibres which terminate on another group of cells in the brain, the Lateral Geniculate Nucleus. The axons of the cells of this nucleus in their turn carry the message up to the primary visual area of the cerebral cortex. From here the connecting fibres pass to other parts of the visual cortex, first to a group of regions still primarily concerned with vision, then to the other parts of the brain.

One can record the activity of all these elements, but it is misleading to think of the receptor, bipolar, ganglion cell, lateral geniculate neurone, primary cortical cell, and secondary cortical cell as successive links in a linear chain. They really form an intricate network with cross-connections at each level, and as the pattern of excitation is passed on from one set of cells to the next it is transformed by both excitatory and inhibitory interactions at the synapses. Because we do not know the anatomy and physiology of these synapses in sufficient detail we cannot possibly describe these transformations in terms of the mechanisms that produce them. In fact, as is often the case when trying to understand electronic apparatus, one can do without knowledge of the detailed modes of action of parts of the circuit, for what one wants to know is the nature of the operations that these parts perform. It is, to use that jargon, the block diagram not the circuit diagram that is of interest.

The most successful experimental approach to the problem of describing the transformations has been the single-unit method. The activity of single cells is recorded one at a time at various points along the visual pathway, while the visual input is manipulated. Some types of input produce vigorous activity, others none at all, and the properties of the nerve cell one is recording from are described by finding out what visual stimuli cause it to respond most vigorously. In this way one can build up knowledge of the variety of visual events that cause activity in different cells at different levels.

As one might expect, the results of recording from the receptors tell one mostly about the biophysics of reception, and the visual event that causes activity can be simply defined: receptors are affected by light of the appropriate wavelength falling on the appropriate part of the retina. But two stages later, from the retinal ganglion cells, one already finds cells which discharge only to a distinctive *pattern* of excitation of the receptors, and which therefore have quite a complex meaning.

This is the type of transformation that is of interest, and the simplest one is inversion. The receptors are all alike in the direction of their responses to light, but some ganglion cells are activated by light, some by the reduction of light. Further, temporal transients are emphasised, for the responses of receptors are quite well sustained whereas those of ganglion cells often depend on a transient increase or decrease of light, the average level appearing unimportant.

As well as emphasising temporal changes, the connections to the cell are so arranged that spatial discontinuities cause stronger activity than uniform illumination. This is done by having opposing connections to a ganglion cell from receptors in the immediate vicinity and from receptors placed a little further away. Uniform illumination excites both sets of receptors, and since their effects are opposed the resulting excitation of the ganglion cell is less than it would have been if only one set had been excited. So, for instance, one can record from a ganglion cell that discharges to a reduction of illumination in one particular region of the visual field, but because of inhibition from remote regions this cell will not discharge if the whole visual field is synchronously dimmed. It is convenient to describe the essential property of such a cell by its "trigger feature", the stimulus that is particularly effective in causing it to discharge; in this case it would be the sudden appearance of a small black spot, or the movement of a black spot on to the part of the visual field served by the particular ganglion cell.

Even at the level of retinal ganglion cells the trigger features may be much more complicated. In animals with well-developed colour vision, such as many species of fish and monkey, one often finds that black and white stimuli are ineffective for a particular ganglion cell and instead the trigger feature is a localised patch of colour. In other species, such as the rabbit and ground squirrel, the most effective stimulus for a cell may turn out to be movement in a particular direction. To discriminate movement, say, upwards from movement downwards demands a mechanism that can distinguish between two classes of spatio-temporal patterns of excitation of the receptors. Furthermore, one finds that the discrimination is correctly performed over a wide range of conditions: it makes little difference if the moving object is light against a dark background or dark against a light background, and the intensities of illumination can also be varied greatly. This invariant property, the ability to disregard changes in non-essential aspects of the scene, is characteristic of the much more complex types of pattern recognition required to identify named objects (such as snakes, or the voice of a friend), though of course direction of movement is a vastly simpler task.

These changes to the physical image make a good deal of intuitive sense when considered both from a subjective or a behavioural point of view. What comes to be more prominently represented in the optic nerve message is also more prominent subjectively, and similarly for features like prey and predators that are important for an animal. It is fascinating to see how information about the factors known to trigger various forms of behaviour is preserved in the optic nerve message, whilst information about other aspects of the physical image is discarded. It is also fascinating to have an opportunity of studying in the retina the physiological mechanisms that achieve some degree of specificity for these behavioural releasers. But these are not the aspects of interest for this essay: we want to discuss what these results imply about the internal language of the brain.

Clearly the optic nerve conveys a complicated message, and since we have a pretty good idea what these messages achieve we can tell something about the language used in the brain. The list of trigger features that single unit recording has built up show that the eye is communicating with an "alphabet" of one or two dozen different symbols, each replicated in many places to cover the whole visual field. The referents of these symbols are the trigger features we have just described, such as the appearance of a small dark spot, or movement of an object upwards in a particular region. Let us compare this with a written sentence, which is a linear string of alphanumeric symbols or punctuations chosen from about 80 alternatives. Now the message from the eye differs from this linear string in being two-dimensional rather than one-dimensional, and it also differs fundamentally in the possibility of combined activity of two or more of the twenty or so varieties of fibre coming from each region of the visual field: letters and words are used in a mutually exclusive fashion, whereas nerve fibres are not. However, this latter statement has to be qualified, for there are subgroups of nerve fibres that *are* only used in a mutually exclusive fashion. For instance, two retinal ganglion cells that pick up from the same retinal region, one discharging for local increases of light, the other for local decreases, cannot ordinarily respond together. There are other infrequent, or ungrammatical, combinations, for instance a set of four directionally selective units that respond maximally to movements up, down, left, and right, cannot normally all discharge at once, and there are similar restrictions among the colour-selective neurones.

Nerves from the eye are apparently not used in the mutually exclusive fashion of letters, but their pattern of use does have restrictions that can perhaps be likened to grammar. Furthermore, in comparing activity at the level of receptors with the message elaborated by the ganglion cells and sent to the brain, there is a clearly defined tendency towards what may be called "economy of impulses". The eyes are used under very varied circumstances and one cannot make any general statements about whereabouts in their dynamic range the receptors are likely to operate. On the other hand, although the ganglion cells are capable of firing impulses at rates up to nearly 1000/sec, very few if any will actually be firing above 100 to 200/sec in any particular circumstance. The situation will change if something in the field of view moves, when a small proportion of fibres will discharge vigorously, or if the eye itself is moved, when a larger proportion will give a burst.But as soon as movement ceases, the level of general activity declines towards the lower end of each cell's possible range of activity, and this must be the usual state. It seems that the brain is economical with its impulses and does not waste them.

This tendency to reduce the average level of firing is even more surprising when one looks at numbers. There are some 10^8 receptors in each human eye, but only 10^6 nerve fibres. Furthermore, in the visual centres of the human brain there are at least 10^9 cells which can be reached pretty directly by the incoming message, so the 10^6 fibres of the optic nerve must be considered something of a bottleneck. It would be reasonable to expect highly efficient usage in which the nerves fired, on average, at half their maximum rate and every possible combination of activity was employed. This would be the way to utilise them most efficiently, but as in the case of muscle control the nervous system seems to avoid combinatorial tricks, and to a large extent the meaning of activity in a given fibre is independent of the context of activity of the other fibres around it. It is actually this feature which has made single unit recording such a powerful tool, for we would not have got far if it had been necessary to know what was happening in half a dozen other, unrecorded, fibres before we could understand the conditions for activity in a single one successfully isolated.

Visual Cortex

When we look at the optic nerve we can be pretty sure that the purpose of the messages is to inform the brain about the retinal image, but when we follow these messages into the brain itself we lose this invaluable knowledge and cease to be certain what the messages we intercept achieve. This is partly because of the vastly greater complexity compared with the retina. Even in the primary visual cortex, the first place the messages reach, there are at least 100 cells for each input fibre, and we really do not understand why there are that number. Some, perhaps the majority, are concerned with relaying the message to other regions of cerebral cortex, and elsewhere in the brain, though as in the retina the information is also sorted out and transformed before being passed on. But the other difficulty is simple ignorance: though many of the facts are agreed to, the interpretation of them is uncertain while it is certain that much remains unknown. I shall therefore give my own interpretation, and this is of course likely to require revision over the coming years.

The image seems to be divided up, like a jig-saw puzzle, into a few thousand pieces which are first analysed separately. The pieces in the centre of the field of view are very small, and they get progressively larger in the periphery. There are an approximately equal number in each annular ring centred on the point of fixation, and each fragment is not greatly different in radial and tangential size. Unlike a jig-saw puzzle, the pieces overlap, each including parts also covered by its neighbours.

It is the job of the primary visual cortex to describe the salient features of each piece, and disseminate this information to other visual centres. Like the retina it does this by having cells selectively sensitive to particular features of the input from the retinal image, and the trigger features of the different neurones provide a set of terms for relaying a description of the image. The most important of the descriptive terms available are those giving the orientation of lines and edges in the fragment of the image, and probably also their size and fuzziness, or spatial frequency content. The images from the two eyes are superimposed and both included in one piece of the jig-saw, so it seems likely that the description will take note of how well lines and edges are lined up in the images from the two eyes, for this information is important for assigning depth to parts of the image. Furthermore, colour must be included, and it is possible that there may be terms for direction and velocity of movement, and perhaps other properties.

This information is relayed to secondary visual cortical regions which specialise in various aspects of the description, looking at the descriptions from several of the original pieces simultaneously. Knowledge of these secondary regions is still very fragmentary, but the cells respond selectively to quite complicated pattern features of the image. For example, many units in the area immediately adjacent to the primary visual cortex are vigorously excited by a line or edge of a particular orientation in a particular part of the visual field, but they also require that the line is terminated in that region. The trigger feature for such a cell is not just a border at a particular orientation and position, but rather discontinuity of a border, that is its change of orientation or termination. One also finds that the positional requirement may be relaxed, without any reduction in other requirements, so that such a unit detects a specific feature, but has acquired some capacity for positional invariance. This is comparable to the invariance for contrast and luminance noted previously in retinal motion detectors, but invariance for the positional dimensions brings out even more strongly the fact that these cells are truly pattern selective.

In other regions one finds further examples of pattern selectivity based upon what has already been achieved in the primary cortex. For instance in one region one finds cells that respond selectively to objects nearer than the plane of fixation while neighbouring cells respond only to more distant objects. These cells receive their inputs from the primary cortex, and the selectivity for depth is based on the requirement that the image be correctly positioned in both eyes simultaneously.

Other regions specialise in colour, yet others in movement. The most complex and specific cells so far described are certain ones in the infero-temporal cortex of monkeys which respond to objects having the characteristics of a monkey forelimb. Other cells in the same region have much vaguer requirements and there are genuine difficulties in establishing and documenting the properties of rarely isolated cells with such highly specific trigger features. It is clear that modern anatomical methods will, with a great deal of work, build up a picture of where these secondary visual regions are and how they interconnect, but I am not sure that current ideas about the functional requirements in image processing are sufficient to enable the physiology to be properly elucidated. There is a double gap in our knowledge: first, we do not know what a single cell or a slab of cortex does or can do in the way of combining, transforming, and storing, information; second, we do not know what operations have to be performed to make sense of pictures. One can hope for some guidance on the second point as the art of computer image processing progresses, but knowledge of the first can really only be built up as we go along. It is part of the problem of understanding the language of the brain, and this will be a good point to summarise where we have got with the visual image.

We saw that the retinal image is signalled to the brain using an "alphabet" of some twenty different types of nerve fibres. In the cortex the first step is to subdivide the image into a few thousand small regions, and then to transmit to other regions a description of each piece of this jig-saw puzzle. The number of optic nerve fibres for each piece is of the order of 100 to 1000, but one can expect that only a small proportion will be vigorously active at any one time, and some combinations of activity are impossible. The number of terms available to provide a description of the pattern of activity of each piece is too large to designate as an alphabet; they would perhaps be better thought of as words from a restricted vocabulary giving the orientation of contours, their fuzziness, direction and velocity of movement, and also information about colour. As with the retinal output, the words are not mutually exclusive, though certain combinations will not occur, and it is also true that the maximum number ever employed is enormously less than the number available. This is because, to an even greater extent than with retinal ganglion cells, the cortical neurones practice economy of impulses and are quiet or only gently active except when their quite specific trigger feature is present. The patterned excitation required to excite any given cell must occur relatively frequently in ordinary use.

The description provided by primary cortex are distributed to secondary regions where presumably they are recombined and another description passed on elsewhere. This is where we lose sight of the visual image, but it does not seem too fanciful to develop the analogy with written language. Marks on paper are grouped first into lines. Collections of these lines are classified as letters of the alphabet. Collections of letters are recognised as words, collections of words as sentences, and so on. One can find an equivalent sequence in spoken language. The serial formation of retinal output, primary cortical output, secondary cortical output, and so on may have some kind of isomorphism with the steps that are needed to extract meaning from written or spoken language. Or to express this another way, the steps required to process language input may be very similar to the steps applied to all inputs, all the time. At all events, setting out the analogy perhaps increases confidence in our belief that the visual image, as we lose sight of it physiologically, has come to be represented by a collection of symbols which have a status not too different from that of a word in spoken or written language.

REPRESENTATION OF LANGUAGE IN THE BRAIN

The next source of information is derived from observations very different from the recording of activity in single units one at a time. Since the middle of the last century it has been known that damage to certain regions on the left side of the cerebral cortex, resulting from traumatic injury, tumours, or vascular accident, causes aphasia, or defective speech. Acquiring this knowledge has been a slow and distressing process. First the actual defect of speech must be analysed and recorded as carefully as possible, which is not at all an easy job when dealing with skills as complex as those involved in language. Then one has to keep a careful check on the progress or remission of the signs until the patient dies, when it may be possible to do a complete neuropathological examination of the brain. One may perhaps be able to make more rapid progress now, for scanning X-ray techniques enormously improve the accuracy of localisation of pathological changes in the brains of living patients.

As with the visual system it is necessary to have some rudimentary knowledge of the anatomy of the cerebral cortex. On each side it is a thin sheet about 2.5 mm thick and 1000 cm^2 in area, but much folded to give the characteristic wrinkled appearance of the surface of the brain. Each square centimetre contains about 5×10^6 cells which are clustered in six layers, but usually have processes extending the full 2.5 mm depth of the cortex. The whole sheet is surprisingly uniform in appearance and structure, though under detailed microscopic examination it can be divided into some fifty distinguishable regions.

Some of these regions are distinct from the rest in receiving direct input from the sense organs of sight, hearing and touch, or making direct connections to the centres controlling muscular movements. These three sensory and one motor region constitute four of the five "primordial zones", which are the earliest to develop. The fifth primordial zone is the "limbic" cortex, which is a region thought to be concerned with the sensations, appetites, and motor responses dealing most directly with the immediate survival of the animal and species, such as hunger, thirst, sexual sensations, rage and fear.

In many animals the primordial zones occupy almost the whole cortex, but in the much larger cortex of humans they occupy only a small fraction and are separated by "intermediate zones", which develop later and are more uniform in detailed structure. These regions are also called "association cortex" and they are thought to be where information from different modalities is combined, or associations with limbic activities or motor actions formed. In one such region, lying close to the auditory primordial zone and between it and the visual and tactile zones, lies a region called Wernicke's area, injury to which results in disturbances of speech. A second association area adjacent to the part of the motor primordial zone controlling lips, face and tongue, also controls speech and is called Broca's area. This was discovered in 1861, 13 years earlier than Wernicke's discovery in 1874.

The most surprising feature of these two speech areas is that, unlike most other specialised regions of the cortex, they are unilateral. In right-handed people they lie in the left cortex, which controls the right half of the body, while in some left-handed people they lie in the right cortex. The disturbances to speech resulting from damage to the two areas are not the same. As might be expected, Wernicke's area, lying between the auditory and visual sensory areas, seems to be concerned with attaching words to objects and events. Broca's area, lying adjacent to the motor cortex, is concerned with the production of speech, though it would be a mistake to suppose that the impairment following damage to this area is simply due to muscular paralysis. The defect seems rather to be an inability to form grammatical sentences. A patient with this form of aphasia comprehends language and can name objects, but he produces halting speech lacking the short connective words and often containing grammatically incorrect endings. By contrast the Wernicke's type of aphasic produces fluent and grammatical speech, but it lacks connected meaning, is full of circumlocutions and paraphrases, and may contain plausible-sounding but incorrect words. The patient also shows a failure to comprehend language, unlike the sufferer from the other type of aphasia, and it is rather as if what was damaged in Broca's type ran freely and out of control in the Wernicke type. In both cases the disorder of the spoken language tends to be paralleled in any written language they produce.

These facts give rise to the following picture of the functional organisation of language. Close to the auditory cortex is a region where the patterned sounds that form words are represented. In the visual association cortex the visual image is analysed into nameable objects, and cross-modal connections between visual and auditory association areas enable the words to be associated with the nameable objects. The region where auditory, visual-auditory, and also tactile-auditory associations are formed is Wernicke's speech area. Connections pass from there to Broca's area, which is responsible for forming a unidimensional, grammatical, string that can then be executed by the motor area.

Geschwind has pointed out the key role of cross-modal associations in this model. In lower animals the main connections from primordial cortex go first to their own association cortex, then to limbic cortex and its association cortex. Each sense can communicate with the region devoted to survival of the individual and species, but the senses communicate hardly at all with each other. The result of this arrangement is that cross-modal transfer is difficult or absent. If an animal learns that a seen object, such as a cross or ring, is associated with reward or punishment, it will have difficulty anticipating reward or punishment when a cross or ring is presented to it to feel sightlessly. It does indeed seem plausible to suppose that this underlies language, for cross-modal transfer requires, not only a high degree of symbolic representation of sensory experience, but also a sharing of symbols between modalities with agreement about their meaning. That is, it requires a developed internal language. Once the internal symbols have been created, associating them with uttered sounds seems a comparatively small step to take.

There have been elaborations on this picture since Wernicke proposed it 100 years ago, and it is particularly fascinating to find that some patients lack particular classes of words, such as those for colour or parts of the body; this may link up with the developing anatomical knowledge of secondary visual areas. But I am unhappy about it partly because, not being a clinician or neuropathologist, I fear that I have done justice neither to the wealth of detail that has been built up during this past century of work, nor to the real exceptions and difficulties with the model that have been brought out. But I am also unhappy because, being a physiologist, I would like to be able to visualise the processes in much more concrete terms. How is a word represented? How is an association formed? How is a group of words organised into a sequential string of letters or sounds? If the overall picture is right perhaps one should not worry about details of mechanism, and yet I am not sure that any overall scheme is acceptable without a good deal more knowledge of mechanism than we have now. It would be extraordinarily misleading to think that you understood the time-keeping principle of a watch if you understood its trains of gears but did not understand the oscillatory device that really makes it keep time.

What is required in order to understand and produce language? Not surprisingly there has been little discussion of this in the literature on aphasia, but computer simulation of these abilities may give new insight.

THE REPRESENTATION OF LANGUAGE IN COMPUTERS

The first 20 years of the computer revolution showed that language was a more complex matter than some people initially thought. Those optimists had supposed that mechanical translation would quite readily be done by computers, which can store the contents of dictionaries and rapidly refer to this store. What they did not appreciate was that translation at an acceptable level requires understanding the meaning, not just of single words, nor yet of the whole sentence that is to be translated, but it also requires knowledge contained in previous sentences, and even general knowledge of the world and its ways that is not referred to at all in the material being translated.

Recently, however, computers have been successfully programmed to "understand" ordinary English, in the sense that they can execute commands, answer questions, and absorb new knowledge presented in ordinary language. In Winograd's programme a computer "sees" a table with coloured blocks of various sizes and shapes on it. This is represented on the computer's graphic display which the operator observes, and the computer of course has internal access to what is displayed there. The following kind of dialogue takes place:

Operator: Pick up a big red block.
Computer: OK (and a suitable block is raised by a "hand" on the display).
Operator: Grasp the pyramid.
Computer: I don't understand which pyramid you mean. (There are several: the computer does nothing.)
Operator: Find a block which is taller than the one you are holding and put it into the box.
Computer: By "it", I assume you mean the block which is taller than the one I am holding. (Does the required actions.)

And so it goes on. It can accept new information. For instance:

Operator: The blue pyramid is mine.
Computer: I understand.
Operator: I own blocks which are not red, but I don't own anything which supports a pyramid.
Computer: I understand.
Operator: Do I own the box?
Computer: No.
Operator: Do I own anything in the box?
Computer: Yes, two things: the blue block and the blue pyramid.

It keeps track of its actions and recalls them if needed. For instance:

Operator: Had you touched any pyramid before you put the green one on the little cube?
Computer: Yes, the green one.
Operator: When did you pick it up?
Computer: While I was stacking up the red cube, a large red block and a large green cube. (This was a previous instruction.)
Operator: Why?
Computer: To get rid of it.

Of particular interest to readers of this book is the fact that it recognises ignorance:

Operator: Is a steeple bigger than every pyramid but not as wide as the thing that supports it?

Computer: Sorry, I don't know the word "steeple".

Winograd's programme required about 80,000 computer words (in a PDP 10) to operate, taking several seconds for each response. Most of the computer's processing and storage facilities were devoted to compiling instructions from the programme for parsing the sentences and finding their meaning. The English vocabulary utilised was tiny, being restricted to only 200 words. Furthermore, the "world and its ways" that was dealt with was a very small one. Nevertheless one has no complaints about the computer's use or comprehension of language, so the claim that it "understands" is not exaggerated. To have the knowledge, vocabulary, and command of English of, say, a person of school-leaving age, would obviously require an expansion of its base of knowledge and vocabulary by orders of magnitude, but Winograd and others seem to think that this is only a matter of degree and would not require radically different operating procedures.

One of the principal features of the programme is the extent to which it is constantly striving for meaning. A word does not have a strict definition, but rather institutes a search for other words in the input stream. This search is based on links between words and propositions about them in the computer's memory, and what is found is crucial in determining the meaning of the word. Out of one word and a search for others emerges a concept, but the model is circular, for the meaning of any concept depends upon the entire knowledge of the system, not just knowledge of the type that would be found in a dictionary. Concepts of this type are hard to visualise or define, but they can be manipulated by the computer, and it is this manipulability which, as Winograd points out, makes them as real as the concept of, say, "force" is to a physicist. The extent to which knowledge must be enmeshed in language is well brought out by the success of this approach.

We now have one working example of a system that can produce and understand external language, while its internal communications are of course conducted in its own internal language. This example cannot fail to help one understand what is required in the knowledge-language system of the brain. As has already been hinted, one can hope that the internal and external languages have some isomorphism in the case of our own brain; this is lacking in a PDP 10, and one may further hope that the daunting complexity of Winograd's programs is largely due to this lack of isomorphism.

CONCLUSIONS

The language for communication between the various higher parts of the brain must use symbols with an agreed meaning, for otherwise one part would not understand what another part said. We do not know what these symbols are, nor how their meanings are agreed to. However, the material discussed in this essay does enable us to hazard a few guesses about the general nature of the language, and here are three of them. (1) Whatever a symbol may correspond to in the external world, internally it is likely to be represented by nerve impulses in a nerve cell or in a family of nerve cells, for we know of no other tokens that can be rapidly exchanged between the parts of the brain. (2) There is nothing to indicate that these internal symbols occur in a single serial string, as do the letters or phonemes in a written or spoken word, and as do the words in a sentence. The cortex is basically a two-dimensional sheet, and activity can certainly occur synchronously at remote places. It is also likely that two different types of symbol can occur synchronously at closely neighbouring points in this sheet. Translation from this parallel, cortical, language into the serial string that must be formed by the speech organs is presumably the function of the grammar of external language, and it must be performed by Broca's speech area. Much of the complexity of computer programmes for dealing with language are concerned with this operation. (3) Nonetheless there are some restrictions on the occurrence of the symbols in the cortical language, for contradictory symbols do not occur together, and there appears to be a general principle of economy of impulses which means that only a very small selection of the possible symbols are used at any one time. The symbols are mutually discouraging, though not mutually exclusive.

This is not a very rich harvest of knowledge to reap from the field surveyed, so our main conclusion is that there is still much to learn about the languages of the brain. But the interface between the internal and external language is the part of our brain's operation of which we are most acutely conscious, and the prospects are good for learning more on this highly interesting topic during the next few decades.

REFERENCES

Some references for the material discussed in this essay are listed below; these have been selected partly as sources of further references. The neurophysiology of muscular control and sensation can be found in any standard textbook of medical physiology, such as Mountcastle's. Levick's review deals with the work on single-unit recording from retinal ganglion cells. The section on visual cortex is based mainly on the work of Hubel and Wiesel. Geschwind's collected papers give a modern viewpoint on aphasia, and Konorski attempts to give a comprehensive picture of the associational capacities of the brain, based partly on similar studies. The section on computer simulation of language is derived from Winograd's book.

Geschwind, N. (1974) Selected papers on language and the brain. Boston studies in the *Philosophy of Science*, Vol. XVI, edited by R. S. Cohen and M. W. Waitofsky, Reidel, Boston.

Hubel, D. H. and Wiesel, T. N. (1974) "Uniformity of monkey striate cortex: a parallel relationship between field size, scatter and magnification factor." *J. Comp. Neurol.* 158, 295-306.

Konorski, J. (1967) *The Integrative Activity of the Brain*, University of Chicago Press, Chicago and London.

Levick, W. R. (1972) "Receptive fields of retinal ganglion cells." Chap. 14, pp. 531-66, in *Handbook of Sensory Physiology*, Vol. VII/2; Springer-Verlag, Berlin.

Mountcastle, V. B. (Ed.) *Medical Physiology*, C. V. Mosby, Saint Louis, 1968 (Part 10, Sensation; Part II, Control of Movement).

Winograd, T. (1972) *Understanding Natural Language*, Academic Press, New York and London.

Richard L. Gregory

Professor of Neuropsychology and Director of Brain and Perception Laboratory, University of Bristol.

Formerly Chairman of the Department of Machine Intelligence and Perception at the University of Edinburgh and Visiting Professor at U.C.L.A., M.I.T. and New York University.

Awarded the CIBA Foundation Research Prize in 1956, the Craik Prize for Physiological Psychology in 1958 and the Waverly Gold Medal in 1960. F.R.S.E., F.R.M.S., F.Z.S. and F.R.S.A.

CONSCIOUSNESS

Why are we so ignorant of the nature of consciousness, though sensation is all we seem to know directly? To many philosophers it is only sensations – of red, pain, tickle and so on – that we know without doubt: yet it is just such "contents of consciousness" that we do not know in other people. Can we, indeed, be sure that other people (or animals, or insects; or computers?) have consciousness? If this cannot be shown – is the difficulty logical or empirical? It might at least be of some help to decide what kind of problem this is.

We believe that consciousness is tied to living organisms: especially human beings, and more particularly to specific regions of the human brain. This notion could be (though of course it might not be) cast in doubt by successful experiments in psychical research: so it does seem to be an empirical question which might be settled by experiment. For the moment we assume that consciousness is always tied to living organisms, and especially to human brains. This in turn generates the question: "What is the relation between consciousness and the matter or functions of the brain?" And what, then, is the relation between physiology and psychology? I shall continue to ask such questions (as headings) and see how far answers might be provided. These will necessarily be tentative and controversial.

1. DO WE *KNOW* THAT OTHER PEOPLE ARE CONSCIOUS?

Philosophers tend to use the word "know" in a strong sense, meaning: "there is no rational doubt", or "It is not possible to doubt". Now one clearly has doubts at specific moments, or in specific situations, as to whether another person is conscious. This can indeed be a serious matter for neurologists, or anaesthetists concerned with states of coma, or with patients undergoing an operation. One may also doubt whether a friend is conscious (or aware) of what one is oneself seeing; for possibly "his mind is wandering" or is even "absent" at that moment. Our concern here is not with temporary absence of consciousness in other people; but rather with its permanent and complete absence. Can we believe that other people are *never* conscious?

Normal life is deeply based on the knowledge, or firm assumption, of consciousness in other people; and that other people's conscious states are in general similar to one's own. This is obvious from the way we protect each other's feelings by being polite; by the way we try to please with food and alcohol, and with the emphasis on the arts of painting and music and the niceties of architecture which are not purely functional. Belief in other people's sensations and emotional states is the starting-point of civilised living. It seems absurd to doubt it. If it is this that Behaviourism denies, Behaviourism as a theory of psychology is an ass. Our belief in the consciousness of other human beings is so strong that – as G. E. Moore and Ludwig Wittgenstein have pointed out – we do not have techniques or language powerful enough to begin to cast doubt on such a firmly held belief. So it is irrational to doubt it.

The same goes for the even more extreme scepticism of solipsism, in which doubt is stated for the existence of *anything*, except the dreamer (as the observer would then be) for again we have no language or techniques sufficiently powerful to weigh against our belief in an external world. Although we cannot experience consciousness in other people, it is irrational to doubt that they have consciousness, and so for normal life and as philosophers we should believe it.

2. DO WE KNOW OURSELVES?

We may be unable to doubt our sensations of red and pain and so on; but does this make it certain that we know ourselves – as origins or centres of sensations? Just how much more do we know of ourselves than of other people? This takes us to Descartes' phrase "I think therefore I am". He might also have said "I experience red therefore I am". But it is not clear how Descartes is arguing from sensations (or of thinking) to the existence of an I as something beyond sensations. He is not claiming he experiences his I (himself) directly: indeed he is claiming that this is some kind of inference from the sensation of thinking. But how does he (or we) know that sensations need "I's"? We may not be quite sure that consciousness needs physical brains (though nowadays we assume this) but it is less sure that an "I" is needed – for we are not at all sure what an I is, beyond sensation, thought, intentions, and so on.

It might (for example) be like saying that a newspaper is more than its pages. We do not doubt the pages, but we may do well to doubt that there is more to a newspaper than its pages – and in the same way doubt that there is an "I" beyond our sensations.

Perhaps we know ourselves very much as we know other people – from reactions or behaviour in various situations. We may also derive the notion of an I, or self, from the syndrome of our sensations, which (as the spots and high temperature of measles) seems to need a unifying concept, or even a substance – consciousness? – having or producing the collection of sensations we know. The psychoanalytical notion (which is indeed quite common in normal thinking) of multiple personalities shows that we do not quite equate *a* mind with *a* body. We may see minds as having cohesion, or unity, from criteria of consistent behaviour. Thus we speak of someone being "not herself" or "beside herself with anger". More subtly, appreciation of loss of integrity, or of honesty, can mean that parts of the personality are lost. But clearly such notions of the unity of personalities (or minds or I's) is not given simply by the "inner" awareness of our sensations: or by the assumption that sensations are linked as a unity in others. This is clear because the "multiple personalities" in a single body may all have (or be supposed to have) similar or identical sensations irrespective of which personality is uppermost. It follows that we do not think of the unity of personalities simply in terms of sensations, or of bodies, but rather in terms of consistency of behaviour; and – though for ourselves only – perhaps also of "syndromes" of sensations. This could be so not only for the analyst looking at such a multiple personality, but also for such a patient considering his several personalities from his single body and brain.

3. HOW IS CONSCIOUSNESS RELATED TO THE BRAIN?

Although the brain has not always been regarded as the seat of consciousness (for example, the ancient Egyptians attributed consciousness to the heart, or the stomach) the significance of the brain was recognised by Greek physicians, and is now universally accepted. The evidence is from changes of sensation by brain damage and stimulation. From the subtle changes of consciousness produced by drugs believed to act on the brain, and other evidence, we infer a close, highly intimate relation between brain and consciousness. But there is a seemingly unbridgeable conceptual gap between the brain as a *physical* object and *mental* consciousness. This is the most baffling problem.

Large gaps are not uncommon in science. There is a large conceptual gap between the voltage across an electric motor and its shaft rotating. This is bridged by theories of electricity and magnetism. But somehow the mind/brain gap seems wider, and much harder to bridge. Why should this be so?

In science we tend to think in terms of functional, or causal significance. This makes it possible to apply inductive procedures (first described by Francis Bacon in *Novam Organum*) for establishing what is necessary for what. By isolating phenomena, it has been possible to describe and map the physical world conceptually with concepts in explanatory theories. One trouble about consciousness is that it cannot (or has not yet been) isolated from brains, to study it in different contexts. So the classical methods of scientific inquiry are not fully available for investigating the brain/mind relation. This is not to say that *nothing* is known. Many correlations have been established between brain function, structure, and consciousness; but we are ignorant of linking principles.

One trouble is that correlations are not in any case sufficient for explanation or for establishing causes. Consider "A causes B". This is not given simply by correlations between A and B. Complete correlation does not tell us whether A causes B, or B causes A, or whether both are caused by some other (hidden) agent, C. The classical example of day following night will serve to illustrate the difficulty. We do not think of day *causing* night, or of night *causing* day. Neither do we quite think of a third agent ("C") doing the causing. We see the causal relation from a *conceptual model* of the solar system, with the Earth rotating and so on. In general, causes and causal directions are given not by correlations but by adequate theoretical models. Such models are accepted and used (if implicitly) in common situations, such as stones breaking windows. We know that it is not due to breaking windows attracting stones – because we have a theoretical model of the world to direct our conceptual cause arrow the other way round: to stones breaking windows rather than breaking windows attracting stones. But sometimes it is extremely difficult to decide on the direction of causes, though correlations are well established. A change of theory (or paradigm) may change the direction of causal arrows without change of data. This is so, for example, with psychosomatic diseases. There are fashions for "physical" or "mental" causes here, and such changing accounts occur also in physics. This uncertainty is endemic to psychology because we do not know how to think about consciousness. Does consciousness causally affect the brain? Is it enough of a *thing* (or force or whatever) to *be* a causal agent? If not, what *use* is consciousness?

The physical sciences have succeeded largely by rejecting consciousness and intelligence from the Universe they describe. Kepler started by thinking of the planets as pushed around by intelligent angels. When he realized that they move in elliptical orbits, guiding intelligence did not seem necessary, and was dropped. This dropping of intelligence as causal in Nature, extended from physics to Darwin's Natural Selection. This is seen as *creating intelligent solutions* to design problems, but not by means of a guiding conscious intelligent entity. The processes are random variation and selection against competition. This generates intelligent solutions without itself being intelligent or conscious. There is indeed no "it" to have intelligence, or consciousness, on Darwin's account. (As there is no valid "I" in "I think therefore I am".) If the brain was developed by Natural Selection we might well suppose that consciousness has survival value. But for this it must, surely, have causal effects. But what effects could awareness, or consciousness, have?

4. IS CONSCIOUSNESS ANY USE?

To physiologists, the muscles move according to signals, which are trains of electrical impulses known as action potentials, travelling at about the speed of sound and varying in frequency from three or four per second up to nearly 1000 per second. All the senses signal to the brain by the same code. What sensation is experienced depends on which part of the brain is stimulated. Each bundle of nerve fibres provides only one kind of sensation, which increases in "subjective" intensity with increase in the frequency of the action potentials. Patterns of firing rates is all the brain receives from the eyes, via the bundle of a million fibres of the optic nerves. Sensations are somehow given – though perhaps very indirectly – from these patterns of firing of peripheral nerves; but sensations can also be produced experimentally by stimulating regions of the brain with weak electric currents. Curiously, though, pain is never elicited by artificial brain stimulation.

When fibres are cut (either "efferent" motor fibres activating muscles *from* the brain, or "afferent" sensory fibres from sense organs *to* the brain) there is typically loss of function. Motor control and sensory stimulation depend on the signals from the fibres. Why, then, do we *need* consciousness? What does consciousness have that the neural signals (and physical brain activity) do not have? Here there is something of a paradox, for if the awareness of consciousness does not have any effect – if consciousness is not a causal agent – then it seems useless, and so should not have developed by evolutionary pressure. If, on the other hand, it is useful it must be a causal agent: but then physiological description in terms of neural activity cannot be complete. Worse, we are on this alternative stuck with mentalistic explanations, which seem outside science. To develop science in this direction we would have to reverse the direction of physical explanations which have so far proved so successful. It might be argued that the "inner" world of consciousness is essentially different from the physical world: but then it seems strange that the physical signals of the nervous system are so important.

Perception is more than the reception of signals from the eyes and other sense organs. Perception completes probable missing features, and it is predictive in time. We do not, however, believe that there are central perceptual processes which are not dependent on physical neural activity. If it could be shown that completion of missing bits, and prediction are *not* entirely neural, but *do* require consciousness, then we should have evidence for mentalistic causal agencies. Excepting telepathy (which is far from proved), there is no such evidence *between* brains, and no evidence *within* brains for non-physical causes.

It is worth pointing out that reactions associated with pain from, say, an accidental burn to the hand with hot water occur *before* the *sensation* of pain, which may be delayed by a second or more. It is therefore difficult to believe that the sensation causes the reflex withdrawal of the hand – for it comes too late.

Much of behaviour cannot, however, in any case, be described adequately in terms of reflexes. Predictive perceptual behaviour is initiated *before* the events for which it is usually appropriate – so it cannot be initiated as a reflex by these events as stimuli. Such behaviour is evidently following an internal simulation of the world, allowing predictive "look ahead" from the brain's simulation, (or map, or hypothesis) of what is happening in the outside world. Perceptual reality – consciousness – may be features of the *simulated* world of the brain on which we act predictively. This may depart from physical reality – as errors, illusions, imaginations and fantasy. Why, though, are we only *aware* of the present, as perceived by our internal simulation – though we *act* on predictions of the immediate future?

Could we say that the internally simulated world (with its fantasy and powers to create new possibilities) *is* consciousness? We might say this; but we are stuck with our previous worry, for presumably *these* high level perceptual processes of the brain, involving memory, guessing and so on, function by physical brain activity. So, again, why is consciousness involved?

We can imagine, and we have begun to build, computers with sensory inputs which function in this kind of way, with internal representations to enrich available information; but we do not believe that these machines are conscious. They remain crude compared to ourselves. Could this be because they lack consciousness? If so, we have no idea how to make machines having fully human performance – or how would we recognise such a machine if it were made – for we do not know how to recognise consciousness except in ourselves.

5. CAN MACHINES BE CONSCIOUS?

Here we meet this same paradox, over consciousness having or not having causal effects. If it has *no* causal effects there seems no reason why machines should not be as perceptually sophisticated as ourselves. If consciousness *is* causally effective we might need to put consciousness into machines. If so, *what features of brain structure or function should be reproduced for machines to be conscious?*

Suppose we made a machine which displayed remarkably human-like perceptual performance, and found that it has signals very like our brain activity during our consciousness. Would this be a clue that the machine is conscious? This seems the best chance of isolating consciousness, and studying it apart from brains. Such machines would serve as "test tubes" used for isolating and examining chemical reactions for examining properties and conditions for consciousness.

If we could establish correlations between physical processes in the brain and states of consciousness (say, by electrical recording from the living brain) and if we found this same kind of activity in machines, would we have evidence for consciousness in such machines? Here I think we must go back to asking what kind of evidence is given by correlations. Correlations are never *sufficient* for establishing causal relations. What is needed is a *unifying account*: a theory providing what we recognise as causal links. Perhaps such experiments might suggest an adequate linking theory. But so far we do not have such a theory: so the causal role of consciousness remains mysterious; and we lack this empirical criterion for knowing a machine is conscious, or for designing conscious machines.

If we should find that consciousness *is* causally important, and that it is associated with characteristic activity which can be detected physically, then we should have evidence for consciousness in other people, for they should have identifiable states related to consciousness. But if consciousness is not causally significant then this empirical bridging is *not* possible. So we have the curious situation that a mentalistic account – which seems "unscientific" – allows empirical bridging from ourselves to other people: while an epiphenomenal account – which seems more "scientific" because it does not trespass upon physiology – does not allow this empirical evidence for consciousness beyond ourselves.

6. ARE "LAWS OF MIND" LAWS OF PHYSICS?

Psychology has traditionally sought "laws of mind" to explain behaviour. There are many terms, such as "motivation", "fear", "hunger" and "shame", which it is quite difficult to conceive as having simple physiological correlates. One can well imagine that the physical state of lack of food is monitored, and signalled to brain regions which activates food-seeking behaviour; and we might describe this in an animal, or another person, to include a sensation like our feeling of hunger. It is more difficult to conceive a physiological state for shame, or guilt, or pride. Is this because these are *social* feelings? Why should these be more difficult to conceive physiologically? Does the difficulty suggest that consciousness is specifically related to social situations, or social needs, or roles? Possibly: but perhaps more likely, this is merely a matter of the complexity or subtlety of social conscious states. It is difficult to conceive of appreciation of music, or of the beauty of a flower in this way: but perhaps appreciation of beauty is social.

The issue is important. It raises the question of how physiology is related to psychology, and whether consciousness can be affected or controlled apart from physiological changes. Even if these are one:one relations, we may still resist physiological explanations in terms of the states of our nervous system for, say, our philosophical opinion or political beliefs. We prefer to think of people "making up their minds" by considering experience, evidence, or arguments – rather than from neural action potentials. We believe that advice can help people in trouble, by making them feel or think differently. For this we consider their habits of thought, and their interests, and fears and so on, rather than physiological states of their brains. Would we retain this attitude if we were less ignorant of physical brain states?

This issue is similar to the issue of how far psychology can be "reduced" to physiology, as chemistry can be "reduced to" physics, even though descriptions in terms of physics can be cumbersome. Are psychological statements shorthand for physiological statements? To consider this, it is useful to look at computers. They have "hardware" processes, described in physical terms, and "software" data processing which is described not in terms of mechanical movements or electronic processes, but rather in terms of the logic of the programmes and logical characteristics of the problem the computer is solving. Just how the computer functions physically is unimportant provided only it carries out the steps of the programmes correctly. Now, errors *may* be due to malfunction of the mechanism, *or* they may be due to logical errors or inadequacy of the programme. Further, a fault may be corrected by changing the computer physically, or by changing program instructions. This looks very like the distinction between treating a psychological problem by drugs or brain surgery, or by "reprogramming" people with words such as advice or analysis. We should note, however, that this distinction does not involve consciousness. We now prefer to think of "mind" as broader than consciousness, and perhaps akin to computer programmes though computers are not conscious. It is at any rate easier – from possible to impossible – to mechanise feeling. But this is complicated by the embarrassing fact that we would not know we had succeeded in creating feelings in a machine – though it beats us at chess.

7. RELATIONS BETWEEN BRAIN AND MIND

It has been suggested that (1) mind and brain are not connected (epiphenomenalism); or that (2) the brain generates consciousness; or (3) that consciousness drives the brain; or (4) that they both work in parallel (like a pair of identical clocks) without causal connection. Clocks, though, rapidly get out of step – so this may be evidence *for* a causal connection.

Another relation is (5) *identity*. It has recently been argued, especially by philosophers in Australia, that brain states and conscious states are *identical*; but that we describe them differently. There are plenty of cases where the same thing is given very different alternative descriptions – so different that we think that the descriptions must be referring to different things though they refer to the same thing. For example, a table may be described as hard and solid and square; but it may also be described as a bunch of atoms and electrons in violent motion. These descriptions appear so different they may *seem* not to refer to the same thing – the table. Could consciousness be *another appearance* of physical brain states? Could they be the same, though described differently? One immediate question, which seems not to have been answered, is: Why, of all physical objects, are only brain states conscious? Or is it supposed that consciousness is widespread among objects? If so we have no knowledge of this. But if not, what makes brains *uniquely* conscious objects?

It is worth pointing out that physical objects cannot be paradoxical, but perceptions can appear paradoxical – so we may question whether brain states can be both *physical* and *paradoxical* – which they would have to be in the (admittedly rare cases) of paradoxical perceptions: such as given by the Penrose impossible figures, some of Escher's drawings, or our impossible triangle object (Fig. 1a, b). As it is generally held that physical objects cannot be logically impossible or paradoxical but perceptions can be: how could such perceptual states of consciousness be *identical* with physical brain states? This may be an argument against the Identity Theory.

Symbols, however, can be paradoxical. For example, $3 \times 4 = 13$ is paradoxical. It is paradoxical in meaning to one who knows how to read numbers. But the numbers *as objects* (marks on paper) are *not* paradoxical. Perhaps this suggests that (6) consciousness is somehow *represented* by brain states, rather as meanings are represented by symbols. If brain states were symbols, at least the difficulty over perceptual paradoxes would disappear. Meaning is in any

Fig. 1a.

Fig. 1b.

case mysterious. Perhaps the best road to follow for understanding consciousness is to explore just *how symbols represent* – to give meaning for men or some machines. Are brains conscious through reading themselves? If so, we should watch out for conscious computers.

REFERENCES

Anscombe, E. and Geach, P. T. (1954) *Descartes: Philosophical Writings*, Ed. and Transl. Nelson, London.
Armstrong, D. M. (1968) *A Materialist Theory of the Mind*, Routledge, Kegan & Paul.
Ayer, A. J. (1940) *The Foundations of Empirical Knowledge*, Macmillan.
Ayer, A. J. (1936) *Language, Truth and Logic*, Revised Edition, Gollancz, 1964.
Ayer, A. J. (1956) *The Problem of Knowledge*, Penguin.
Boden, M. A. (1972) *Purposive Explanation in the Behaviour*, Harvard.
Boring, E. G. (1963) *The Physical Dimensions of Consciousness*, Dover.
Borst, C. V. (Ed.) (1970) *The Mind/Brain Identity Theory*, Macmillan.
Broad, C. D. (1929) *The Mind and its Place in Nature*, Kegan Paul.
Gazzinago and Blakemore, C. (Eds.) (1975) *Handbook of Psychobiology*, Academic Press.
Gregory, R. L. (1970) *The Intelligent Eye*, Weidenfeld and Nicolson.
Hook, S. (Ed.) (1960) *Dimensions of Mind*, N.Y.U.
James, W. (1890) *Principles of Psychology*, (Chapter 9), Henry Holt.
Kenny, A. (1973) *Wittgenstein*, Penguin.
Mach, Ernst (1950) *The Analysis of Sensations*, Dover.
Penfield, W. and Roberts L. (1959) *Speech and Brain Mechanisms*, O.U.P.
Skinner, B. F. (1974) *About Behaviourism*, Jonathan Cape.
Watson, J. B. (1913) Psychology as the behaviourist views it. *Psychol. Rev.* **20**, 158-77.
Wisdom, J. (1946) Other minds. *Proceedings of the Aristotelian Society* Supplementary **20**, 122-147.
Wisdom, J. (1934) *Problems of Minds and Matter*, C.U.P.
Wittgenstein, Ludwig von (1969) *On Certainty*, Blackwell.
Wittgenstein, Ludwig von (1953) *Philosophical Investigations*, Blackwell.
Wooldridge, D. E. (1963) *The Machinery of the Brain*, McGraw Hill.

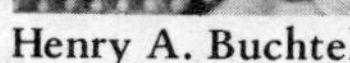

Henry A. Buchtel

Giovanni Berlucchi

Henry A. Buchtel

Research Fellow at the Institute of Human Physiology, University of Parma, Italy.

IBRO/Unesco Fellow at the Institute of Physiology of the University of Pisa, 1969-71 and Clinical Neuropsychologist at the National Hospital for Nervous Diseases, London, 1972-75.

Research interests are in the reorganisation of nervous connectivity as a function of experience or damage, with special reference to the visual system.

Giovanni Berlucchi

Professor of Human Physiology in the Medical School of the University of Pisa and research worker in Neurophysiology at the National Research Council of Italy Laboratory in Pisa.

Research interests are centred on the neurology of vision.

LEARNING AND MEMORY AND THE NERVOUS SYSTEM

"Where or how does the brain store its memories? That is the great mystery." This statement, taken from Boring's (1950) classic work on the history of experimental psychology, is still valid today, despite a quarter of a century of intensive work. However, while speculation at that time might have been along the rather simple lines of electrical fields or reverberating circuits in nervous tissue, it is now clear that the answer to Boring's question will be immensely complicated and will involve the collaborative work of researchers and theorists in many scientific fields.

Research into the brain mechanisms underlying learning and memory has suffered the usual fate of scientific problems receiving concentrated attention: a simple question has generated a myriad of hypotheses, theories and conceptual approaches which need to be tested, unified and extended in order that we may eventually construct a coherent picture of the responsible system or systems. We shall attempt to predict the most probable direction of the research advances which will allow this unification and extension to occur, but at the outset we must admit that we are painfully aware of the multiple pitfalls of prognostication. The popular research of today may lead nowhere; the next step may well depend on advances in entirely unrelated fields about which today's researchers are ignorant. This essay, then, may have value only in providing a clear picture of some of the biases present today – we suspect that in 10 years our predictions will probably be seen as hopelessly short-sighted.

The terms learning and memory have always provided difficulties of definition and we must begin by stating how we intend to use them. We use the term learning to refer to a process which allows an organism to modify its responsiveness or to acquire a new pattern of behaviour as the result of interactions with the environment. A more careful definition requires the exclusion of factors which are capable of modifying behaviour and which may interact with learning, but which must be distinguished from learning: growth; maturation; aging; fatigue; receptor adaptation; changes in arousal, attention or motivation; and disease. Memory can be seen as the relatively permanent manifestation of this change in responsiveness, or as a state or process which allows the retention of information, loosely called "knowledge".

TYPES OF LEARNING

How many kinds of learning and memory are there? This question must be confronted at the outset since the possibility exists that for every kind of learning and memory there might be a fundamentally different kind of underlying mechanism. Physiological psychologists and other brain scientists working in this area have mostly designed their experiments to study habituation and the two well-known kinds of conditioning. These three cases of learning will be referred to several times in this article and for those who are not familiar with them, we shall provide here a brief description of their defining characteristics.

Habituation is classically defined as a reversible decrease in the strength of a response as a result of repeated stimulation (see Thompson and Spencer, 1966, for review). It can be seen as a kind of gating out of information which is insignificant to the organism.

The two kinds of conditioning are type S and type R. Type S conditioning, also called classical conditioning or Pavlovian conditioning, typically involves learning to respond to a previously neutral stimulus due to the pairing of that neutral stimulus with another stimulus which elicits the response to be learned. Type R conditioning, also called respondent conditioning or Skinnerian conditioning, typically involves the increase or decrease in frequency of a particular act as a function of the consequence of the act – basically, if the act is followed by something good, it increases in frequency; if the response is followed by something bad, its frequency decreases.

As will be described in detail below, we are well on the way to understanding the physiological nature of habituation, and we shall argue as a working hypothesis that the basis of conditioning will be in the same general category as that of habituation, although certain details of the underlying morphological and chemical changes will undoubtedly differ. It should be emphasised, however, that habituation and conditioning represent only a part of the field of learning and memory. In higher animals at least, there is another set of phenomena of equal or more importance which will have to be taken into account. These phenomena will be expanded upon below, but we may anticipate that discussion and state here that our working hypothesis of a unitary class of mechanisms underlying learning and memory applies to them as well. Thus, we predict that the problem of different kinds of learning and memory will influence developing models only in terms of particular details, not in general terms.

LOCUS OF LEARNING

Where does learning occur? The simple answer is that learning occurs in the nervous system. Thus the question has very little meaning in a general sense, considering the fact that practically every part of the nervous system is capable of exhibiting at least some kind of learning (see Berlucchi and Buchtel, 1975, for review), and considering the unexpected discovery that virtually all animals, including those of only one cell, are capable of some form of learning. All that seems necessary is some means of receiving information and of expressing the response being studied.

The problem of localising learning and memory has attracted considerable attention in animals with well-developed brains since specific kinds of learning and memory can be shown to depend on activity in different parts of the brain. In man, for instance, learning and memory for many kinds of material have been shown to require an intact limbic system and/or the tissue adjacent to the primary sensory receiving areas. But no matter how precise this kind of localisation becomes, it is clear that not all kinds of learning are going to be attributed to particular cortical or subcortical structures. Therefore, although clinical assessment of brain damage has profited from these kinds of localisation studies, it would hinder our search for the basic mechanisms underlying learning and memory if our examination were limited to learning which depends on one particular brain system simply because it is involved in the acquisition and holding of information which we find most interesting (usually verbal or able to be verbalised; and usually having access to consciousness, at least in humans).

BASIS OF LEARNING

Still, the question remains, do habituation, conditioning and other kinds of learning share a common neural basis? As already indicated above, we suspect the answer to this question to be "yes", both on the grounds of parsimony and because unified principles have been found for other biological (and non-biological) functions. Of course, the techniques for demonstrating this conclusion will have to evolve considerably in the coming years. In any case, a theory of the neural basis of learning and memory presupposes a more general theory of nervous function which is still lacking. It may be said, however, that at the present time most students of nervous function subscribe to the neuron theory (see below), although there is another set of alternatives which have attracted adherents and it may be useful here to describe the two approaches in their historical setting.

Since about the time of Tanzi, whose theory of learning will be outlined below, there have been two competing general theories of neural organisation, each one enjoying periods of relative popularity. One approach has been to consider nervous tissue as diffusely organised into neural nets, electrical fields or hologram-like patterns. The other approach has been to emphasise the precise connectivity of neurons which has been so well demonstrated histologically. This latter approach, the neuron theory, has resisted attempts to refute it while the various alternatives have failed to withstand a detailed scientific analysis (for the most recent alternative, the hologram analogue, see Pribram. 1971).

REACTIVITY AND PLASTICITY

For a general theory of function one must discover general properties of neurons, and the most widely accepted of these is that of reactivity. Reactivity is the capacity of nervous tissue to respond to direct stimulation or to stimuli applied to the sense organs, and is based on the excitability of single neurons and on the propagation of electrical impulses along the nerve cell's membrane. It also depends on the transmissibility of excitatory and inhibitory influences across synapses. The biophysical bases of neural excitability are known in considerable detail and the laws governing them have been shown to have general validity across different animal species and throughout different portions of any given nervous system.

Neural plasticity, a term which was coined by J. Konorski (1948), is used to refer to another property which, along with reactivity, is thought to be basic to nervous tissue. Plasticity is the ability of the nervous system to modify its reactivity as a result of previous activation. It is believed by many to be the specific cause of the changes in behaviour subsumed under the terms learning and memory, and to be different from transformations in neural structure and activity associated with the phylogenetically determined ontogenesis of the organism. Unlike neural reactivity, neural plasticity is poorly understood because heretofore it has been almost exclusively studied by inference from behaviour rather than by direct analysis of the neural mechanisms involved (for review see Vital-Durand, 1975). Accordingly, the nature of the events and processes encompassing neural plasticity is largely obscure, and current attempts at clarifying it are based more on speculation than on experimentation. Nonetheless, the literature on this subject is huge and it would be impossible to review it even cursorily within the present article. We shall discuss selectively a limited number of experimental and theoretical contributions which we regard as being historically important for our present understanding of this subject, or as having heuristic value for the future development in this field of research. Our guiding idea will be that it is still justified to consider neural plasticity as a rather uniform class of mechanisms, provided that we follow Konorski in limiting the term plasticity to defining the neural processes essential for learning and memory. While modifications in nervous organisation almost certainly accompany growth, maturation, aging, recovery from brain damage, and adaptation to unnatural distortions in sensory input, and while they may in principle share some features with the neural processes of learning and memory, there are reasons to believe that they differ in important respects. Plastic changes underlying learning and memory may reasonably be accounted for by functional modifications in synaptic transmission without *major* structural alterations of the nervous tissue. After lesions and during development, on the other hand, striking changes in neural structure such as the establishment of new connections by sprouting and growth of the nervous elements (see Hirsch and Jacobson, 1975) have been observed. And this brings us to a consideration of the putative changes thought to underlie plasticity.

THE SYNAPTIC HYPOTHESIS

Following along the lines of Pavlov's thinking, Konorski (1948) suggested that the plastic neural changes in establishing a conditioned reflex (type S conditioning) consist in the formation of new connections between the neural receiving areas of the conditioned and unconditioned stimuli (the previously neutral, and the initially effective stimuli, respectively). However, contrary to Pavlov, he dismissed the possibility that previously non-existent nerve pathways could appear during the relatively short time needed for the appearance of the conditioned reflex. Rather, he favoured the hypothesis of the transformation of already existing, but only potentially functional, connections between reflex centres. During the training session, these connections become endowed with full functional power. Such a transformation would have to take place at the synapses, perhaps due to the increase in size or number of synaptic contacts between pre-synaptic axons and post-synaptic dendrites or cell bodies. Konorski gave priority to Cajal (1909) for putting forward this morphological and functional hypothesis of neural plasticity but in fact the synaptic hypothesis of learning and memory was originally suggested by the Italian psychiatrist Tanzi, at a time when the word synapse had not yet been coined. Tanzi's theory was later referred to by Cajal in his great textbook of neurohistology (1909) but it was probably too far advanced for its time since it aroused little interest.

The theory was introduced in a review of contemporary neurohistology (Tanzi, 1893), written to support the neuron theory against the "diffuse nerve set" hypothesis entertained by Golgi. Tanzi ventured to advance a "timid" but remarkably lucid "physiopsychological" hypothesis of learning and memory, in which he proposed that each external stimulus acts on the nervous system by producing, in addition to a temporary modification, a longer lasting impression which may become permanent. According to the theory, neurons activated by the stimulus tend to grow in length, thereby reducing their distance from neurons located in their vicinity and conjoined in the same activity. Reduction of the distance between the terminal arborization of the axon of the expanding neuron and the dendrites or soma of the next neuron, brings about an increase in the capacity for transmitting what Tanzi called the "nervous wave" from the first neuron to the second. Such a reduction of the interneuronal distance and the resulting facilitation in transmission are only temporary when a single activation occurs, but they produce a relatively permanent facilitated associative link between neurons when there is massed, repetitious activation, thus providing a basis for memories and learned motor skills.

As mentioned above, Tanzi's hypothesis quickly disappeared from the neurological literature and it was only after more than half a century that Konorski (1948) and especially Hebb (1949) revived interest in the possible importance of modifiability of synaptic transmission.

Hebb apparently did not know of Tanzi's monograph, but the 1949 theory of cerebral organisation is remarkably similar to that of Tanzi in several ways. As Tanzi had done so many years before with the diffuse nerve net theory, Hebb refuted the "field theory" of the nervous system which was then prevailing because of the influence of Gestalt psychology, and constructed a model of the nervous system entirely founded on the neuron theory. Within that framework, he postulated a dual trace mechanism for memory, with a temporary maintenance of information committed to a reverberatory activity within neural circuits, and a relatively permanent incorporation of the same information into a structural change. The proposed structural change is more or less the same as that envisaged by Tanzi. To quote Hebb: "When an axon of cell A is near enough to excite a cell B and repeatedly or persistently takes part in firing it, some growth process or metabolic change takes place in one or both cells such that A's efficiency as one of the cells firing B, is increased" (1949, p. 62).

MODERN DEVELOPMENTS

Since the publication of Hebb's book on the organisation of behaviour (1949), several attempts have been made to confirm or refute the synaptic theory of learning and memory, on both experimental and theoretical grounds. Although the present state of knowledge is very far from allowing any firm conclusion, it is fair to say that the hypothesis has withstood opposition very well, the main reason for this success being its superior clarity and parsimony relative to all other alternative hypotheses. If one were to summarise the position currently held by most students of the neurology of learning and memory, the following two statements would probably suffice:

1. The general plan of the connections within the nervous system is laid out by heredity alone, but proper functioning of the connections must be maintained or brought out by interactions with the environment, especially during the maturation period.
2. Specific environmental influences resulting in learning act by modifying interneuronal communication at one or more of the synapses already existing in the connection system produced by heredity and maturation.

The first statement is corroborated by a host of neurological and behavioural data (see Hirsch and Jacobson, 1975); experimental evidence for the second statement is somewhat scanty and much less conclusive, although research activity in this field is growing steadily and is continuously providing new contributions.

The work of Kandel and his colleagues on the mechanisms of habituation and dishabituation in the mollusk *Aplasia* comes close to the ideal experimental conditions for demonstrating the neural substrates of learning and memory (for reviews see Kandel, 1974; Kupfermann, 1975). Its main features are: (a) the neural pathway subserving the behaviour under study is almost completely identified, due to the relative simplicity of the nervous system of *Aplysia*; (b) the changes in the activity of this identified pathway which accompany and probably cause the changes in observed behaviour, have been localised in specific synapses along the pathway; and (c) electrophysiological investigations have indicated the nature of such synaptic changes and such findings can be used as the basis of further morphological and chemical analyses.

The learning in the case of these *Aplysia* experiments is habituation of a gill withdrawal response following repeated tactile stimulation of the siphon. Dishabituation is caused by strong exteroceptive stimulation of a different part of the body. Habituation is not due to receptor adaptation or neuromuscular fatigue, but rather can be correlated with a decline in the production of action potentials by an identified gill motor neuron located in the abdominal ganglion. This reduced motoneuronal activity is in turn contingent on a decrease in the excitatory efficiency of the sensory neurons which are aroused by the tactile stimulus and which drive the motoneuron. The change in synaptic efficiency has been shown to be presynaptic in nature, and to involve a decrease in the amount of synaptic mediator released with each action potential. A long-term retention of the electrophysiological change (at least 24 hours) has been demonstrated, and it is virtually certain that this retention is related to the retention of behavioural habituation. Dishabituation is caused by heterosynaptic facilitation of the presynaptic terminal which had become the seat of habituation. A schematic representation of this simple system is shown in Fig. 1.

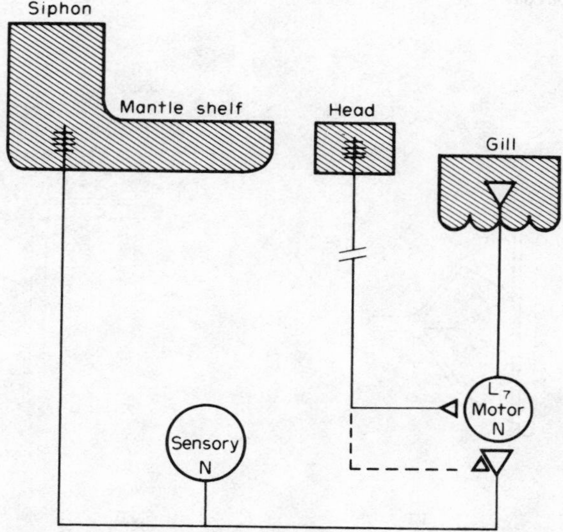

Fig. 1. Simplified scheme indicating receptors and effectors, and nerve circuitry involved in the habituation and dishabituation responses studied by Kandel and collaborators in *Aplysia*. Tactile stimulation of the siphon brings about a withdrawal of the gill; repetition of the stimulus leads to habituation, i.e. suppression of the gill withdrawal. Dishabituation, i.e. reappearance of the gill withdrawal response to siphon stimulation, is produced by exteroceptive stimulation of the head. Habituation appears to depend on a reduced efficiency of transmission at the synapse between the sensory neuron and the motor neuron; this reduced synaptic efficiency is purely presynaptic in origin. Dishabituation appears to depend on facilitation of the previously habituated synapse; such a facilitation is due to an influence exerted on the sensory terminal presynaptic to the motoneuron by a sensory pathway from the head. (From Kandel, 1974.)

As mentioned above, one can predict that by continuing these studies, researchers will soon unveil the precise morphological and chemical synaptic modifications which subserve behavioural habituation and dishabituation. These discoveries will in turn suggest approaches to the study of similar and more complex learning phenomena in vertebrates. However, one might object that these findings in *Aplysia* have no great significance for the neurology of learning and memory in general, both because they concern a very rudimentary kind of learning, and because the neural mechanisms of habituation and dishabituation in mollusks may be quite different from those of higher animals and indeed of other invertebrates. An easy counter-objection is that similar though not identical changes in neural activity may in principle

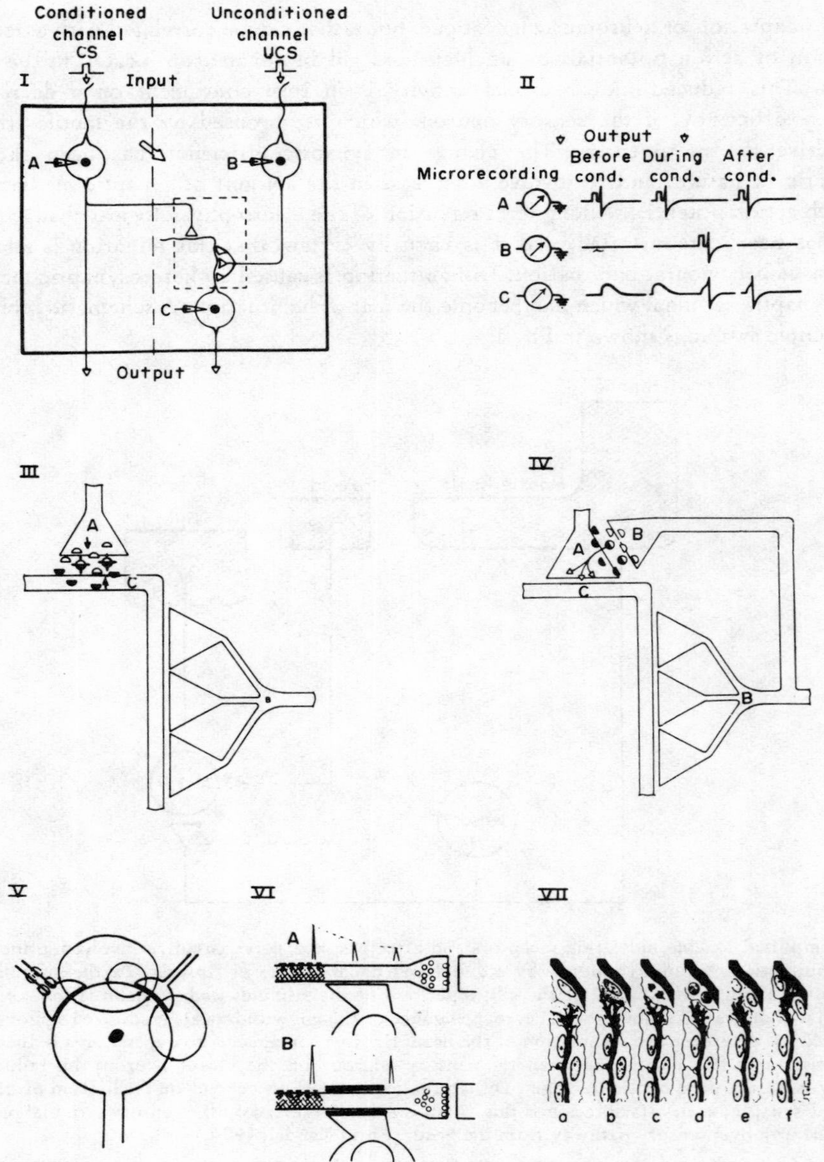

Fig. 2. (From Berlucchi and Buchtel, 1975.)

I and II: classical conditioning of a single neuron. Experiments of this kind have been performed on identified neurons in the abdominal ganglion of the mollusk *Aplysia* (see Kandel and Spencer, 1968; von Baumgarten, 1970). The conditioning stimulus is an electrical shock applied to neuron A. The action potential produced by this stimulus (see IIa) causes a post-synaptic excitatory potential in neuron C which is below threshold for the emission of an action potential (see IIc). The unconditioned stimulus is an electrical shock applied to neuron B, and the action potential which is produced (see IIb) is sufficient to produce an action potential in C (see IIc). If A and then B are repeatedly stimulated in that order, neuron C becomes "conditioned" to respond to stimulation of A with the response that it originally gave only to stimulation of B, that is, an action potential (see IIc). This change in responsiveness in neuron C may last for several minutes or even hours, and is extinguished by stimulating A several times without the matching stimulation of B.

be invoked for the explanation of more complex forms of learning. An answer to the second part of the objection is that our present good understanding of neuronal excitability in mammals has developed largely as the result of studies based on previous findings on the giant nerve fibres of another invertebrate, the squid.

Although the experimental evidence concerning possible synaptic mechanisms of conditioning is minimal, and again largely restricted to invertebrates (Kandel and Spencer, 1968), a number of theoretical models have been developed which attempt to account for the behavioural facts of conditioning in terms of conjectural neural circuits. Some examples are shown in Fig. 2. Other logically possible nerve cell nets capable in principle of producing learning have been extensively explored, but with no reference to the actual mechanisms involved (Burke, 1966; Brindley, 1969). While supporting the view that conditioning, in addition to habituation, may involve changes in synaptic transmission, we must underline the importance of temporal factors in conditioning. The fact that conditioned and unconditioned stimuli must be presented in close temporal proximity is clearly anticipated in all models of learning based on synaptic modification. What is virtually never taken into account is the

---

Fig. 2 (cont.)

III and IV: possible synaptic changes of an essentially chemical nature underlying the neural conditioning illustrated in I and II. These parts of the figure are expanded views of the area outlined by a broken line in I. It is hypothesized that when an action potential arrives at the axon terminal, the synaptic mediator released by A (open half-circles) reacts with a substance released by the post-synaptic neuron C when it is discharged by B (hatched half-circles). The combination of the pre-synaptic and post-synaptic substance would induce changes in the pre-synaptic and/or post-synaptic membrances, thereby increasing synaptic efficiency. The reaction is possible only when A and B fire almost simultaneously, with the action potential in A preceding that in C. In IV an identical interaction is supposed to occur between A and B, resulting in an increased synaptic effect of A and C by means of hetero-synaptic facilitation. The schemes in I, II. III and IV have been taken from von Baumgarten (1970); the synaptic models of conditioning illustrated in III and IV are essentially the same as suggested by Doty (1969).

V: possible synaptic changes, essentially electrical in nature, suggested by Milner (1960) to underlie neuronal conditioning. Milner hypothesized that the synaptic terminal on the right fires the neuron (unconditioned stimulus) and that the current flowing through the membrane invades the synaptic terminal on the left. It is assumed that an action potential running along the fibre of this terminal (conditioning stimulus) dies out before reaching the terminal itself (because, for example, of the increase in membrane surface area at the terminal), the current from the depolarized post-synaptic neuron may help the action potential to get past the critical point. This hypothesis, however, does not account for the temporal paradox of conditioning. In order to do so, it should also postulate some relatively permanent change in the membrane of the initially ineffective synaptic terminal as a result of the combination of the two currents.

VI: glial hypothesis of synaptic changes in conditioning according to Roitbak (1970). It is assumed that the pathway of the unconditioned stimulus and that of the conditioned stimulus converge on common neurons. Action potential travelling to the common neurons via the pathway of the unconditioned stimulus are a signal for mobilization of oligodendrocytes in the area of the neurons in common. The fibre terminals of the conditioned stimulus pathway are thought to be bare of glia in the region of their contact with the common neurons (as in A). Due to the shunting of the depolarizing current prior to conditioning, the action potential dies out before reaching the terminals, and the small amount of current which does arrive releases only a small amount of mediator (open circles in A). It is further assumed that if the bare terminals have just been active, the oligodendrocytes mobilized by the unconditioned stimulus grow around them. This provides a better membrane insulation and causes a large amount of mediator-releasing current to reach the terminals (B).

VII: glial hypothesis of synaptic changes in learning according to Pribram (Karl H. Pribram, *Languages of the Brain*, 1971, by permission of Prentice-Hall, Inc., Englewood Cliffs, N.J.). The original hypothesis simply assumes that RNA released by the excited neuronal terminal induces cell division in the glial cell, thus making space for the growth of the axon toward the neuron beyond. The hypothesis can be applied to conditioning if it is assumed that the glial changes are possible only when the axon terminal and the neuron beyond the glial cell become active together in a fixed temporal sequence.

so-called "temporal paradox" of conditioning (Doty, 1969): the initially neutral stimulus in type S conditioning must come before and not after the unconditioned stimulus; the act or response in type R conditioning must come before and not after the reinforcement or punishment. Attempts to produce learning when the temporal relations are reversed, the so-called "backwards" conditioning, have never been convincingly successful, nevertheless the Burke and Brindley models cited above, and all but one of the models shown in Fig. 2, would permit backwards conditioning. Undoubtedly, the explanation of the temporal factors in conditioning will be made in chemical terms rather than in morphological terms, and in the following section we shall discuss the probable direction of work aimed at discovering the chemical modifications during learning.

CHEMICAL FACTORS IN LEARNING AND MEMORY

We may introduce this section by raising a question often encountered in psychology and clinical neurology: Why are recent memories more subject to disruption than well-established memories? As described earlier, the labile phase of memory may be due to a reverberating process of the type proposed by Hebb (1949) on the basis of the histological drawings of Lorente de Nó (1938). (Strangely enough, Lorente de Nó apparently dismissed the possibility of reverberation: see p. 236 of Delafresnaye, 1954.) Or purely chemical modifications of short duration may be shown to be responsible (Huttunen, 1973; Mark, 1974). The relatively insensitive phase of memory may have a structural basis as suggested by Tanzi, or it may be shown to depend in part on chemical changes at the synapse. In fact, a combination of the two is most likely. At the moment, the chemical changes which are assumed to occur during learning are being studied by fractionating and quantitatively analysing cellular constituents, by radioactively labelling cellular metabolites and precursors, and by inhibiting the synthesis of nucleic acid and protein. The studies are essentially of two types. The first concerns the internal and external environment of the cells during learning, without any presumption that the transitory or permanent changes may be transferred physically to other cells in the same animal or in another animal; the other type of study is orientated toward discovering chemical products of the learning process which in some way contain the acquired information and are capable of producing the new pattern of behaviour in a naïve animal. The latter experiments are perhaps more fascinating than the former, at least superficially, but they are still very controversial and cannot always be replicated. However, some of the findings of apparent inter-animal transfer are convincing, especially those which show a savings in the time taken by the naïve, receiving animal to learn a new task which had previously been learned by the donor; and those experiments in which the transferred behaviour is a passive avoidance.

Unfortunately, the work on transfer of information from one animal to another is rather difficult to relate to electrophysiological studies of synaptic modifiability whereas the results of studies restricted simply to the chemical changes in the cell and synapse during learning are easy to translate into terms of spatial or temporal firing patterns of cells. One may conclude, as has Mark (1974), that the transfer studies are ahead of their time and cannot at the moment be fitted into physiological and psychological models of the learning process. Whether they will ever help in understanding learning and memory is still debatable and we feel that the restricted amount of evidence on a solid footing after so many years of work is an ominous sign. Perhaps what is lacking is a better technique for transferring the information that is thought to reside in a chemical code. Certainly there are indications from other kinds of studies that chemically coded learning does not transfer easily from one part of the brain to another within a given animal: the work by Sperry and colleagues (Sperry, 1961; Gazzaniga, 1970) on patients and animals with surgical disconnection of the two cerebral hemispheres has shown that the two halves of the brain can act independently and that each half can possess information to which the other hemisphere has no access.

HIGHER FORMS OF LEARNING

Habituation and conditioning have been regarded as the forms of learning most amenable to neurological analysis. They have in common the repetition of stimuli and the practice of responses. Yet learning in higher animals, and especially man, is often of a different nature in that it may occur following the presentation of a single stimulus, and in the absence of primary and secondary reinforcement (see Hebb, Lambert and Tucker, 1971). How is one to account for these important aspects of learning in neural terms?

A crucial notion for understanding these modes of higher learning is that of set (Gibson, 1941), a psychological concept which Sperry (1955) has already attempted to deal with from the neurological point of view. He underlines the possibility that central patterns of cerebral organisation, relatively independent of sensory and motor events, may instantly link up different responses with different stimuli so that, in man for instance, virtually any element or elements of the motor repertoire can upon instruction be emitted without prior practice in response to any stimulus capable of exciting the sense organs. Sperry suggests that this dynamic central adjustment of sensorimotor coordination may cooperate with previously existing structural changes in allowing long-term retention of information, but he does not elaborate on the neuronal mechanisms probably involved in the working of the central set. Once again, it is possible and parsimonious to look at changes in synaptic transmission occurring at some strategic place within the nervous system as the substrate for the immediate channelling of sensory commands into motor responses, on instruction. Two entirely hypothetical mechanisms for such quick synaptic change are shown in Fig. 3.

SENSORIMOTOR INTEGRATION

Analyses of brain function based on electrical stimulation, localised lesions and electrophysiological recordings have all led to a distinction between sensory centres and motor centres. This distinction is particularly clear-cut for the cerebral cortex of higher animals. Current neurophysiological analyses in terms of single neuron activity have reinforced the distinction by showing the existence of neurons which code selected features of the sensory environment (Barlow, 1972) or discrete parameters of motor output (Evarts *et al.*, 1971). Yet this distinction may reflect the experimental conditions more than a real separation of input and output functions within the brain. As advocated by Sperry many years ago (Sperry, 1952), *sensorimotor* integration is the end to which the working of the brain is directed. All accounts of central neural organisation based on a drastic separation of sensory and motor operations are bound to be unsuccessful.

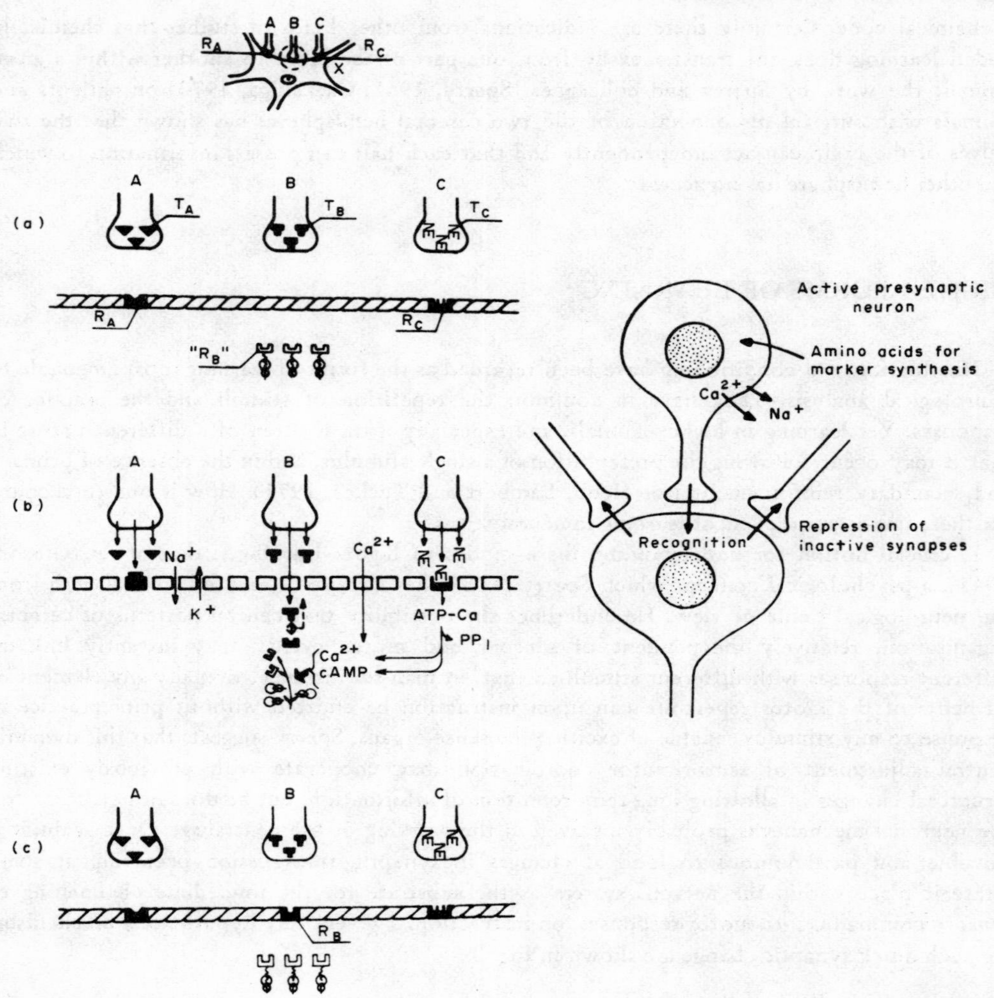

Fig. 3. Models of rapidly modifiable synapses.

Left: Huttunen's model of a neuron (X) post-synaptic to three terminals (A, B and C) which release different synaptic mediators (T_A, T_B and T_C). Synapses A→X and C→X are thought to work effectively in a stable fashion, while synapse B→X is the modifiable synapse which functions only conditionally (see top drawing). When the synapse B→X is not working, this is because the post-synaptic membrane of neuron X contains receptors for T_A (R_A) and T_C (R_C) but not for T_B (R_B). However, R_B molecules, presumably proteic in nature, are consistently synthesised inside neuron X and incorporated into a polisome complex near the unfunctional synaptic contact (see (a)). When neuron X fires in response to A and C is also active, R_B molecules are liberated from the polisome complex because T_C (which is supposed to be norepinephrine, NE) has a releasing action on the polisome complex, exerted through calcium ions (Ca^{++}) and cyclic adenosynmonophosphate (AMP). R_B molecules thus liberated are destroyed by proteolytical enzymes unless a certain amount of T_B, released by a concurrent or previous action potential in B, enters the X membrane and helps them to be incorporated into the post-synaptic membrane (see (b)). The synapse B→X will become therefore functional because of the simultaneous activity of A, B, C and X. (From Huttunen, 1973.)

Right: Mark's model of quick synaptic facilitation and repression. An active pre-synaptic neuron accumulates calcium and sodium, one or both of which improve synaptic transmission. Increase of sodium

There are now some studies showing the existence in the brain of neurons which fire only when certain stimuli impinge on the sense organs *and* certain responses to such stimuli are made or contemplated (Mountcastle, 1975). There is little or no activity when the stimulus does not evoke a response or a preparation to respond, or when the motor act is carried out in the absence of the stimulus. Studies of these cells have so far been performed in relation to innate responses such as reaching or eye-tracking, but it should be possible to extend them to learned associations between different kinds of sensory stimuli and different kinds of motor acts. The demonstration that behavioural learning proceeds in parallel with the emergence in certain neurons of activity strictly specific for the learned sensorimotor integration would undoubtedly provide a model for testing the synaptic hypothesis of learning in complex nervous systems and complex learning situations. Synaptic control of these cells could indeed be tested by electrophysiological methods (i.e. by electrically stimulating their inputs and recording extra- or even intra-cellularly from them) before, during and after learning. Fine anatomical and neurochemical studies could then be started. Admittedly, these electrophysiological and morpho-chemical analyses will be difficult and require methods more sophisticated than those available at present, but in principle the approach would be feasible.

One of the major difficulties of this daunting task will be to locate such cells since they may be few in number and intermingled with other, non-plastic cells. A good place to begin may be the cortical areas which are usually disregarded by cortical physiologists because their activities show little or no association with sensory or motor processes (the "uncommitted cortex" of Penfield, 1966). Cells in these areas would perhaps become more interesting were they studied in relation to sensorimotor associations.

But cortical cells important for learning need not be those which are relatively insensitive to sensory stimulation and motor activities before the acquisition of a sensorimotor association. In fact, a learned sensorimotor association could require the suppression of unwanted responses, just as well as or in addition to the appearance of a new response. An animal can be trained to lift its right forepaw in response to a vertical bar of light and its left forepaw in response to a horizontal bar of light. After it has learned this association, it can be trained to unlearn it and learn the opposite stimulus-response combination. We know that there are cells in the visual cortex responding selectively to vertical or horizontal visual stimuli (Hubel and Wiesel, 1968) and in the case of the learning described immediately above these cells must presumably guide the neurons directly involved in the control of the paw muscles. Other cortical cells respond *unselectively* to both vertical and horizontal visual stimuli and at first sight it seems logical to disregard them for the purposes of the analysis of the learned response. Yet the fact that a particular movement can be attached at one moment to the vertical stimulus and at another moment to the horizontal stimulus, argues for the importance of neurons receiving information about both vertical and horizontal visual stimuli, and capable of channelling the information into the appropriate motor output.

Because of the complexity of the nervous system and the constraints of learning situations, changes in neural activity appearing in temporal relation to behavioural effects of conditioning can only rarely be categorized as causes or effects (Olds *et al.*, 1972), however we believe that eventually this approach will turn out to be successful, especially if single-cell analysis can be

---

Fig. 3 (cont.)

within the neuron potentiates the work of the pump which extrudes this ion from the cell interior, thereby leading to increased accumulation of amino acids within the cell. These amino acids are rapidly incorporated into an antibody-like protein (marker) which is carried to and recognised by the post-synaptic neuron. This latter will repress transmission from pre-synaptic terminals bearing different markers. (From Mark, 1974.)

freed from the distinction between sensory and motor cells, and if more attention can be paid to patterns of sensorimotor integration.

REFERENCES

Barlow, H. B. (1972) "Single units and sensations: a neuron doctrine for perceptual psychology?" *Perception*, 1, 371-94.

Berlucchi, G. and Buchtel, H. A. (1975) "Some trends in the neurological study of learning." In Gazzaniga, M. S. and Blakemore, C. (Eds.), *Handbook of Psychobiology*, pp. 481-98, Academic Press, New York.

Boring, E. G. (1957) *A History of Experimental Psychology*, Appleton-Century-Crofts Inc., New York.

Brindley, G. S. (1969) "Nerve net models of plausible size that perform many simple learning tasks." *Proc. Roy. Soc., Lond.*, Series B, 174, 173-91.

Burke, W. (1968) "Neuronal models for conditioned reflexes." *Nature*, **210**, 269-70.

Cajal, S. R. (1909-11) *Histologie du Système Nerveux de l'Homme et des Vertébrés*, Paris, Maloine.

Delafresnaye, J. F. (Ed.) (1954) *Brain Mechanisms and Consciousness*, Blackwell, Oxford.

Doty, R. W. (1969) "Electrical stimulation of the brain in behavioral context." *Ann. Rev. Psychol.*, **20**, 289-320.

Evarts, E. V., Bizzi, E., Burke, E. R., DeLong, M. and Thach, W. T., Jr. (1971) "Central control of movement." *Neurosci. Res. Prog. Bull.*, 9, 1-170.

Gazzaniga, M. S. (1970) *The Bisected Brain*, Appleton-Century-Crofts Inc., New York.

Gibson, J. J. (1941) "A critical review of the concept of set in contemporary experimental psychology." *Psychol. Bull.*, 38, 781-817.

Hebb, D. O. (1949) *The Organization of Behavior*, Wiley, New York.

Hebb, D. O., Lambert, W. E. and Tucker, G. R. (1971) "Language, thought and experience." *Mod. Lang. J.*, **55**, 212-22.

Hirsch, H. V. B. and Jacobson, M. (1975) "The perfectable brain: principles of neuronal development." In Gazzaniga, M. S. and Blakemore, C. (Eds.), *Handbook of Psychobiology*, pp. 107-37, Academic Press, New York.

Hubel, D. H. and Wiesel, T. N. (1968) "Receptive fields and functional architecture of monkey striate cortex." *J. Physiol.*, 195, 215-43.

Huttunen, M. O. (1973) "General model for the molecular events in synapses during learning." *Perspectives in Biology and Medicine*, 17, 103-8.

Kandel, E. R. (1974) "An invertebrate system for the cellular analysis of simple behaviors and their modifications." In Schmitt, F. O. and Worden, F. G. (Eds.), *The Neurosciences. Third Study Program*, pp. 347-70, The M.I.T. Press, Cambridge, Mass.

Kandel, E. R. and Spencer, W. A. (1968) "Cellular neurophysiological approaches in the study of learning." *Physiol. Rev.*, **48**, 65-134.

Konorski, J. (1948) *Conditioned Reflexes and Neuron Organization*, Cambridge University Press, Cambridge.

Kupfermann, I. (1975) "Neurophysiology of learning." *Ann. Rev. Psychol.* 26, 367-91.

Lorente de Nó, R. (1938) "Synaptic stimulation of motoneurons as a local process." *J. Neurophys.*, 1, 195-206.

Mark, R. (1974) *Memory and Nerve Cell Connections*, Clarendon Press, Oxford.

Milner, P. M. (1960) "Learning in neural systems." In Yovitz, M. C. and Cameron, S. (Eds.), *Self-organizing Systems*, pp. 190-204, Pergamon Press, New York.

Mountcastle, V. B. (1975) "The view from within: pathways to the study of perception." *Johns Hopkins Med. J.*, 136, 109-31.

Olds, J., Disterhoft, J., Segal, M., Kornblith, C. and Hirsch, R. (1972) "Learning centers of the rat brain mapped by measuring latencies of conditioned unit responses." *J. Neurophys.*, 35, 202-19.

Penfield, W. (1966) "Speech, perception and the uncommitted cortex." In Eccles, J. C. (Ed.), *Brain and Conscious Experience*, pp. 217-37, Springer, New York.

Pribram, K. H. (1971) *Languages of the Brain*, Prentice-Hall, Englewood Cliffs, N.J.

Roitbak, A. I. (1970) "A new hypothesis concerning the mechanisms of formation of the conditioned reflex." *Acta Neurobiol. Exp.*, 30, 81-94.

Sperry, R. W. (1952) "Neurology and the mind-brain problem." *Amer. Scientist*, **40**, 291-312.

Sperry, R. W. (1955) "On the neural basis of the conditioned response." *Brit. J. Anim. Behav.*, 3, 41-4.

Sperry, R. W. (1961) "Cerebral organization and behavior." *Science*, **133**, 1749-55.

Tanzi, E. (1893) "I fatti e le induzioni nell'odierna istologia del sistema nervoso." *Riv. Sper. Freniat. Med. Leg.*, 19, 419-72.

Thompson, R. F. and Spencer, W. A. (1966) "Habituation: a model phenomenon for the study of neuronal substrates of behaviour." *Psychol. Rev.,* 73, 16-43.

Vital-Durand, F. (1975) "Towards a definition of neural plasticity: theoretical and practical limitations." In M. Jeannerod and F. Vital-Durand (Eds.), *Aspects of Neural Plasticity*, pp. 251-60, Paris INSERM.

Von Baumgarten, R. J. (1970) "Plasticity in the nervous system at the unitary level." In F. O. Schmitt (Ed.), *The Neurosciences, Second Study Program*, pp. 260-71. Rockefeller University Press, New York.

F. H. C. Crick, F.R.S.

Kieckhefer Distinguished Research Professor, The Salk Institute, San Diego, California, U.S.A.

Awarded the Nobel Prize for Medicine (jointly) in 1962, the Royal Medal of the Royal Society in 1972 and the Copley Medal in 1975. Honorary Fellow of Churchill College and Caius College, Cambridge. Fellow of University College, London.

DEVELOPMENTAL BIOLOGY

Our ignorance of developmental biology has the following curious feature. We understand how an organism can build molecules (even very large molecules) in great variety and with great precision, although the largest of them is far too minute for us to see, even with a high-powered microscope; yet we do not understand how it builds a flower or a hand or an eye, all of which are plainly visible to us. We understand much of what goes on inside a small cell, such as a bacterial cell, whose dimensions are no more than a few times the wavelength of visible light, yet we are ignorant of many important processes in our own cells, and indeed in the cells of most animals and plants. How these cells unite together to form tissues, the tissues to form organs, and the organs to form the whole organism, we can describe but we cannot yet fully comprehend. True, much can be followed with the optical microscope, which allows us to see cells, to see when they divide and when and where they move, but this is not enough. The reason is simple. Life is engineered at the molecular level. To explain what we see, we must understand what we cannot see. And thus, paradoxically, the nearer we are to atoms the more easily we can grasp what is really happening, in spite of the formidable experimental difficulties.

The aim of developmental biology is to explain as fully as possible how an egg and a sperm are made, how they come together to form the fertilized egg, how the cell divides and divides again and again to form a small hollow ball of cells – the blastula; how this undergoes a complicated series of internal movements to form the gastrula, how the various cells change their shape and character to form tissues – and so on until the mature animal itself is built. But developmental biology is wider than this, since it encompasses plants as well as animals, both the familiar higher organisms like the apple tree and the giraffe, as well as those more lowly and less familiar such as the sea urchin and the slime mould.

To appreciate what we yet have to discover, we must first outline what we already know, and this is no easy task since our knowledge has grown enormously in the present century, and especially during the last 25 years. Leaving aside most of the hard parts, living organisms are mainly made from organic molecules, and most of these molecules the organism must synthesize for itself. The raw materials are obtained in various ways. In plants by using photosynthesis to fix the carbon dioxide in the air and from salts taken up by the roots from the soil. In animals by breaking down food into small molecules which are then used as building blocks or as a source of energy. Each simple step of chemical synthesis, typically turning one small molecule into a closely related one, is catalysed by a special catalyst, specific for that step alone, and each of these catalysts (or enzymes, as they are called) is a *large* chemical molecule, typically having tens of thousands of atoms, of the family known as protein. That is, it is made by stringing together certain smaller molecules into long chains. Of the smaller molecules (known as amino acids) there are twenty different types used to build protein and it is the precise order in which they are strung together which makes a particular protein what it is. Moreover, since proteins are made with great precision and are, as a family, both versatile and subtle in their chemical properties, they are used for other important functions: to build structures, to act as signals and so on. A small bacterial cell may possess several thousand *types* of protein, our own cells rather more. Not surprisingly different sets of cells have somewhat different batteries of proteins. A muscle cell has many special proteins needed to construct the contractile apparatus. A red blood cell is stuffed with the protein haemoglobin, which is an oxygen carrier. To a large extent, a cell is what it is because of its proteins. Thus we need to know how proteins are synthesized and how each type of cell produces just the proteins it requires.

Here again, our knowledge has grown enormously. The production of each protein is controlled by a different gene. The essential part of each gene is made of nucleic acid which acts as a molecular message, specifying with great precision the amino acid sequence of that particular protein. We know how working copies are made of this message, how they are read by special reading heads (called ribosomes) each an assembly of large molecules. We know the genetic code; that is, the dictionary which relates the four-letter language of the nucleic acid to the twenty-letter language of the protein. And in addition we know how genes are copied so that each daughter cell may have an exact replica of the genetic instructions stored in its parent. Such, in the briefest outline, is our understanding of the fundamental processes of gene structure, gene replication and gene action.

The first major gap in our knowledge is that while we understand a gene in a very simple organism, such as a bacterial cell, we are uncertain exactly what is in a higher organism. Our chromosomes appear to have much more DNA (deoxyribonucleic acid) than we would expect. Also they have much more protein associated with them than we find associated with bacterial chromosomes. The primary gene product appears too large for a simple messenger function and much of it is broken down rather quickly and never leaves the nucleus of the cell – why, we do not yet know. Nor do we have any example of how, in detail, genes are controlled in higher organisms. However, we are sure that, as in bacteria, proteins are coded by sequences in the DNA of the chromosomes, so we think we understand the essential features of all these processes. It is the baroque superstructure which has so far eluded us; but this is an area of intensive research and one can reasonably expect that the picture will clear dramatically within the next 5 or 10 years.

There are other aspects of the cell which we now understand much better than we did even 5 years ago. Every cell is bounded by a membrane which acts as a barrier to molecules on the outside getting in and, equally important, to many small molecules on the inside which would otherwise leak out. Special proteins act as gates across this external membrane, allowing desirable small molecules to be pumped into the cell, and various unwanted molecules to leave. There are also transducer proteins which, when activated by special chemical signals on the outside, will produce a chemical change inside the cell. In addition, there are membranes *inside* the cell – the membrane of the nucleus is one example – and there are devices whereby packets of molecules, wrapped in a small piece of membrane, can be released from the cell.

Basically a membrane is made of a lipid bilayer which has hydrophobic (water-hating) groups on its inside and hydrophilic (water-loving) groups on its two outer sides. We now know that the bilayer is fairly fluid and we are beginning to understand something about the proteins in it. We need to know much more, especially how the components of the membrane are synthesized and assembled together. Again this is a very active field of research.

Within each cell molecules and organelles are moved about. Cells can change shape, divide, send out processes. How are these movement produced? The molecular tools used appear to be fibrillar protein molecules, long thin aggregates of molecules assembled from protein subunits. We know four major types of these: two, actin and myosin, are closely related to the two major proteins in muscle which slide over each other to produce muscular contraction; one, tubulin, is also found, with other proteins, as the basic structural element in the cilia and the flagella of higher organisms; and one, the protein of neurofilaments, is only just being characterised. There are probably others – certainly other molecules found in muscles are also found in ordinary cells – but it is possible that these four are the only major factors as far as cell movement is concerned. They occur in many animal cells but we still do not know if they are all present in plants. (Plant cells usually have rigid walls and thus move much less than animal cells, although their cytoplasm often flows rapidly within the cell.) Again, we may expect rapid progress in identifying these molecules, locating them in the cell, and finding how they interact. There is one major unsolved problem. How do these fibrillar elements, which produce movement, themselves get shunted to the right place in the cell, at the right time, to produce just the right movement? For example, how does the mitotic spindle form in the correct orientation? And how does a muscle fibre assemble all its components to produce a highly ordered contractile machine? The answer may come from studies on the fibrillar molecules themselves and how they interact or it may involve some other principle. Here we know so little we can only wait and see what will turn up.

But by far the largest area of our ignorance is not what happens inside a cell but what happens between cells. Here our knowledge is fragmentary, since what we know is quite inadequate to explain what we see. Each cell in a multicellular organism is obviously influenced by the general chemical environment in which it is bathed. It must be supplied with sugars, amino acids and other small molecules it needs, since our own cells are not as versatile as many bacterial cells and, not being able to synthesize some of these small molecules for themselves, are dependent upon an external supply. But in addition there are specific chemicals which are in effect instructions. It is not their metabolism that matters, but the controlling effects they have on many sets of metabolic processes inside cells. Many such molecules are known – the sex hormones are good examples of this – and something is being learnt about how they act at the molecular level. But the interactions of cells are regulated in other less familiar ways. In a sheet of cells – an epithelium – a cell can often be shown to know, in some way, where it is in the sheet. It has what is called "positional information", as can be demonstrated in special cases by moving it somewhere else in the sheet and seeing how it behaves. It also knows which way round it is in the sheet. For example, if it produces a bristle, that bristle will often lean over in a particular direction. Cells are probably influenced quite strongly and specifically by their contacts with neighbouring cells. We can see that they often form special junctions, of several different types, when they come into contact and we know some molecules can move from cell to cell across some of the junctions. It is a measure of our ignorance that we do not know whether "positional information" is mediated by small molecules, by cell to cell contact, or by some other process such as signals, analogous in a loose sense to the signals passing down a nerve cell, passing over the tissue. Then again, for a tissue to grow to a fixed size and shape, there must be some control of the divisions of its cells and in many cases over the direction of these divisions. Here again, when we realise what a complicated job has to be done, we can see that our ignorance far exceeds our knowledge. At least we know that it is not always done by a strict lineage mechanism. In some simple organisms – for example a soil nematode – it has been shown that a particular cell in a developing animal will always produce exactly the same descendents. It will divide twice; one of the four grand-daughters will always die and the other three each develop in a special way. If this mother cell is killed in early development, all the three special grand-daughter cells will be absent. No other cell will take over the function of the dead cell. But in other organisms – the fruit fly, for example – killing one cell of a group, at an early stage in development, will merely mean that the others divide rather more often so that in the adult the final number of descendants of that group will be roughly the same as usual.

Perhaps the most challenging problem in the whole of developmental biology is the construction of the nervous system of an animal. Many years ago it was shown by Sperry that if a newt's eye was removed, so that the optic nerve from its eye to its brain was broken, then even if the eye was replaced upside down the optic nerve would regenerate from the retina, grow towards the brain and connect up again. After a period the animal could see again with this eye but it always saw upside-down (newts not being clever enough to adjust to such a situation). In other words the new connections had been made "correctly" except that the eye did not know that it had been inverted. Such experiments, and many sophisticated variants of them, have been repeated in recent years by several groups of workers. All their results show that fairly precise processes are at work to make the correct, rather intricate, connections needed between one set of nerves and another but exactly what these mechanisms are we do not yet know. We do not even know just how precise they need to be.

I have spoken as if all these problems were the same in all multicellular organisms but this ignores the immense variety of Nature. To a biologist an elephant and a field mouse are very similar – they are both mammals. It would be surprising if in these two cases the general problems of development did not have similar answers. Again one might guess that the other vertebrates – the fishes, the amphibia, the reptiles, the birds – would not be very different. But what about the higher invertebrates such as the octopus and the squid, or the vast array of insects, to say nothing of various worms, jellyfish, sponges . . . right down to the amoebas. Are all these unknown processes likely to be just the same in plants as in animals? And will there be major differences between the flowering plants and the lower members of the plant kingdom such as the fungi? We can only guess at the answers. Because living organisms look so different it does not mean that their basic biochemical mechanisms are not the same. The great families of the nucleic acids and the proteins are put together in the same general way in all organisms. Even though individual proteins usually differ somewhat in different species they are often remarkably similar – not surprisingly since they have usually evolved from a common ancestor. The general uniformity of biochemistry throughout Nature is one of the most astonishing discoveries of this century. But we must not press it too far. The hormones of plants are very different from those of animals. Moreover, plants and animals are, to some extent, built on different principles. On a broad scale a tree is only statistically defined. The exact branching pattern can be very different from one tree to another, even if they are identical genetically. On this scale a higher animal is made with greater precision – two eyes, four legs and so on. But then the flowers of a particular plant are often constructred rather precisely while, on the other hand, certain tissues in animals – for example the branching patterns of the bronchioles in the lungs – appear as loosely defined as the branches of a tree. Certainly there must be a variety of mechanisms at work. Some are likely to be the same, or very similar, throughout Nature; others may be less widely distributed.

Developmental biology, then, is an area of the greatest biological importance where our ignorance is more striking than our knowledge. At the visible level, as seen in the optical microscope, we can describe an organism in terms of its cells – itself a major advance over the knowledge of earlier times. In many cases we can see these cells grow, divide and change shape. At the molecular level we understand some of the most fundamental biochemical processes and we are rapidly extending our knowledge. But how an organism constructs a hand, with its thumb and four fingers, with all the bones, the muscles and the nerves, all assembled and correctly connected together, that we cannot yet explain with the force and finality which characterises a successful piece of science. We have a profound and reasoned belief that it can be done, but it will take some time, much hard work and no little imagination to do it successfully.

H. S. Micklem

Reader in the Department of Zoology at the University of Edinburgh.

Spent several years on the scientific staff of the Medical Research Council and worked at the Pasteur Institute in Paris. Has published extensively in the fields of immunology and haematopoiesis.

IMMUNOLOGY

The immune system is perhaps the most intensively studied aspect of vertebrate physiology. Although there are many gaps in our understanding of it, the broad outlines of what it does, and how, are well established. More precisely that is true of the mouse and to a less extent of man, rat, chicken, toad and a few other species. We assume (with some reason and supporting evidence) that other vertebrates behave in the same way.

I do not wish to concern myself much with the missing pieces in the jigsaw. Such is the number, energy and ingenuity of immunologists that most of these will certainly be filled in before very long. We are, as Jerne[1], put it, "waiting for the end". The questions I wish to pose, therefore, relate not to How, but to Whence, Why and Whither. From what precursors did the immune system evolve? Why did it evolve – or, more precisely, what selective pressures favoured its evolution? (To the question Whither we will return later.)

The immune system, as it is normally understood, exists only in vertebrates. It is present (with some variations) in all extant vertebrate species.

The first vertebrates appeared some 450 million years ago having evolved, it is presumed, from a chordate progenitor – perhaps something similar to the tadpole larva of today's tunicates (sea-squirts). The earliest forms lacked jaws. This deficiency was soon made good, and the first jawed fishes, from which the vast majority of extant vertebrates are descended, came onto the scene in the late Silurian period. Only a small line of animals maintained the jawless state – the Agnatha, represented today by the lamprey and the hagfish. These fishes are notable for their somewhat unsavoury parasitic habits, but what is particularly interesting about them is that they, too, possess lymphocytes and are capable of rejecting allografts (grafts from other individuals of the same species) and making specific antibodies. This means that recognisable rudiments of the immune system must have existed in very early vertebrates, before the divergence of the jawed from the jawless lines.

The three features of the immune system which are generally thought most clearly to differentiate vertebrates from invertebrates are: (i) the family of antibody molecules, (ii) the ability to recognise and reject allografts even from closely related individuals, and (iii) "memory" for previously encountered antigens. Not a trace has so far been found, in any invertebrate group alive today, of possible relatives of the antibody family, or of any functional analogues of comparable specificity. In other respects, the functional dichotomy between vertebrates and invertebrates appears somewhat less sharp than it did 20 years ago. Then, invertebrates were believed to lack any but the most non-specific defences. Some species, especially among the arthropods, had been shown to have bactericidal substances in their body fluids, but the specificity of these substances for particular organisms was not notably greater than that of carbolic acid. Many arthropods, and other invertebrates, were also known to possess phagocytic cells. In fact, arthropods and protostomes in general (annelid worms and molluscs being the other large protostome phyla) are not a propitious group in which to search for the precursors of vertebrate immunity. If we are to believe the evidence of comparative morphology and embryology, they diverged from our own ancestors soon after the dawn of multicellular life. It is, therefore, all the more interesting that the last few years have shown many invertebrate groups, including protostomes, to have some capacity for recognising and reacting against "non-self". This effectively disposes of the (anyway remote) possibility that discrimination between self and non-self is an accidental, functionally unimportant, peculiarity of vertebrates. Recognition of foreign-ness and its obverse, self-recognition, may in fact be universal among animals, and even plants, albeit often in a crude form.

In some groups, it is easy to discern possible advantages in the facility for recognition. In flowering plants, it may be used to impose self-sterility and hence promote outbreeding and genetic mixing. In amoebae it may prevent one end of the organism from devouring the other, like a dog eating its tail. Many symbiotic relationships appear to involve a more or less precise recognition of one partner by the other. In some colonial and quasi-colonial species the density of individuals might be adversely affected if colonies were able to intermingle freely. Examples may be seen among the phylum Coelenterata. One of the groups in this phylum, the marine Alcyonaria, consists of species which form branching colonies, tree-like in form. These are composed of numerous small feeding polyps which capture small organisms and organic particles from the surrounding water. The polyps intercommunicate by means of channels of living tissue and are supported by a calcareous or horny skeleton. Since the individual polyps do not exist independently, the alcyonarian is not so much a colony as an individual with many mouths whose number and density can reflect the availability of food. This balance between supply and demand would be upset by the intrusion of another colony, to the detriment of both. It is therefore not too surprising to find, as Jacques Theodor has described, that a destructive interaction, apparently more akin to suicide than to murder, occurs at the site of contact between branches of different colonies. A similar incompatibility has been reported between colonial sea anemones.

In these instances, recognition is plainly advantageous, and perhaps vital, for the survival of the species. Other examples, uncovered by the interference of Experimental Man, are less easy to explain in such terms. Of what benefit, for example, is it to an earthworm resident in Strasbourg to be able to reject skin grafts from its cousins (of the same species) from Bordeaux? Having rejected one graft, the worm will often manifest some memory of the insult, by rejecting a second graft from the same source more rapidly. This is an impressive performance, but hardly one which the worm is likely to have to put on frequently. It is a commonplace of evolutionary theory that characters will only spread in a population if their presence confers some reproductive advantage. Once the character is installed, however, it may linger on, or be modified for other uses, long after the selective pressure that encouraged its spread in the first place has vanished. Viewing the earthworm's capabilities in this light, one might be tempted to regard them as a residue, inherited perhaps from some far-off coelenterate ancestor.

This simple explanation will not do. The reason is that vertebrates have, quite independently, evolved a system of individual markers (cell surface antigens) and a concomitant capacity for recognising and reacting against non-self, far more complex than the earthworm's. There are so many variations within any species that it is doubtful whether any two individuals, aside from identical twins and (superficially at least) some experimentally inbred strains, are completely compatible. There are again two sides to the coin: one side represents the antigenic complexity of each individual; the other represents the finely tuned recognition system without which the antigenic complexity would be indetectable. These seem in a strict sense to be two sides of a single coin. This is suggested by the close genetic linkage between at least one of the complex loci which determine the histocompatibility ("H") antigens, and the "immune response" genes which partially govern the individual's ability to recognise non-self.

Thus recognition of self and non-self poses two problems. (1) Why does it exist at all? (2) Why is it so exquisitely discriminating in vertebrates?

There are in fact other markers of individuality which may approach the H-antigens in precision. In man they include fingerprints, facial characteristics and smell. Of these, the last two, at least, serve an obvious function in communication and differ in that respect from the H-antigens. From a molecular point of view the most interesting comparison is, as Lewis Thomas[(2)] suggested, with smell. The field of scent discrimination is relatively sparsely mapped. It is not, for example, clear whether scents are ever strictly individual-specific in a chemical sense, or whether the apparent specificity observed is due to the combination of a quite small number of olfactants in critical concentrations. The distinction is important, because in the latter case the apparent parallelism between scent discrimination and cell surface antigen discrimination would be illusory. What does seem to be clear is that trained dogs are capable of tracking an individual human for considerable distances without being diverted by the scents of other individuals. It is therefore possible, if not probable, that scents are truly as individual-specific as H-antigens. As Thomas pointed out, we may have much to learn from a comparison of events at the surface of olfactory receptor cells and of lymphocytes.

Lymphocytes are one of the success stories of modern immunology. Less than 20 years ago it was possible for O. A. Trowell[(3)] to summarise his review of "The Lymphocyte" in the following terms:

> The small lymphocyte seems a poor sort of cell, characterised by mostly negative attributes: small in size, with especially little cytoplasm, unable to multiply, dying on the least provocation, surviving *in vitro* for only a few days, living *in vivo* for perhaps a few weeks . . .
>
> It must be regretfully concluded, however, that the office of this Cinderella cell is still uncertain. There is, in fact, little to add to the statement of Rich (1936) that "The complete ignorance of the function of this cell is one of the most humiliating and disgraceful gaps in all medical knowledge".

It is encouraging, for anyone who finds himself dismayed by our present ignorance of Nature's laws, to contemplate the change that our vision of the lymphocyte has undergone since the publication of Trowell's review. Lymphocytes have emerged as the vertebrate's agents of discrimination not only between self and non-self, but also between innumerable varieties of non-self. There are at least two functionally distinct types, each highly heterogeneous in specificity. On present evidence they may well use two quite separate classes of molecule as their tools of recognition. One of these certainly consists of antibody molecules (of which more shortly), while the other (less well characterised) may be related to the more primitive recognition mechanisms found in other animal groups.

We know little about the evolutionary origin of lymphocytes. The search is now concentrated among the tunicates (supposedly our nearest non-vertebrate cousins) and other deuterostomes. Tunicates possess cells which appear morphologically similar to lymphocytes and which, like the thymus-derived ("T") lymphocytes of vertebrates, are active in allograft rejection. However, as Trowell noted, lymphocytes are not possessed of very striking morphological attributes, and much more work will be needed to establish a homology between vertebrate lymphocytes and any cell so far recognised in non-vertebrates.

Still more obscure are the evolutionary pressures which called the fine individual cell-surface specificities of vertebrates into being. Of the several explanations which have been put forward, none looks particularly cogent. (1) Diversity might be advantageous to a species by reducing its susceptibility to parasites and infectious micro-organisms. The proposition is that parasites would be unlikely to be able to adapt to all the varieties of its "host" species and therefore some members of the latter would be unaffected. However, vertebrates as a class are rather heavily parasitised; and parasitologists as a class have tended to be particularly sceptical about the efficacy of the vertebrate immune system. (2) Variability ensures that cancer is not contagious. This is true, but seems unlikely to have much relevance to our early aquatic ancestors. (3) Detection of inter-individual variation may be just a by-product of the capacity to detect deviant, potentially cancerous, cell surface patterns within the body ("surveillance"[(4)]). There is some experimental support for the surveillance hypothesis, since individuals who congenitally lack lymphocytes, or whose lymphocytes have been partly destroyed by experimental means, are prone to develop certain cancers. However, most of these cancers belong to rare types and involve components of the immune system itself; the incidence of the commoner kinds of cancer is not increased. Recent evidence demonstrates that even if surveillance by lymphocytes is a reality, it is not the only defence against cancer development. (And are invertebrates more cancer-prone?) (4) A related proposition is that surveillance operates not merely against potential tumours, but against any intrusion of foreign genetic material, notably from viruses. There is no doubt that T-lymphocytes are active against virus infections, but that is no proof that this activity is their primary *raison d'être*.

In summary, we really have no firm idea what the vertebrate histocompatibility antigens are for, or what encouraged the evolution of T-lymphocytes capable of recognising them.

The mystery surrounding the origins of vertebrate immunity acquires a new dimension when one begins to consider the molecular structure of antibodies. A unique picture has emerged. Each molecule consists of four (or certain multiples of four) polypeptide chains. Half or more of each chain has an amino-acid sequence which is constant from molecule to molecule; but at the amino-terminal end, the chain has a variable sequence over a stretch of more than 100 amino acids. The variation is not random and free from constraints. On the contrary, some positions are constant, some carry one of perhaps two or three amino acids, while a few are "hypervariable". The variability endows antibody molecules with affinity for an undetermined, but certainly very large, array of antigens. The structural evidence suggests that the polypeptide chains as they exist today evolved by a process of successive duplications from an ancestral gene. A subject of much controversy has been the nature of the generator of diversity (sometimes abbreviated to GOD) in the variable regions. Basically, the question is how much of the genetic information for the different variable regions is carried in the germ-line of the species (i. e. is generated and selected during evolution) and how much is generated anew in each individual, presumably by somatic mutation? Even when this is resolved, two questions are likely to remain. Firstly, how did the primordia of today's complex of antibodies evolve in or before the Ordovician period? What, for example, was the function of the postulated pre-antibody fragment? Secondly, what selective advantage promoted the evolution of antibody variability?

The first question can hardly be approached until or unless a homologue of part of an antibody molecule is identified in non-vertebrates, presumably in lower chordates. The answer to the second question might be supposed to be clear, since we have observations on a whole sub-phylum on which to base it. However, it is not in reality quite so simple. The immune system may now be indispensable to vertebrates – it certainly is to man – but it is not strikingly efficient. Many micro-organisms – viruses, bacteria, protozoa and fungi – as well as helminths can get the better of it. Discrimination between self and non-self is not, it turns out, functionally absolute: potentially self-destructive lymphocytes do exist and probably require checks and balances to stop them from running amok. Finally, the immune system frequently reacts inconveniently and even dangerously to antigens present in, for example, pollen, bee venom and even food. If we accept that evolution discourages the spread of disadvantageous characters, it is difficult to explain the existence of these allergic reactions and, in particular, the special class of antibody which is responsible for many of them. The disadvantages must be (or have been) outweighed by so far unidentified advantages.

It is something of a luxury to be able to adopt a negative viewpoint and simply point to uncompleted areas of the jigsaw. Finding the missing pieces, and fitting them in when found, are much more taxing occupations. I shall end by briefly discussing another, quite different, area of uncertainty.

The first "golden age" of immunology benefited the human race in a dramatic way. Some diseases such as smallpox and diphtheria have been almost eliminated. Immunization with killed or modified micro-organisms provides protection against many others. Few people dispute that immunology has been enjoying a second golden age for the last 20 years or so. What practical prospects does it hold out?

At present, there is a danger of immunological research becoming largely self-generating and self-satisfying – problem-solving undertaken for its own sake. Much of it is a series of skilful micro-dissections, fascinating in themselves, but uncertain in their long-term outcome.

There are several fields in which we may reasonably hope for some practical returns. They range from cancer therapy, contraception and allergy to (more doubtfully) measures effective against protozoal infections such as malaria and trypanosomiasis. However, we have yet to see returns on any significant scale and they may, as the example of molecular biology shows, prove unexpectedly elusive. Certainly they will have to be great if they are to match, per man-year of effort, those of the first golden age.

More uncertainty, both scientific and social, lies in the relationship between the immune system and ageing. Ageing is in fact another field about which there is an impressive corpus of ignorance. There is no doubt that the immune system, in common with other physiological systems, loses some of its efficiency in aged individuals. It also seems to lose some of its constraints, so that autoimmunity becomes more frequent. Immunological research may, then, be expected to alleviate some age-related diseases. The proposition has also been made, quite seriously, that the immune system contributes fundamentally to the whole process of vertebrate ageing, perhaps providing a mechanism of post-reproductive self-destruction. If that is so, the implications of immunological knowledge could become more disturbing.

REFERENCES

1. N. K. Jerne, *Cold Spring Harbor Symp. Quant. Biol.* 32 591-603 (1967).
2. L. Thomas, *Prog. Immunol.* 2, edited by L. Brent and E. J. Holborow, Vol. 2, pp. 239-47 (1974).
3. O. A. Trowell, *Int. Rev. Cytol.* 7, 235-93 (1958).
4. F. M. Burnet, *Immunological Surveillance*, Pergamon Press, Oxford, 1970.

Arthur E. Mourant, F.R.S.

Former Director of the Blook Group Reference Laboratory of the U.K. Medical Research Council. Fellow of the Royal College of Physicians and the Royal College of Pathology.

Although mainly a human geneticist has a strong interest in geology and is a past Vice-President of the U.K. Minerological Society.

WHY ARE THERE BLOOD GROUPS?

It is well known that everyone belongs to one of four main blood groups, O, A, B, and AB, and that in a blood transfusion the groups of donor and recipient must be carefully matched in order to avoid a harmful and possibly fatal reaction. Subdividing these four groups into eight are the principal Rhesus or Rh groups, Rh-positive and Rh-negative. Mismatchings with respect to these can also give rise to transfusion reactions, but their better known effect is that of occasionally causing haemolytic disease of the newborn (jaundice and anaemia) in the infants of couples where the wife is negative and the husband, and hence the child, is positive. Incompatibility between mother and foetus is also one of the causes of spontaneous miscarriage early in pregnancy.

The types within each of these two systems are inherited according to fairly simple genetical rules. In addition to these two there are about a dozen other systems of blood groups, each system subdividing humanity into two or more types, so that the number of possible combinations runs into millions. Incompatibility between the types within any one of these systems can give rise, under appropriate circumstances, either to transfusion reactions or to haemolytic disease of the newborn, but ABO and Rh incompatibilities account for about 99 per cent of all such cases.

The tests for the blood groups are carried out, basically, by mixing the red cells of the blood of the person to be tested with a drop of an appropriate human or animal serum and looking under the microscope for agglutination or clumping together of the red cells.

Haemolytic disease and transfusion reactions, as well as the test results observed in the laboratory, are the effects of reactions of certain complex and highly specific molecules, usually of protein, present in the membrane, only a few atoms thick, which surrounds the red cell: these substances, called antigens, interact with other molecules, of proteins called antibodies, present in the surrounding serum.

The blood of a healthy adult human being contains about 5×10^{15} red cells, and every one of these carries on its surface molecules of some dozens of different blood group substances, each of which can give rise, under appropriate circumstances, to haemolytic disease of the newborn, and to transfusion reactions.

How is it that everyone carries all these substances, almost the only known functions of which are to cause diseases? They are passed on from parents to offspring according to precisely known genetical rules and, according to modern views on evolution, must have some selective advantages which counterbalance their known harmful effects. What can these advantages be, and how can we hope to find out?

We get one hopeful clue from a study of the varieties of human haemoglobin, the red iron-containing pigment which occurs inside the red-cell, and is responsible for its power of carrying oxygen from the lungs to the other parts of the body where it is needed. Haemoglobin is a protein with a very complex structure which has perhaps been more intensively studied than that of any other molecule in the whole of chemistry and biochemistry. Among its principal components are four chains of linked molecules of amino acids, or polypeptide chains, two identical ones known as alpha chains and two identical beta chains. Both of these chains are subject to genetic mutation, but nearly all the variants have such harmful effects that they are rapidly eliminated by natural selection.

However, there is one more persistent mutation of the beta chain which is responsible for the variant known as haemoglobin S, or sickle-cell haemoglobin, to distinguish it from haemoglobin A, or normal haemoglobin. About 99 per cent of all people have nothing but haemoglobin A, but there are parts of Africa where a substantial part of the population, up to 40 per cent, have haemoglobin S. The S variant is dependent on a small hereditary change in the beta chain, which causes the molecule, when it gives up its oxygen to the tissues, to become insoluble in the watery medium inside the red cell, so that it forms crystals which give the red cell a peculiar shape, which has been compared to a sickle.

Persons who are heterozygous, having received the variant gene from only one parent, have a mixture of haemoglobin A and haemoglobin S, and have almost normal health, but those who are homozygous, having received the abnormal gene from both parents, have virtually only haemoglobin S, and most of them die in infancy of sickle-cell anaemia. Thus in a population with 40 per cent of heterozygotes, like several African tribes, the laws of genetic equilibrium tell us that about 4 per cent of all babies must be homozygotes, destined, under African conditions, to an early death. One would expect such a process to reduce the frequency of the variant rapidly from one generation to the next – yet it persists. Around 1950 this problem was the subject of much research and it was finally Dr. A. C. Allison who showed that the heterozygotes, with genes for both A and S, are substantially more resistant to malaria than normal A individuals. Thus in a malarial area homozygous AA individuals tend to die in infancy of malaria; SS individuals die of sickle-cell anaemia, while the AS heterozygotes are resistant to both diseases, and pass both genes on to the next generation.

It has since been amply proved that, as expected on this hypothesis, there is a close relation in African and some other populations between the intensity of endemic malarial infection and the frequency of the haemoglobin S gene.

This condition of equilibrium between the environment and the genes is known as balanced polymorphism, and research as to the cause of the persistence of the polymorphism of the blood groups has aimed, so far without success, at finding a similar relation between blood-group frequencies and certain endemic widespread diseases, not necessarily malarial, or even infectious.

The frequencies of the blood groups of all systems vary widely from one population to the next, throughout the world, so that it is even more difficult to account for their distribution than for that of the haemoglobins. For many years now, and especially since 1950, many workers have been looking for differences in the frequency of certain blood groups, especially those of the ABO system, between the general population of healthy persons and classes of people suffering from certain common diseases. Some most interesting patterns are now emerging, but nothing like the simple balanced one presented by the haemoglobins.

It was hoped, indeed, that such associations, more especially with infections of well-defined geographical distribution, would, through natural selection, explain a large part of the world distribution pattern of the blood groups. However, in contrast to the orderly associations found for some classes of diseases, the infections tend to show erratic results, with some investigators finding associations which are denied by other investigators. In the specific examples which follow, mention will be made chiefly of blood groups A and O, since groups B and AB are much rarer, so that statistically significant associations are much more difficult to establish.

One disease which has been particularly baffling has been smallpox, once a major killing disease in large parts of the world. Some workers have found a close association between blood group A and susceptibility to smallpox, while others have denied this. One is almost forced to the conclusion that an association exists, but that differences in methods of selection of patients and in laboratory techniques account for the different results. Tuberculosis and leprosy similarly give very variable results, and diphtheria shows no tendency at all to association with any blood group. Malaria seems to be associated with group A, and not either positively or negatively with group B as some workers had expected to find.

Typhoid, paratyphoid and scarlet fevers are pretty clearly associated with group O.

Virus diseases tend on the whole to associate with group O. In particular, this is true of poliomyelitis, and among those infected, those of group O seem to be more likely than those of group A to go on to paralysis.

At the present state of knowledge, however, we certainly cannot account for the world distribution of the blood groups through selective mortality from infections.

In strong contrast are the various forms of cancer. Here group A persons are, on the average, about 10 per cent more susceptible to these diseases than those of group O, but the increased tendency varies from a very firmly established 21 per cent in cancer of the stomach to almost nil in cancers of the breast and lung, to name three where there are data on many thousands of patients.

To take two other classes of common diseases, or symptoms of disease, there is a striking contrast between thrombosis (clotting) and haemorrhage, with group A persons tending to clot and group O persons to bleed. This is at least partly explained by the observation that one of the main clotting agents of normal blood is present in greater amounts, on the average, in A persons than in O persons.

Another very consistent group of diseases comprises the ulcers of stomach and duodenum, which are distinctly commoner in O than in A persons, and this tendency is even more marked in those who have ulcers which bleed.

Thus the infectious diseases, which are killers of the young, and might therefore be expected to have a natural selective effect on blood group frequencies, show, on the whole, an erratic picture of associations, while the cancers and intestinal ulcerative conditions, which are diseases of middle and old age, and so should have little selective effect on the composition of the next generation, give clear-cut associations. Somehow it looks as though selection by disease processes *is* affecting the blood-group-distribution pattern, but the unifying hypothesis, which might also answer the question as to the natural function of the blood groups, remains for the present out of reach.

A small clue to the possible solution is given by the blood-group secretor factor. Some people do and some do not secrete into their saliva, gastric juice and other body fluids, a substance closely related to the blood-group antigen of their red cells. Secretion is under separate genetic control from that which determines the blood groups themselves, and the frequencies of the secretor gene differ from one population to another. Two classes of diseases both show a marked deficiency of secretors. These are gastric and duodenal ulcers, and diseases associated with infection by bacteria known as haemolytic streptococci, especially rheumatic fever, acute nephritis and rheumatic heart disease. In the case of the ulcerative conditions, we think we know part of the answer. A special form of the enzyme alkaline phosphatase is normally present in the intestine of all persons, and appears in the blood of some persons, but almost solely in those who are of blood group O or B, and are non-secretors. Thus there seems to be some relation between the permeability to this enzyme of the walls of the stomach and duodenum, and the tendency of these walls to ulcerate. The reason for the relation of the secretor factor to haemolytic streptococcal disease is more obscure but perhaps arises from similarities or differences between the bacterial and the blood-group antigens.

A completely new field of associations between hereditary blood antigens and diseases has recently been opened up by reserarch on the transplantation of kidneys and other organs. It had long been known that the white cells of the blood carry antigens distinct from those of the red cells, but tests for them were much more difficult, so that for many years they remained a rather obscure curiosity. Then it was shown that compatibility between organ donor and recipient with respect to these antigens was a major factor in determining the compatibility and survival of the grafted organ. This brought about widespread research on them, leading to improvements of technique, to the elucidation of the complex genetics of these antigens, now known as the histocompatibility antigens, and to the demonstration of complex world distribution patterns. Then, very recently, it was found that a whole class of well-known but ill-understood diseases showed very close associations with one or another of these antigens. These were the auto-immune diseases, in which certain organs or tissues are attacked by the body's own antibodies. Such diseases include ankylosing spondylitis, multiple sclerosis, one form of diabetes, coeliac disease, and a number of skin diseases. There are two main possible explanations for this relationship. One, at present the less favoured, is that the histocompatibility antigens, being present in the affected organs, directly influence their susceptibility to attack by the body's own antibodies. The other, and more widely accepted, theory is that in the course of evolution those genes controlling the histocompatibility antigens, and those determining the tendency or lack of tendency to auto-immunity, have migrated into proximity with one another on one particular chromosome, so that in any given population certain genes of the two classes tend to be inherited together.

A suggestion is now gaining ground that, not only in multiple sclerosis, but also in other auto-immune diseases, the susceptibility of a particular tissue to attack by the body's own antibodies is due to the combination of an inherited tendency, and the infection of the tissue by a virus. In investigating this, studies of the associations with the histocompatibility antigens mark one of many promising new lines of attack on these hitherto intractable and inexplicable diseases.

A further ray of light on both cancers and auto-immune diseases is projected by Sir Macfarlane Burnet's suggestion that there is an antithesis in the genetic make-up of individuals between susceptibilities to these two broad classes of disease. According to this theory the tendency to cancer is due to the failure of the body's antibody-producing mechanism to police effectively and to destroy small incipient cancers. On the other hand, the tendency towards auto-immune disease is due to too efficient policing, resulting in the destruction of some of the body's normal tissues. In this connection it is instructive to look in a very broad way at the associations of diseases of these two classes with blood groups on the one hand, and histocompatibility antigens on the other. There is no doubt that in general cancers show a significant association with blood group A, and auto-immune diseases with certain histocompatibility antigens. There is on the other hand a suggestion, which needs much further work if it is to be substantiated, that cancers are basically unassociated with histocompatibility antigens, and that auto-immune diseases are associated with blood group O.

If this were indeed so, then at least the blood-group end of the picture would be explained if it could further be shown that, deep in their structures, and not readily accessible to the body's antibody policing system, both cancers, and those organs which tend to be affected by auto-immune disease, contained an antigen similar to that of blood group A. Then group A persons, constitutionally lacking the anti-A antibody, would be marginally less efficient, in attacking and destroying incipient cancers, than persons of group O, who have circulating anti-A antibody. This same anti-A antibody would, however, make group-O persons more liable than group A persons to attack their own normal tissues.

FURTHER READING

1. W. C. Boyd, *Genetics and the Races of Man*, Blackwell Scientific Publications Ltd.; Boston, Little Brown & Co., 1950.
2. A. E. Mourant, A. C. Kopeć and K. Domaniewska-Sobczak, *The Distribution of the Human Blood Groups*, 2nd edition. London: Oxford University Press, 1976.
3. A. E. Mourant, A. C. Kopeć and K. Domaniewska-Sobczak, *Blood Groups and Diseases*, Oxford University Press (in the press).
4. R. R. Race and Ruth Sanger, *Blood Groups in Man*, 6th edition, Blackwell Scientific Publications, Oxford, 1975.

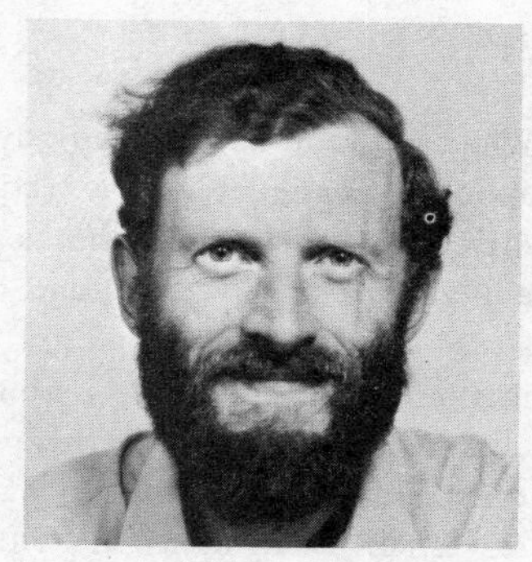

Peter J. Grubb

Fellow and formerly scholar of Magdalene College, Cambridge, and Lecturer in Botany in the University.

Early research concerned the physiology of bryophytes, especially the uptake and translocation of mineral ions. Has become increasingly committed to ecology, his chief research interests being plant-soil relations and the regeneration of vegetation. In 1970-1 made comparative studies of rain forests in New Guinea, Malaya, Japan and New Zealand.

LEAF STRUCTURE AND FUNCTION

When we look at leaves in any one part of the world we tend to take it for granted that they should be assignable to a certain range of sizes and shapes. Those who have written about the differences in leaf size between areas with different climates have concentrated on trying to explain the *relative* sizes and have never come to grips with the problem of explaining *absolute* sizes. To take an example, in the lowland tropical rain forest the areas of most tree leaves are in the range 30-200 cm^2, while in forests and scrub communities of areas with a Mediterranean climate they are mostly in the range 2-30 cm^2. Botanists have not lacked for ideas to explain the occurrence of smaller leaves in the drier climate, but they have never explained why the leaves in both areas are not an order of magnitude bigger or smaller. Presumably the explanation is to be sought in terms of the mechanical properties of the cell wall materials, the absolute permeabilities of the cuticular sealing materials so far evolved by plants and so on, but no convincing analysis has yet been made of the rationale of the absolute leaf areas and thicknesses with which we are familiar. Can we imagine landscapes with leaves much bigger than those of lowland tropical rain forest or smaller than those of Mediterranean scrub? What are the ultimate limitations to size? Several tropical trees (especially when juvenile) and a few epiphytes have undivided leaves up to 1.5 m long and 30 cm wide. In Malaya the palm *Johannesteijsmannia altifrons* has undivided, diamond-shaped leaf blades up to 3.5 × 1.8 m (Fig. 1). Why not bigger still?

A related problem is understanding the factors that determine the optimal distribution of organic materials between the leaves and the twigs that support them. For a wide range of leaf sizes it is found that about 60-70 per cent of the shoot dry matter is in the leaves. Why not more or less?

Even the more familiar kind of discussion concerning *relative* sizes is full of uncertainty. The general trends in size are clear, but the explanations of them are not. Thus within the tropics there is a marked decrease in mean leaf area for trees and shrubs between the lowlands and the subalpine zone. There is a similar decrease among evergreen species to be found on passing from the lowland tropics to both the dry areas with Mediterranean climates and the moist areas in the cool temperate zones. The basic point yet to be established is whether the leaf size is adapted primarily to performance during extended intervals of time or to survival during critical periods, e.g. intense summer drought or severely drying winter winds. It has been shown that the trends in leaf size outlined above would be expected if the key determining factor were the maximization of the ratio between the amount of carbon dioxide absorbed in photosynthesis and the amount of water evaporated from the leaves. But is this what matters to plants? Is minimization of leaf death (and loss of absorbed mineral nutrients) during drought periods more important? In some cases the two selective forces may act in opposite directions and it may be possible to determine which selective force has been more effective. In other cases, e.g. in Mediterranean vegetation, they act in the same direction and, until we have much more information on CO_2/H_2O ratios for plants growing naturally, we cannot evaluate the two supposed selective forces.

Leaf shape in evergreen flowering plants in all parts of the world is remarkably constant; the vast majority of species have the leaves more or less ovate, oblong or obovate. Why? Why do we not find more dissected leaves or round leaves? Presumably the narrow range of shapes found is a compromise between what is optimal for light interception by a whole canopy and what is optimal for the temperature and water relations of the individual leaves, but no quantitative analysis has been presented. The leaves of deciduous trees and shrubs, in contrast, have greatly variable shapes – consider the range shown by the many oaks, maples and hawthorns. Herbs similarly have a wide range of shapes, which represent different kinds of lobing. It has been

Fig. 1. The palm *Johannesteijsmannia altifrons* in rain forest in Malaya. (Photo by T. C. Whitmore.)

shown that such lobing must increase the effectiveness of leaf-cooling by both evaporation and conduction-convection. But is this effect the most important one? Why do we not see these features in trees of tropical or temperate rain forests? There are two differences between evergreen and deciduous leaves that are possibly relevant. Firstly, the deciduous leaves are usually much thinner, perhaps half as thick on average. Secondly, their walls are generally much thinner. It is known that the lamina tissue between the water-supplying veins has a relatively low conductance for water in leaves generally and it is tempting to ask whether this conductance is lower in thin-walled deciduous leaves than in thick-walled evergreen leaves. Is the lobing and toothing of deciduous leaves primarily an adaptation whereby the mean distance travelled by water between the major veins and sites of evaporation is reduced – so minimising desiccation-damage during stress? This issue cannot be resolved until critical determinations have been made for a sufficient number of species of the conductance of the leaf tissues for water, of the heat-balance of leaves in the field and of the relations between the rates of photosynthesis, respiration and water loss at different temperatures.

It is worth remembering that leaves have not always been the shapes they are now. In the Coal Measures, 200 million years ago and more, many of the dominant plants (all belonging to families now extinct) had strap-shaped leaves superficially rather like those seen today in the Dragon's Blood tree (*Cordyline*) and the Joshua Tree (*Yucca*). There were also many kinds of seed-ferns, plants with leaves like modern spore-bearing ferns. Why did these groups not evolve leaves like those of modern-day trees? Was their inability to do so one of the causes of their decline? If the leaf shapes of modern-day trees are indeed superior, in what functional sense is this true? Trees with leaf types that seem odd to most temperate-zone botanists are very successsful in parts of the tropics now, most notably the palms with their comb-like and fan-like fronds and the screw pines (pandans) with their huge strap-shaped leaves, rather like monstrous sedge leaves borne in great tufts at the ends of the branches. Any rationale of

(a)

Fig. 2. Cordate (heart-shaped) leaves seem to be particularly common in climbing plants. Two examples are shown here: (a) *Philodendron erubescens* (Araceae) and (b) *Aristolochia cathcartii* (Aristolochiaceae). (Photos by P. Freeman.)

optimal leaf shape ought to take into account these minority types and the types found in former ages.

In any one particular community at the present time leaf-shape is often correlated with life-form (tree, shrub, herb, etc.). One intriguing correlation, most often seen in the tropics, is particularly worthy of mention. That is the development of heart-shaped leaves in climbers belonging to many different families, e.g. Araceae, Aristolochiaceae, Dioscoreaceae, Menispermaceae, Piperaceae (Fig. 2). It appears that no convincing explanation of this peculiar and striking feature has been published.

Two other mysterious features peculiar to the tropics are worthy of note. Firstly, the development of "drip-tips" – long acuminate tips seen on the leaves of very many tropical lowland rain-forest plants, especially the lower-growing trees and shrubs. Rain certainly does drip from the tips, and there is some evidence that the leaf surface dries more slowly after rain on leaves from which the tips have been cut off. If the drip tips do facilitate surface-drying, what is the advantage? Do many of the leaves concerned have pores (stomata, cf. Fig. 4) on the

Fig. 2(b)

upper surface and is inward diffusion of carbon dioxide for photosynthesis significantly hindered by a water film on this surface? Or does the benefit to the leaf arise from the hindrance to development of parasitic fungi and of epiphyllous liverworts and lichens? The epiphylls can shade the leaf, take mineral nutrients from it and perforate the cuticle so making it less water-tight. Is the danger from parasitic fungi or epiphylls the more important? Any explanation should take account of the fact that drip-tips are rare or absent not only in tropical dry forests but also in forests high up on the tropical mountains, shrouded in fog for long periods.

An even more remarkable feature is the habit of many lowland rain-forest trees in developing limp, pendent young leaves, which contain little chlorophyll and are brightly coloured – red, pink, white or even blue. When such leaves cover a crown, they make a remarkable sight. Some time after they reach their full size they turn green, stiffen and assume normal angles of inclination. What is the advantage of this habit of producing limp, colourful leaves? Are they unpalatable and do they exhibit warning coloration to predators, as certain caterpillars do? Could it be that the young leaves of some species are highly unpalatable and that other colourful species are mimicking them? Detailed studies are needed on the losses to predators at this early stage and in later stages, particularly with species in which different varieties have red or green young leaves.

In recent years it has become fashionable to explain many leaf features in terms of reducing palatability: the latex channels of dandelions and spurges, the resin ducts of pine needles, the

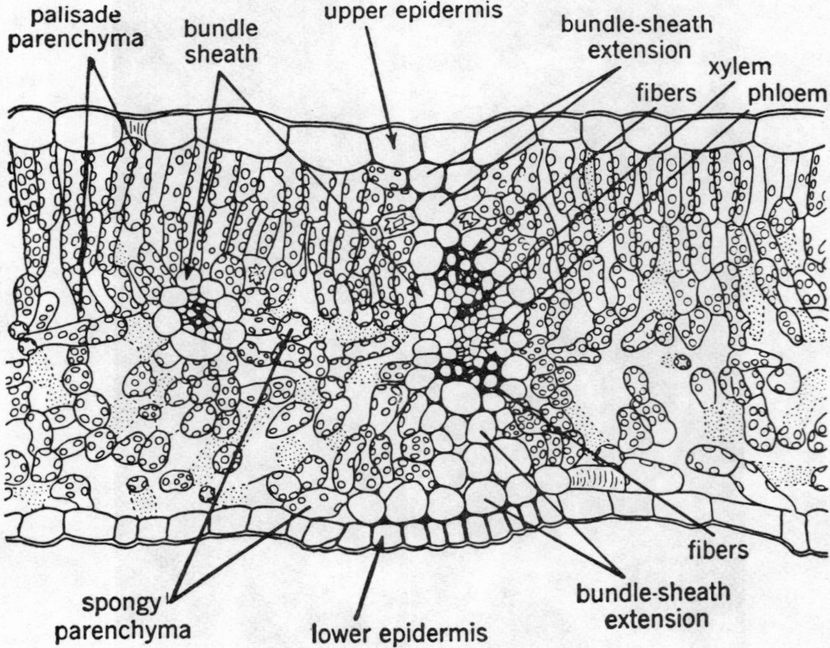

Fig. 3. Part of a transverse section of a pear leaf to show (i) the columnar, tightly packed cells of the palisade layer, which contrast markedly with the more rounded, less tightly packed cells of the spongy mesophyll, and (ii) the cells of the bundle-sheath extension forming a continuous water-conducting bridge between the conducting cells of the vein and the two epidermides. (Reproduced from *Plant Anatomy* by K. Esau with permission of John Wiley & Sons Inc.)

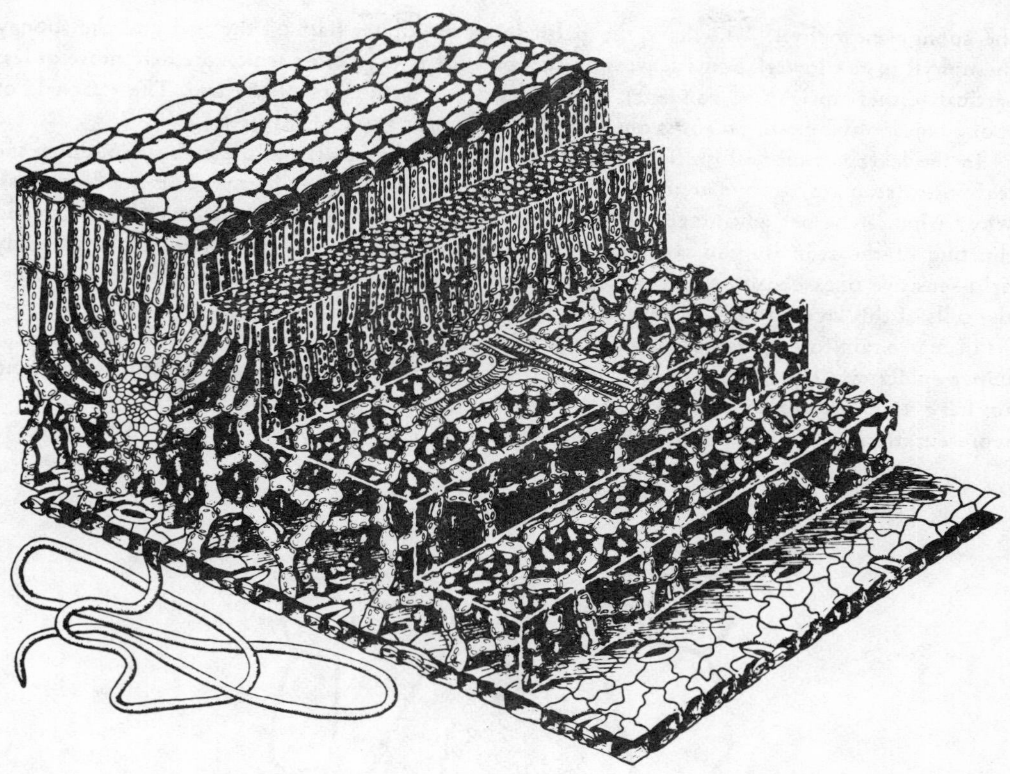

Fig. 4. A three-dimensional view of the internal structure of an apple leaf. Note particularly (i) the contrast between the tightly packed palisade cells (in the upper half of the leaf) and the widely separated cells of the spongy mesophyll (in the lower half), and (ii) the pores (stomata) confined to the lower epidermis. (Reproduced from *Introduction to Plant Anatomy* by A. J. Eames and L. H. MacDaniels. Used with permission of McGraw-Hill Book Company.)

essential oils of mints, the tannins and calcium oxalate crystals of very many leaves and the hairs of many species, especially those which have dense hairs on young, otherwise apparently highly palatable leaves but few or none on the adult. However, it is hard to see how one may obtain critical evidence bearing on this idea of adaptation by unpalatability and the idea is in danger of remaining plausible but untested indefinitely. Similarly the nectaries found on the leaves of many tropical plants are supposed to attract ants, which destroy or discourage virtually all other herbivores, but there is extremely little evidence. In this case patient observation and recording could contribute a lot to understanding.

One of the first points made to any student about the internal structure of a leaf is the fact that in most cases the central tissue, the mesophyll, is divided into two more or less distinct layers: palisade mesophyll and spongy mesophyll (Figs. 3 and 4). Yet there is no accepted explanation of this fact. The palisade consists of smaller cells and has a higher ratio of free surface area to cell volume. It has been shown to contain 67-86 per cent of all the chloroplasts in the mesophyll of a variety of herbaceous species. The leaves in which the mesophyll has the highest proportion palisadiform are those of Mediterranean-climate evergreens. Why are not more leaves like them? What is the advantage of the large cells and large intercellular spaces of the spongy mesophyll? Why have the palisade in the upper half of the leaf and the spongy mesophyll in the lower? Some leaves, mostly but not always those which are held more or less vertical (either upright or pendent), have palisade on both sides of the leaf. The rationale of spongy mesophyll development is one of the most puzzling of leaf features.

In the leaves of most flowering plants only the mesophyll cells and the cells surrounding the leaf-pores (stomata) are green and photosynthesise. Most of the epidermal cells are colourless. Why? What is the net advantage? It cannot be related simply to the mechanism of opening and shutting of the stomata and some possible need for the stomatal guard cells to be the only light-sensitive ones because in many leaves there are no stomata on the upper surface and yet the cells of this layer are still colourless.

In many rain-forest trees the leaves have a hypodermis – an extra colourless layer below the upper epidermis (Fig. 5). In one sample of tropical lowland tree species 25 per cent were found to have this feature and in several tropical montane and subalpine samples the level of representation is 40-60 per cent. The function of this layer has been supposed to be either to

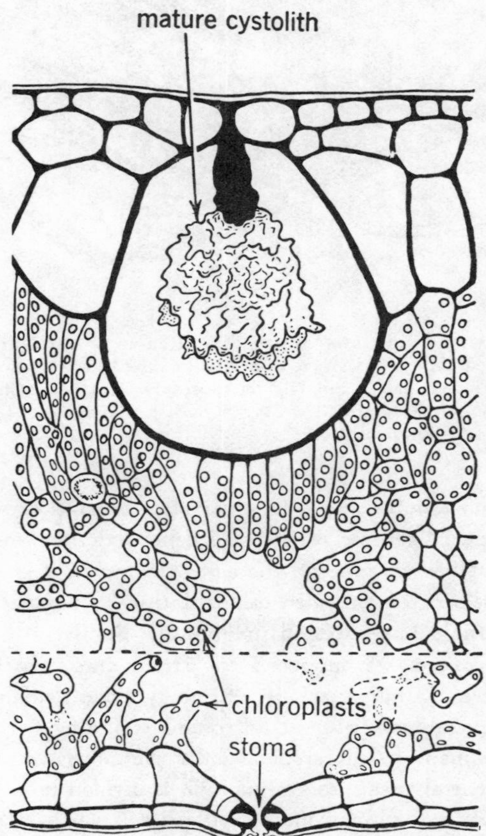

Fig. 5. Part of a transverse section of a leaf of the India Rubber plant, *Ficus elastica*, to show the thick walls of the cells in the hypodermis, the layer just below the upper epidermis. This section also shows a cystolith, a siliceous concretion borne on an inward projection of a cell wall – another leaf structure of unknown function, possibly a deterrent to herbivores. (Reproduced from *Plant Anatomy* by K. Esau with permission of John Wiley & Sons Inc.)

filter out excess ultra-violet light or to act as a reservoir of water for use during drought. There seems to be no experimental evidence for either view, and something of an anomaly is presented by the fact that a hypodermis is common in rain-forest trees but decidedly uncommon in trees of Mediterranean climates certainly subject to drought. If the hypodermis is indeed adapted to give up water to the mesophyll, let us say, in moderate and transient drought stress, why does it so often have relatively thick walls? Surely thin, collapsible walls would be more obviously appropriate, such as are found in the multi-layered hypodermis of some succulent, epiphytic *Peperomia* species.

Another problem concerning water relations within the leaf arises over the distribution of bundle sheath extension tissue. In most flowering plants the vascular tissues of each vein are surrounded by parenchyma cells of a generalized type, either colourless or with few chloroplasts. In a majority of soft-leaved species similar cells form bundle-sheath extensions connecting the vascular bundles to the upper and lower epidermides (Fig. 3). There is a good deal of experimental evidence that water is conducted effectively along the walls of these bundle sheath extensions to the epidermides and that this is the main route of supply for the palisade. What is puzzling is that only about a quarter or less of leathery-leaved species have such extensions – they are lacking in many familiar plants like ivy or cherry laurel and there is a notable resistance to water movement between the vascular bundles and the epidermides. Is it the case that the ancestors of such plants simply have not had the genetic potential to develop bundle sheath extensions or has there been no strong selection for such a development in leathery leaves – even a selection against it? In many leathery leaves which have bundle sheath extensions, such as the holm oak and bay, the cells concerned are thick-walled and seem to be primarily adapted to provide support.

One of the most remarkable discoveries in plant physiology in the period 1965-70 concerned plants with a particular kind of bundle sheath – those in which the sheath cells are densely packed with organelles, particularly chloroplasts (Fig. 6). Examples are maize, sugar cane, sorghum and various species of *Atriplex*. Such plants had been known for some time to be

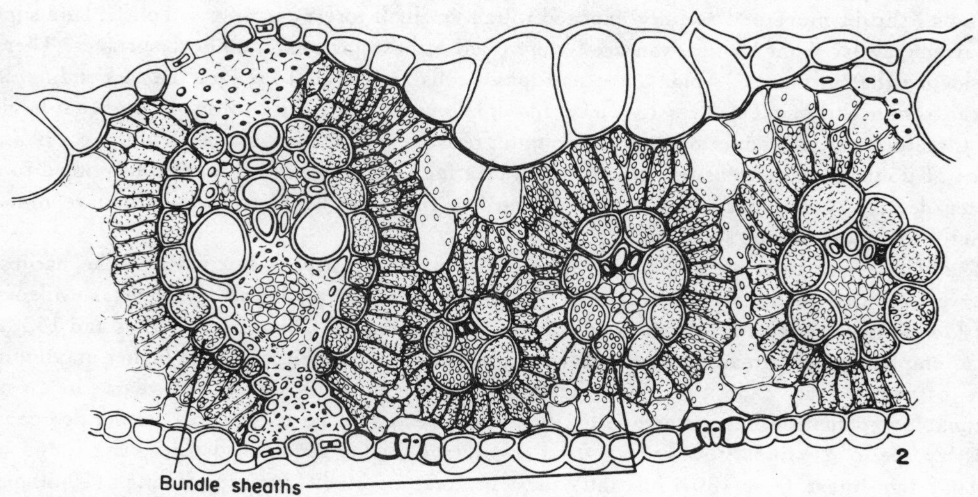

Fig. 6. Part of a transverse section of a leaf of a C_4 grass, *Hyparrhenia hirta*, to show the bundle sheath cells densely packed with chloroplasts. Compare this section with that in Fig. 3, where there are no chloroplasts in the bundle sheath cells. (Reproduced from *Plant Anatomy* by A. Fahn with permission of Pergamon Press.)

exceptional in a number of ways. Their maximum rates of photosynthesis per unit leaf area are particularly high. They can reduce the CO_2 concentration in the leaf to a much lower level than that ever found in the leaves of other plants. They have much higher ratios of CO_2 absorbed to H_2O lost. They have the ability to photosynthesise actively at 30-50°C and indeed the optimum temperature is usually in this range, whereas that for most other plants is in the range 15-30°C. The new discovery was that the chemical pathway of initial fixation of carbon dioxide was quite different from that of most plants with superficially similar leaves. Radioactively labelled carbon dioxide was found to be fixed first in 4-carbon compounds such as malic acid rather than the 3-C compound, phospho-glyceric acid, found in most plants previously tested. Hence the newly characterised group became known as C_4 plants, as opposed to C_3 plants. Possibly the most unexpected feature of these plants is that the C_4 pattern of fixation occurs in the mesophyll, *not* in the bundle sheaths; this can be shown in macerated leaf tissue suitably sorted into different fractions. The C_4 compounds, into which the carbon from CO_2 is fixed in the mesophyll, are translocated to the bundle sheaths, where CO_2 is released from them within the cytoplasm and immediately reincorporated by the "normal" C_3 pathway into organic compounds. As yet no convincing rationale has been produced for this spatial separation of the C_4 and C_3 reaction systems. There are many related problems, mainly requiring investigation at the level of molecular biology. How is it that the C_3 reaction system can function so well at 30-50°C in the bundle sheaths of C_4 plants and not in the mesophyll of C_3 plants? How is it that the enzymes of the C_4 fixation system can reduce the CO_2 concentration inside the leaf to a level so much lower than is ever found in leaves of C_3 plants?

Two other fundamental features of leaf-form which are not understood in terms of adaptation are the distribution and frequency of stomata. Why are the stomata of most tree and shrub species confined to the lower surface? At first sight it seems reasonable to suggest that it is because the leaves are then less liable to infiltration by rain; if the intercellular spaces become filled by water, the diffusion inward of carbondioxide becomes impossibly slow (about 10,000 times slower than with air-filled spaces) and the leaf becomes useless as a photosynthetic organ. But if that explanation holds, why are stomata generally found on both surfaces of the herbs and small shrubs more or less fully exposed to rain in small forest-clearings, on cliffs, land-slips and river gravels? What is the advantage to any plant of having stomata on both surfaces? There is evidence that in many C_3 plants the mesophyll cells are unable to reduce the carbondioxide concentration within themselves to a level much below, say, one-third to one-half of that in air and therefore a small reduction in the minimum resistance to diffusion inward in the gas phase seems of doubtful value. Possibly the plants of clearings, cliffs and so on need to be cooled to a greater degree by evaporation of water during hot sunny spells. If so, why not have more numerous stomata on one surface?

The whole question of adaptation through stomatal density is fascinating. The earliest observers pointed to the very low densities found in succulent plants of dry areas as evidence that low densities were advantageous on balance in dry climates. Then, in the 1920s and 1930s, it was emphasised that many steppe plants had higher stomatal densities (and higher maximum rates of water loss) than their close relatives in moister areas. In recent years the most reasonable explanation has seemed to be that the leaves could be cooled to a greater degree in hot, dry spells. Yet it is now known that, on average, higher stomatal densities are found in lowland rain-forest trees (30-45 m tall) than in trees of Mediterranean climates (10-20 m). Furthermore, it was long ago shown that in temperate deciduous woodland, the mean densities were successively lower in trees, shrubs and shade-tolerant herbs. Now it appears that in rain forest the highest densities are found in some of the tallest trees. There is abundant evidence that stomatal density is greater, other things being equal, in leaves subjected to greater water stress during development. Leaves that are higher on a plant are further from the water supply and, with a given degree of exposure, are bound to suffer more water stress. However, the differences in stomatal density in plants with different life-forms outlined above, although they parallel changes that can be induced during development of the individual, do appear to have a genetic basis. What is the advantage to taller plants in having higher stomatal densities? There is no obvious advantage in a taller plant as such maintaining a higher rate of transpiration. The ecological interpretation of stomatal density will remain elusive until we have experimental determinations of the rates of photosynthesis and water loss per unit leaf area for a large number of species in the field. As yet we cannot be sure that the maximum rate of water loss in different species is strongly correlated with stomatal density and we have almost no information on the correlation with maximum rate of photosynthesis.

One particularly interesting question is the degree to which greater stomatal pore-size can compensate for lesser density. In sizeable samples of species there is, as might be expected, a negative correlation between stomatal size and stomatal density, but there is no clear linear relationship between density and either stomatal length or the square of that length.

One peculiar feature of stomata sadly neglected experimentally is the accumulation of wax in the pore. This feature seems to be common, at least in certain groups of plants. In some cases the material is porous and there is abundant reflection of light from internal surfaces so that each little plug appears as a white dot. Very little indeed is known about the development, function and significance of these plugs.

So far we have treated a leaf as though it were never appreciably distorted but distortion certainly occurs not as only a result of drought or animal-damage but also when leaves are exposed to wind. Large, soft, flexible leaves are probably distorted sufficiently by quite moderate breezes to induce an appreciable bellows effect, whereby air is sucked into and out of the leaf via the stomata. The possibility has been appreciated for a long time and the occurrence of slight cuticularisation on the walls of the mesophyll cells in a number of plants has been related to it. But what is its quantitative importance? What proportion of all gas exchange in such leaves is by mass flow in and out rather than by diffusion? There appears to have been no quantitative study of the bellows effect. It might be very considerable in windy districts such as many coastal areas in north-west Europe. Are the rigidity and small size of the leaves found in at least some wind-swept communities, e.g. *Diappensia* on ridges blown free of protective snow in arctic regions, primarily an adaptation to prevent the bellows effect? How small may a soft leaf be and yet remain susceptible to excess desiccation by the bellows effect?

Finally we may consider those leaves which move themselves. In some cases the advantages seem obvious, as with the grass leaves that roll up during drought and expose a lesser area for evaporation, or the leaves of the Venus fly-trap that close dramatically on their prey. In other cases the advantage is not at all clear. This is particularly true of the leaves which show "sleep movements". Such are found in a number of plants with leaves made up of several leaflets that fold together each night and open during the day (Fig. 7). In Europe this behaviour is shown by the wood-sorrel (*Oxalis acetosella*) and in the tropics it is seen in the leaves of many tree species in the pea family. No convincing rationale of this behaviour has been forthcoming.

It should be clear that students everywhere – in the tropics and in temperate lands, in desert country and in rain forests – can still contribute greatly to our knowledge of the relationship between structure and function in leaves. Every kind of aptitude is required. Not only are the traditional kinds of expertise in plant physiology still badly needed, but also a whole spectrum of expertises in related disciplines, ranging from an ability to observe and record the activities

Fig. 7. Shoots of *Oxalis vulcanicola* to show the kinds of changes in leaflet inclination that occur during "sleep movements": (a) leaves in a "waking" state, (b) leaves in a "sleeping" state. In fact the leaves respond to environmental changes in a very complex fashion; for example, leaves may assume the "sleeping" position by day if exposed to bright sunlight. (Photos by P. Freeman.)

of ants and other animals on leaves to an ability to interpret the molecular structures of enzymes.

FURTHER READING

The best guides to the older ideas about some of the problems reviewed are the two following books:

Haberlandt, G. (1914) *Physiological Plant Anatomy*, Macmillan, London.
Skene, McGregor (1924) *The Biology of Flowering Plants*, Sidgwick & Jackson, London.

Relevant modern contributions are the following:

Black, C. C. (1971) "Ecological implications of dividing plants into groups with distinctive photosynthetic production capacities." *Adv. Ecol. Res.*, 7, 87-114. A very valuable introductory account of C_4 plants.

Grubb, P. J., Grubb, E. A. A. and Miyata, I. (1975) "Leaf structure and function in evergreen trees and shrubs of Japanese Warm Temperate Rain Forest. I. Structure of the lamina." *Bot. Mag., Tokyo*, **88**, 197-211. This paper incidentally shows the sorts of differences that emerge when leaves from different rain forests are compared.

Longman, K. A. and Jénik, J. (1974) *Tropical Forest and its Environment*, Longman, London. Deals with drip-tips, limp and brightly coloured young leaves and sleep movements.

Montieth, J. L. (1973) *Principles of Environmental Physics*, Arnold, London. Chapters 7 and 11 include treatments of heat losses from leaves of different size and shape.

Parleya, J-Y., Waggoner, P. E. and Heickel, G. H. (1971) "Boundary layer resistance and temperature distribution on still and flapping leaves. I. Theory and laboratory experiments." *Plant Physiology, Lancaster, Pa.* 48, 437-42. One of the few papers dealing with flapping leaves.

Parkhurst, D. F. and Loucks, O. L. (1972) "Optimal leaf size in relation to environment." *J. Ecol.* **60**, 505-37. Analyses a model based on CO_2/H_2O ratios.

Sherriff, D. W. and Meidner, H. (1974) "Water pathways in leaves of *Hedera helix* L. and *Tradescantia virginiana* L." *Journal of Experimental Botany*, **25**, 1147-56. This paper provides experimental evidence for the role of bundle-sheath extensions in water conduction. See also later papers by the same authors in the same journal.

Anthony Charles Neville

Reader in Zoology at the University of Bristol.

In 1961 discovered daily growth layers in insect exoskeletons, making it possible for the first time to determine accurately the age of an insect.

Fellow of the Royal Entomological Society of London since 1958 and Vice-President 1973-4.

SYMMETRY AND ASYMMETRY PROBLEMS IN ANIMALS

1. WHAT IS ANIMAL ASYMMETRY?

When symmetrical body form first appeared in the evolution of the animal phyla, the early examples were radially symmetrical, like a sea anemone. This was associated with a sessile way of life. It is useful for an animal which sits mostly in one place to be able to sense the approach of danger from all (radial) directions. Then muscle began to be exploited as the main means of mobility, replacing rows of beating cilia, and locomotion consequently became faster. It was then an advantage to become streamlined and so bilateral symmetry of external form became a dominant evolutionary theme. A bilaterally symmetrical animal can be mirror-imaged about its midline. During subsequent evolution, this bilateral symmetry has often been lost, so that what superficially appears to be a bilaterally symmetrical animal is seen on closer inspection to be asymmetrical; such an animal cannot be divided exactly into two halves. Asymmetry may be subtle (e.g. the zig-zag pattern down the back of an adder) or obvious (e.g. the spiral coiling of a snail). The examples may be structural, or behavioural (e.g. earthworms can distinguish right from left in maze experiments). Often asymmetry has evolved as a special functional adaptation (e.g. insect jaws which work like left-handed scissors in some species, right-handed in others).

Biological structures are formed by a hierarchy of assembled sub-units, each of which forms the construction unit for the next size level of organisation. Examples, in ascending order of size and complexity, are small molecules, macromolecules, microfibrils, fibrils, fibres, organelles, cells, tissues, organs, whole organism. This means that symmetry and asymmetry can coexist independently at different levels in the structural hierarchy of an individual animal. Thus all the amino acids from which proteins are made are L-amino acids. My left ear contains proteins with L-amino acids and so does my right ear. Taken at this level of structure, therefore, all animals which appear from the outside to be bilaterally symmetrical, are really asymmetrical. Again, whereas evolutionary pressures led to streamlining of exterior form in the mackerel, its internal viscera are not bilaterally symmetrical and neither are the stripes on its back. In adult beetles, the orientation of the first layer of fibrous chitin laid down after hatching from the pupa follows the bilateral body symmetry. But subsequent layers are laid down asymmetrically. In flatfish, reversal of external asymmetry usually occurs without reversal of internal viscera. It is concluded that separate control mechanisms must exist for the various levels of asymmetry.

Some examples of asymmetrical adaptations in animals will next be presented, as a prelude to the discussion of major unsolved problems in developmental biology which they reveal.

2. SOME TELLING EXAMPLES

Hermit Crabs

Hermit crabs live inside empty shells of molluscs. Primitive ones are bilaterally symmetrical as they live inside homes which are not asymmetrical (e.g. scaphopod shells; pieces of bamboo tube). More advanced forms show striking corkscrew twists, evolved in adaptation to the spiral shells of gastropod molluscs. They twist the same way as a right-handed corkscrew (dextral), correlated with the dextral twist in the majority of gastropods. The robber crab *Birgus* shows a striking sequence of symmetry changes. When it leaves the sea it is bilaterally symmetrical; it then emerges onto the land and becomes dextrally asymmetrical, inhabiting mollusc shells. Finally, air-breathing lungs are developed and the crab leaves its mollusc shell to become free-living on land, redeveloping bilateral symmetry.

Fiddler Crabs

Crabs and lobsters show a marked size and shape difference between the large claws on the two sides of the body. The larger one is adapted for crushing, the smaller one for picking up pieces of food. In the extreme example of the fiddler crabs (*Uca* spp.) the large claw may weigh 20 times more than that of its partner, amounting to 70 per cent of the weight of the whole body. Only males develop this large claw, which is used in courtship, both for display and stridulation. Chance injury determines whether the large claw develops on the right- or left-hand side. If a claw is lost in a young male, another regenerates. In this case the surviving original claw grows on (it has a start) to become the large one.

Patterns

I find it surprising that relatively few animals have evolved asymmetrical camouflage patterns, since this should help to conceal a bilaterally symmetrical animal better than a bilaterally symmetrical pattern. (Camouflage on warplanes is often asymmetrical.) Examples of asymmetrical pattern are the fly larva *Thaumalea verralli*, some frogs (e.g. *Dendrobates tinctorius, Rana temporaria*), european salamander, common viper and mackerel.

Flatfish

The flatfishes are the most asymmetrical vertebrates. In the plaice, for example, the bilaterally symmetrical larva settles on its left side on the sea bed when about 30 days old. A spectacular metamorphosis then occurs, at the end of which both eyes lie on the top (right) side. Cryptic coloration is found only on the right side and the fish now swims with the right side uppermost. There is a difference in the form of the scales of the two sides as well as in the sense organs. An experiment reported last century needs repeating: when young flatfish were prevented from settling on their left side before metamorphosis, and made to keep swimming, they retained their bilateral symmetry.

Hearing in Owls

In Tengmalm's owl (*Aegolius funereus*) the skull is markedly asymmetrical; the right ear points obliquely upwards, whereas the left ear points obliquely downwards. The ear apertures are covered with sound-transparent feathers. This asymmetry permits judgement (after learning) of high-frequency sounds in the vertical plane without the need to tilt the head. This could not be done if the ears were symmetrical. The maximum sensitivity for sound frequencies between 6000 and 10,000 c/s is obliquely upwards for the left ear, obliquely downwards for the right ear. The asymmetry is reversed for frequencies between 10,000 and 16,000 c/s. Young owls learn to interpret the asymmetrical information from both ears by tilting their heads alternately to left and right in the vertical plane, often through more than a right angle (Fig. 1).

Fig. 1. A young Tengmalm's owl (*Aegolius funereus*) learning to pinpoint sounds in the vertical plane by tilting its head. (From Å. Norberg, *Svensk. Naturvetenskap,* 26, 89-101 (1975)).

They can then pinpoint the direction of sounds of potential prey rustling in the vegetation or in the snow. Several species of owls can locate prey in total darkness. Barn owls can pinpoint sound to one degree of accuracy in both vertical and horizontal planes.

Humans

Human faces and skulls are asymmetrical. This is illustrated in Fig. 2, in which symmetrical faces are deliberately reconstructed by photography from each half of a real face. It is more difficult to identify an individual thus reconstructed. Perhaps asymmetry of features decreases the chance of one face being confused with another. Police "Identikit" pictures look more realistic when asymmetry is built in.

Siamese twins are not identical, but are really mirror-images; e.g. one has the heart on the left, the other on the right.

The human brain behaves as two independent halves; each half can be taught opposite solutions to simple problems. Evidence comes from patients with brain damage, patients in which the nerves connecting the two halves have been surgically severed, and from electrical recordings from each half of normal brains. In a right-handed human, the left half of the brain

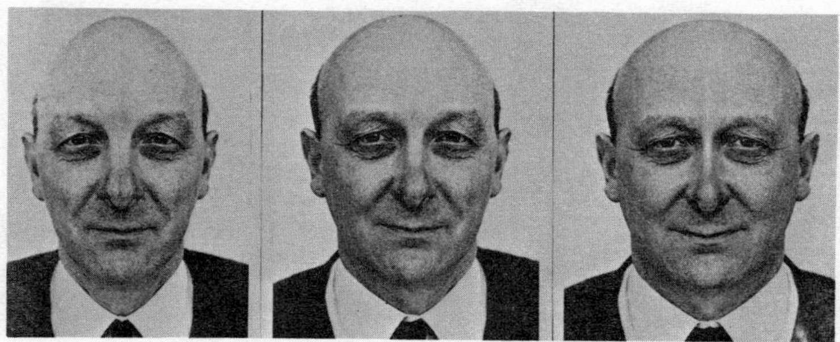

Fig. 2. Landlord of White Bear Hotel, Bristol. Normal face, in the centre. Left shows face photographically reconstructed from two left halves of the face; right shows similar reconstruction from two right halves.

controls analytical thinking, mathematical and verbal functions, whereas the right half is concerned with arts, crafts, music and dancing. This is reminiscent of C. P. Snow's famous division into two cultures, since half of the brain is usually dominant in each individual.

A surprisingly strong asymmetrical bias in locomotion is revealed when humans are blindfolded. Such subjects are told to walk, swim, row a boat or drive a car in a straight line. Although they are convinced that they progress in a straight line, the subjects walk in spirals with diameter as small as 15 metres and swim in circles of diameter 6 metres. There is no correlation between sense of spiralling and left- or right-handedness. As well as this unconscious asymmetry, humans also possess a conscious sense of asymmetry. I can distinguish my right from my left-side with 100 per cent accuracy. A computer cannot do this, or at least only if it is built or programmed by humans so to do.

3. THE EVOLUTION OF BIAS

We have seen that in fiddler-crabs chance injury determines whether the large claw develops on the right or the left side of the body. Also, in the primitive flatfish *Psettodes* there is an equal number of left and right asymmetrical individuals. There are, however, many examples where there is *not* an equal chance of left or right asymmetry. Table 1 gives some examples of what I have termed *bias* in asymmetry (Neville, 1976), expressed as percentages, and covering a wide selection of animals. It is clear that some species are able to select left from right during development, and in some cases to make a correct choice nearly every time. Figure 3 shows dextral (very common) and sinistral (very rare) examples of a snail (*Cepaea nemoralis*), in which bias approaches 100 per cent. In flatfish there is an evolutionary trend towards greater bias percentages (Hubbs and Hubbs, 1945).

4. PROBLEMS POSED

Consideration of animal asymmetry reveals some challenging unsolved problems in developmental biology. How can an animal build a bilaterally symmetrical body using asymmetrical components (e.g. L-amino acids, D-sugars, right-handed α-helix)? In the inorganic

TABLE 1. Examples of Left or Right Bias in Asymmetry
(Adapted from Neville, 1976)

Examples	Species	%
Sinistral shell	*Limnaea peregra*[1]	0.001
Sinistral shell	*Buccinum*[1]	0.1
Large L. chela	*Homarus americanus*[2]	53
L/R wing folding	*Gryllotalpa gryllotalpa*[3]	0.05
L/R wing folding	*Acheta domesticus*[3]	0.05
L/R wing folding	*Acheta assimilis*[3]	4-10
L/R wing folding	*Acheta pennsylvanicus*[3]	4-10
L/R wing folding	*Acheta rubens*[3]	4-10
L/R wing folding	*Teleogryllus commodus*[3]	10
L/R mandibles	*Periplaneta americana*[3]	100?
L. abdominal torsion	*Clunio marinus*[3]	48-55
Eyes on L. side	*Solea*[4]	0.001
Eyes on L. side	*Pleuronectes platessa*[4]	0.001
Eyes on L. side	*Psettodes*[4]	50
Gut reversal	*Cottus gobio*[4]	0.2
More fin rays in R. pectoral fin	*Leptocottus armatus*[4]	60-81
Gut reversal	*Triton*[5]	2-3
L. gular plate larger	*Gopherus agassizii*[6]	90
Lower mandible crosses L.	*Loxia curvirostra*[7]	58
Extra digit on L. foot	*Gallus domesticus*[7]	90
L. handedness	*Homo sapiens*[8]	5

Key: 1. Mollusca; 2. Crustacea; 3. Insecta; 4. Osteichythes; 5. Amphibia; 6. Reptiles; 7. Aves; 8. Mammalia.

Fig. 3. Sinistral (left) and dextral (right) examples of the snail *Cepaea nemoralis*.

world, bilaterally symmetrical crystals form spontaneously with the expenditure of the minimum amount of energy. Since the building units of organisms are asymmetrical (molecules, organelles, cells, etc.), it has been argued that the achievement of bilateral symmetry in a whole organism must involve higher energy expenditure. For instance, in the replication of two-stranded DNA (giving right-left symmetry of base pairing), two enzymes seem to be required (ligase and replicase) instead of one. Hence at this molecular level, bilateral symmetry is expensive in energy terms.

The problem recurs in reverse. Given a bilaterally symmetrical animal, how can it evolve to give rise to an asymmetrical form? DNA codes for proteins which control an organism's shape and metabolism. Since the DNA code occurs identically in every somatic cell of the body (i.e. not counting sex cells), how can it be interpreted differently on the two sides of the midline, in a non-erratic manner?

This problem becomes even more challenging when the question of bias is considered. Not only must we postulate differential interpretation of the genetic code on the two sides of the body, but we have to suggest that the code "knows", often with great accuracy, what to build on a specific side of the body. (A trivial explanation of bias could be that there is a difference in mortality rate between left and right individuals of a species. However, extensive experiments with huge numbers of the snail *Limnaea peregra* did not support this explanation.)

Furthermore, asymmetrical bias can vary geographically. If bias were to be affected environmentally we might have expected it to be reduced (i.e. to approach 50 per cent). However, this is not so. For example, in the Pacific flounder (*Platichthys stellatus*), the bias to sinistrality at metamorphosis is 49–60 per cent off North West America; 68 per cent off Alaska; and 100 per cent off Japan. There is also evidence of geographical variation in asymmetrical bias in the twisting direction of male abdomens of the fly *Clunio marinus*. Clockwise twisting dominates in Norway and is about equal with anti-clockwise twisting in Brittany, but anti-clockwise dominates in north Spain. Again, bilateral gynandromorphs occur in common blue butterflies (*Polyommatus icarus*), with half the body male (blue wings) and the other half female (brown wings). Such examples are more common in Ireland than in England. Similarly, in the case of the snail *Cepaea nemoralis*, sinistral examples (Fig. 3) are frequently found on sand dunes at Bundoran and Donegal in the Irish Republic; they are extremely rare in most other places.

5. POSSIBLE EXPERIMENTAL APPROACHES

On the face of it these problems of asymmetry and bias appear intangible. There are, however, several possible approaches. Old experiments on the prevention of metamorphic asymmetry and bias in flatfish larvae by not allowing them to settle, deserve repeating and extending. Eggs and developing eggs of gastropod molluscs and flatfish could well be subjected to centrifuge experiments and to nuclear transplantation experiments, to try to establish the intracellular location of asymmetry. The genetics of asymmetry is a promising field. We already know that sinistrality and dextrality in the snail *Limnaea peregra* are genetically controlled; also hybrid flatfish produced by crossing sinistral *Platichthys stellatus* with dextral *Kareius bicoloratus* comprise equal numbers of dextral and sinistral individuals. Again, in mice, the luxate mutation affects one side rather than the other. Certain mutagen drugs are now known which induce biased asymmetry. For example, acetazolamide and dichlorphenamide produce specific deformities in the right forelimb of rats. In domestic fowl chicks, the cornea of the eye has collagen fibres arranged in a multiple-ply structure which has a clockwise rotation in both eyes. This is therefore an asymmetrical property. The recently synthesised drug 6-diazo-5-oxo-L-norleucine (DON) induces reversal of this rotation in both eyes. It is hoped that these examples may stimulate new thoughts and experiments on this widespread and unsolved problem in development.

Further Reading

Bochner, S. (1973) "Symmetry and asymmetry." In *Dictionary of the History of ideas*, edited by P. P. Wiener, Vol. 4, pp. 345-53. Scribner's Sons, New York.

Gardner, M. (1970) *The Ambidextrous Universe. Left, right and the fall of parity*, Pelican Books, pp. 1-276.

Gazzaniga, M. S. (1967) "The split brain in man." *Scient. Amer.* **217**(2), 24-9.

Hubbs, C. L. and Hubbs, L. C. (1945) "Bilateral asymmetry and bilateral variation in fishes." *Pap. Mich. Acad. Sci., Arts and Letters,* **30**, 229-310.

Neville, A. C. (1976) *Animal Asymmetry*, Inst. Biol. Studies in Biology No. 67, Arnold, London.

Oppenheimer, J. M. (1974) "Asymmetry revisited". *Am. Zool.* **14**, 867-79.

Walker, I. (1972) "Biological memory." *Acta Biotheoretica* **21**, 203-35.

K. A. Bettelheim

Lecturer in Medical Microbiology at the Medical College of St. Bartholomew's Hospital, London, and Member of the Royal College of Pathologists. Now Serologist, National Health Institute, Wellington, New Zealand.

Interests include the use of immunology to study microbial enzymes, automation and the study of *Escherichia coli.*

BACTERIAL PATHOGENICITY

Bacteria comprise a very large percentage of the biomass of this planet and the majority of them are an essential link in the chain of life on earth. Many of the others, though not essential, perform many important tasks maintaining the balance of nature. Only a very small number of species are associated with humans. These can be subdivided into three interrelated groups: Commensal bacteria, Opportunist pathogens and True pathogens.

Commensal bacteria constitute the normal bacterial flora of the healthy human body. They live on the skin, on the mucous membranes of the upper respiratory tract and the intestinal and female genital tracts. They may also exist in smaller numbers in other sites. They obtain their nutritional requirements from secretions, sloughed-off cells and food residues. Generally they are harmless and incapable of invading other tissues, or if small numbers do penetrate into internal tissues, they are rapidly dealt with by the body's defence mechanisms. If these mechanisms are somehow impaired or if commensal bacteria have been able to gain access to internal sites due to accidental damage or some form of manipulation, where sterile precautions were not strictly adhered to, then they can cause disease and become opportunist pathogens. They do not normally have any direct means of gaining access to the healthy human body and causing disease.

True pathogens are partially or completely parasitic organisms, which are adapted to overcome the normal defence mechanisms of the host and are capable of invading the tissues. This invasion or their subsequent growth within the tissues may cause a number of poisonous or other tissue-damaging substances to be formed. These substances include the various toxins, aggressins, etc., which may be the prime cause of the symptoms associated with the disease. In a number of other instances it appears to be only the ability of the pathogen to grow at a certain site for the disease symptoms to develop, whereas often both factors may play a role. Unfortunately for many of the common human pathogens no adequate explanation for their pathogenicity has yet been forthcoming.

The earliest and simplest mode of explaining bacterial pathogenicity was by means of toxin production. In these cases it was considered that if a substance produced from the pathogen could be isolated and it was found that it reproduced the disease symptoms in the absence of the pathogen in experimental animals or human volunteers then the pathogenicity of that organism could exclusively be ascribed to the production of that substance, which became known as a toxin. Furthermore, treatment with specific antagonists to the toxin would relieve the disease either in patients or experimental animal situations. A number of pathogens do seem to fit into that picture and they will be discussed first.

Botulism is caused by ingestion of food contaminated with *Clostridium botulinum* or a toxin produced by them. The disease, which also affects animals, can be reproduced fully with the toxin. The toxin, of which six serological types are known, is one of the most lethal substances known to man, 1μg contains more than 200,000 minimum lethal doses (MLD) for a 20-g mouse. The symptoms caused by ingestion of the toxin are paralysis, often affecting the eyes first followed by the neck muscles and the muscles controlling respiration. Of the six serological types, type A toxin has been extensively purified and shown to be a protein with a molecular weight of 900,000 D. It contains no lipid, carbohydrate, nucleic acids or unusual amino acids. Toxic polypeptides can be separated from non-toxic fragments of the toxin by dialysis and it is believed that a similar mechanism permits the passage of the toxic fragments through the intestinal walls. The toxin produces complete paralysis of cholinergic nerve fibres probably by inhibiting the release of acetyl choline, by mechanisms, which are still obscure, and whose elucidation may profoundly affect our knowledge on the mechanisms of nerve action. The toxin is only fixed at peripheral nerve endings, not the brain or other parts of the central nervous system. The amino acid tryptophan appears to be closely associated with the active centre of the toxin and injecting serotonin into mice before toxin will reduce the activity of the toxin. The only effective method of treatment is with antitoxin.

Clostridium tetani is another organism which produces a potent toxin, the action of which completely mimics the disease produced by the organisms. It is very difficult to ascertain the minimum lethal doses because the material gets denatured and absorbed onto glass in dilute solution but they are of the same order as botulinum toxin. Also there is some variation in susceptibility of different animal species. There is only one type of tetanus toxin known and it is a simple protein dimer with a molecular weight of 150,000 D containing no carbohydrate or lipid. The role of the toxin to the bacteria is obscure particularly as toxin production in culture can be suppressed without affecting bacterial viability. Also the bacteria are widespread in soil and can really be considered more as opportunist pathogens because they require external skin damage to gain entry. The toxin does not act on isolated body fluids or cells and injection of even 500,000 MLD gives no directly observable pathological lesions. It is only fixed by nervous tissues, attaching itself to a component of the membrane of the nerve cell known as ganglioside. There it blocks neuromuscular transmission by blocking the movement of sodium and potassium ions across the membrane, probably by affecting the adenyl cyclase system.

A very similar mechanism of action has been proposed for a quite different organism, namely *Vibrio cholerae*, the causative agent of cholera. During the course of an untreated case a large number of symptoms, which ultimately lead to death, can be observed. It is now established that all these symptoms can be entirely attributed to local alterations of fluid and electrolyte movement in the small bowel, resulting in fluid and electrolyte loss from the plasma, giving hypovolaemic shock. The technique of the ligated intestinal loop test has been developed to correlate this enterotoxic activity and fluid loss. From this it has been established that cell-free culture filtrates from *V. cholerae*, will mimic the activities of the bacteria, and cause the release or stimulation of intracellular lipases, producing effects similar to those induced by cyclic adenosine monophosphate (cAMP). From these active culture filtrates has been isolated a protein of molecular weight 84,000 D, which is considered to be the active cholera toxin or cholerogen. This protein, which is reversibly dissociable into smaller subunits, appears to have as its primary or possibly sole function the activation of adenyl cyclase, which it does indiscriminately regardless of cell type. Thus a number of cell types can be used to assay toxin activity. Of the cholerogen fragments, a non-toxic protein of molecular weight 56,000 D has been isolated. This protein can inhibit the action of complete cholerogen by occupying the same receptors on the susceptible cell membranes as the complete toxin, indicating the role of this fragment. Another fragment of molecular weight 28,000 D must be present for a preparation to have toxin activity. Thus stimulation of the adenyl cyclase system of nervous tissue by tetanus toxin causes tetanus, while a similar effect on the intestinal epithelial cells by cholerogen produces cholera. Both toxins act first by binding to the membrane component ganglioside, but otherwise the tissue specificity and evolution of these toxins is obscure.

Although *Escherichia coli* is a common commensal of the bowel of humans and domestic animals, circumstantial evidence based mainly on epidemiological studies has shown that certain serotypes of *Esch. coli* can cause a variety of diseases of man and domestic animals. These include infantile gastroenteritis, traveller's diarrhoea, *non-Vibrio cholerae*-cholera-like diarrhoea as well as a number of similar diseases of animals. To date, two types of enterotoxin have been described. Of these a heat-stable, dialysable acid-resistant non-antigenic material has been shown to cause diarrhoea in test animals. Otherwise nothing is known of its composition or mode of action. The other material with enterotoxin activity, which has been isolated from some strains of *Esch. coli*, is a heat-labile protein of molecular sizes varying from 20,000 to above 5×10^6 D. It is antigenic and serologically related but not identical with cholerogen. Its mode of action also appears to be very similar to that of cholerogen, being capable of stimulating the adenyl cyclase systems of a variety of cells. It has recently been suggested, although not demonstrated, that the production of this enterotoxin might be coded for by a plasmid derived originally from *V. cholerae*. The possibility that toxigenic activity can be transmitted from a pathogen to a commensal must be seriously investigated particularly in view of the current controversy on certain types of genetic experiments. So far these two toxins have not always been isolated from strains epidemiologically linked to disease production. Whether these strains produce some types of toxin unlike the ones discussed or whether the disease is in fact due to some other vector and the isolation of certain strains of *Esch. coli* is purely coincidental and has not been established.

The fact that toxigenicity can be transmitted to non-toxigenic bacteria has been established in the case of *Corynebacterium diphtheriae*, the causative organism of diphtheria. Toxigenicity has been shown to be dependent on bacteria carrying lysogenically the corynephage β. This is a DNA virus, on whose twin helical length of DNA of 2.2×10^7 D has been mapped the gene which is transcribed to produce the toxin. This *tox* gene does not seem to serve an essential viral function. Toxin has been synthesized *in vitro* by a protein synthesizing system, using isolated phage DNA as template. Iron ion concentration also appears to play an important though unexplained controlling role in toxin production, thus iron-rich cells will only produce *ca.* 10 per cent of the toxin produced when iron is limiting. The toxin is a protein of molecular weight 62,000 D and 130 mg/kg body weight is the lethal dose to man and animals. It has been demonstrated that it is fixed by most body cells and if added to tissue culture cells it blocks protein synthesis. A fragment of the toxin, known as fragment A, which has a molecular weight of 24,000 D and is released from the complete toxin by trypsin and dithreitol, has been shown to catalyse the transfer of adenosine diphospho-ribose from diphosphopyridine nucleotide (NAD) coenzyme to an enzyme known as elongation factor 2 (EF2, aminoacyl tranferase I, translocase II). As a result of this action EF2 is inactivated. As EF2 is the soluble enzyme involved in messenger ribonucleic acid (mRNA) directed growth of polypeptide chains, the action of fragment A of the toxin is to inactivate ribosomal protein synthesis. It has been suggested that the other component of the toxin, fragment B, is required to fix the toxin to the cell, thereby permitting fragment A to gain access to the cell. Only protein synthesis directed by 80S eukaryose ribosomes is affected by diphtheria toxin.

Although some information is available on the pathogens and their modes of action on the organisms just discussed, it will be realised that in no case is it complete. For the other organisms to be discussed now far less is known; sometimes small tantalizing pieces of information are available, but these are insufficient to provide a complete picture. A number of these organisms including *Staphylococcus aureus* and *Clostridium welchii* are responsible for a number of apparently quite diverse diseases as well as being frequently encountered as commensals.

To date from preparations of *Clostridium welchii* at least twelve antigens with toxic activity, which might also be enzymes, have been defined. These have been given Greek letters to distinguish them. Three haemolysins, which also affect other cell membranes, have been distinguished. These are the α-, δ- and θ toxins. So far the biochemical properties of only the α-toxin have been established and it has been shown to be lecithinase type C. The κ toxin is a collagenase, the λ toxin a proteinase, the μ toxin a hyaluronidase and the ν toxin a DNase. The β-, γ-, δ-, ϵ-, η-, and ι-toxins all have lethal activities in experimental animals. In some cases the specific tissue target is known but not in all. A number of them also have necrotic activities. There are also present such components as a circulation factor which inhibits phagocytosis and a bursting factor, causing oedema. Of the toxins mentioned the α- and the θ-toxins as well as probably the κ-toxin are the most important in causing gas gangrene. The other toxins probably have as their main function the disturbance of metabolism, particularly the lipid metabolism, causing embolisms, or the liberation of host enzymes from lysozomes. There is a great variation in the amount and type of enzyme present in different toxin preparations from different strains of *Cl. welchii*, and this probably accounts for the different types of disease these organisms are supposed to cause.

To different strains of *Staphylococcus aureus* have been ascribed perhaps the largest number of different diseases, without any satisfactory explanation being available for them, although many toxic factors have been isolated. In order of importance the diseases ascribed to these organisms are wound infections, food poisoning, osteomyelitis, pneumonia, meningitis, endocarditis and enterocolitis. Three haemolysins have been characterised of which the α-haemolysin (toxin) probably acts as some type of membrane damaging agent, affecting many cell membranes, including lysing platelets and releasing the contents. It also has a paralysing action on smooth and skeletal muscles and in animals can cause increased vascular permeability. The α-toxin also appears to have effects on the nervous system because intravenous injection into rabbits gives an altered electroencephalogram with the thalamus being the most affected region. The β-haemolysin (toxin) appears to be a lecithinase whose substrates tend to be the membrane components sphingomyelin and phosphatidyl choline. It, too, damages membranes, including platelet membranes. There are contradictory reports on its lethality for experimental animals. The δ-haemolysin (toxin) releases organic phosphorus compounds from the phospholipids of erythrocytes and also appears to have the properties of a surface active protein. Polymorphs, lymphocytes and macrophages are uniformly sensitive and lysed by it. It also acts as an antibiotic against other gram-positive bacteria. Apparently the ability to produce this toxin is also dependent on the presence of a phage, probably by mechanisms similar to those operating for the toxigenicity of *Corynebacterium diphtheriae*.

The ability by strains of *Staphylococcus* to produce the enzyme staphylocoagulase is frequently used as an indicator for pathogenicity. Current nomenclature considers all staphylococci which produce coagulase as *Staph. aureus*, while those not producing it are grouped among the species *Staph. epidermis*, *Staph. albus*, etc. Nevertheless, many of these coagulase-negative staphylococci produce some of the other toxins discussed and have been associated with a pathogenic role, although they are also more frequently associated with the normal flora of the human body, particularly the skin and nose. In blood coagulase reacts with a plasma factor which probably results in the formation of thrombin. The coagulase-thrombin reaction then serves as a defibrinogenating and systemic haemostatic agent, acting somewhat like snake venom.

Other enzymes, which staphylococci have been shown to produce under certain conditions, and which have tissue-damaging properties, include a hyaluronidase, phosphatases, RNases, DNases, proteinases and lipases.

It has recently been shown that staphylococci growing *in vivo* produce a leucocidin, which kills white blood cells, in far larger amounts than *in vitro*. As the host's production of antileucocidin antibodies seems to be its main defence to staphylococcal infections this substance may be the main pathogenic factor of staphylococci.

Staphylococci have also been shown to produce enterotoxins in some foods and media. The laboratory studies on these substances have been hampered by the fact that the only reliable assay method involves feeding experiments on monkeys. They have been shown to be present when enterotoxigenic staphylococci grow in cooked meat and cheese, and impure enterotoxins in contaminated food can survive up to 2 hours at 100°C. The mode of action of these enterotoxins has not been established and it is not known whether they enter the circulatory system.

Thus in the case of the staphylococci the presence of a wealth of potentially toxic materials does not yet seem to have clarified how these organisms act as pathogens.

It appears that the initial or suppurative infections caused by strains belonging to the taxonomic group A of the genus *Streptococcus*, such as acute pharyngitis with its numerous complications, scarlet fever, erysipelas, etc., are largely dependent on the factors that permit the strains to invade the human body and establish themselves there. The most important of these properties seem to be their anti-phagocytic properties, which appear to be mainly due to their possessing of a capsule of hyaluronic acid and of a protein, known as M protein, in their cell wall. Those streptococci which can be ingested by phagocytes are destroyed by them and therefore are not virulent. Some streptococci also produce a leucotoxic factor, with the enzymic property of an NADase which kills phagocytes and thus these organisms also escape destruction. Streptococci also produce a number of haemolysins, some of which are highly toxic for leucocytes. The erythrogenic toxin, which is associated with scarlet fever, is produced only by those strains which carry a specific bacteriophage. It is not known whether the rash is formed as a result of some direct action of the toxin or is due to a secondary hypersensitivity reaction of the host.

It is believed that streptococci can cause some type of immune response that causes the host to develop antibodies to some components of his own kidney resulting in the development of acute glomerulonephritis. The causes of rheumatic fever and the role of streptococci in it are not clear at present. One of the streptococcal haemolysins, which is oxygen sensitive (streptolysin O), has been shown to have cardiotoxic properties. When streptococcal proteinases are injected into the blood stream of experimental animals they produce endocardial lesions and the M protein, when introduced into the blood stream, becomes deposited in the endocardium, where it becomes available for a hypersensitivity reaction.

Although pathogenic strains of *Diplococcus pneumoniae* are commonly carried in the throats of normal adults, pneumonia is a relatively rare disease. It is assumed that under normal conditions the host defence mechanisms prevent a disease situation from arising. This only happens when the defence mechanisms are impaired such as through a viral infection occurring in the upper respiratory tract. The development of pulmonary oedema contributes to the infection because the fluid is a suitable growth medium for *D. pneumoniae*. It has been shown that the severity of the disease is proportional to the population density of the micro-organisms in the lesion and to the size of the capsule produced around each organism. Although numerous different antigenic types of capsule are produced by these organisms they do not appear to differ in pathogenicity. No specific toxins have been shown to be formed by them although they produce haemolysins like the streptococci.

Shigella disenteriae type 1 (Shiga's bacillus) is a less common cause of dysentery than most of the other types of *Shigella*, but the disease is often more severe, being accompanied by profound toxaemia ending in collapse. Only this serotype of the genus *Shigella* produces a protein toxin of molecular weight 82,000 D with a dose range for animals of the order of the tetanus and botulinum toxins. Its main mode of action appears to be in blood vessels, mainly those in the spinal chord where it causes oedema and the swelling causes pressure. This damage to the spinal tissue led earlier investigators to consider that the *Shigella* toxin was a neurotoxin.

In animals it causes the caecum to show gelatinous oedema and haemorrhagic necrosis. The specific cause of the lesions produced in dysentery, caused by this or other shigellae, such as superficial ulcerations covered by pseudomembranes in the terminal ileum and colon have not been defined.

Strains of *Salmonella* are the principal causes of enteric fevers including typhoid fever, acute gastroenteritis and food poisoning. The organisms causing enteric fevers can penetrate the epithelial cells of the intestine, perhaps by a type of phagocytic process, while the other salmonellae cannot. This ability to penetrate epithelial cells has not been sufficiently studied but is presumably based on some material present on the bacterial cell surface or excreted by the bacterium to which the epithelial cells are sensitive. Thus these hypothetical substances would be the prime factors determining whether a strain of *Salmonella* causes enteric fever and can gain entrance to the blood stream or whether it merely remains in the gastrointestinal tract causing gastroenteritis of varying severity. There must be some contributing host factors because *Salm. typhi*, which causes enteric fever in man, only causes gastroenteritis in mice, while *Salm. typhimurium* has a reverse effect. There are similar examples of host specificity for other serotypes of *Salmonella*. During the course of enteric fever the bacteria are able to persist and multiply in mononuclear phagocytes, in which they are protected from specific antibody and phagocytic bacteriolytic activity. During recovery these monocytes acquire the ability to destroy the Salmonellae. The lipopolysaccharide somatic and capsule-like antigens have been shown to be related to virulence and from animal experiments it has been suggested that most of the symptoms are due to circulating breakdown products of the bacterial cell walls, which have endotoxic activity, but this does not seem to apply to human disease. No enterotoxins or similar substances have been isolated from Salmonellae causing gastroenteritis.

In this discussion of some of the more common pathogenic bacteria it can be seen that toxins and similar substances are in many cases important contributing factors to the virulence of these organisms. This does not appear to be always the case. It seems surprising that the many techniques of biochemistry and molecular biology have not been employed to study these problems of the interaction of bacteria and humans, because many of the results described were established many years ago, and the importance of gaining knowledge of the mechanisms of bacterial virulence in order to combat human disease cannot be overstressed.

M. M. Lehman

Professor of Computing Science at the Department of Computing and Control, Imperial College, London, and an industrial consultant on software engineering and total computing systems.

Combined teaching activities with research into evolutionary trends and other properties of software systems, the software engineering process and computer installations.

HUMAN THOUGHT AND ACTION AS AN INGREDIENT OF SYSTEM BEHAVIOUR

1. SYSTEMS SCIENCE

Significant sections of the scientific community are increasingly becoming involved in Systems Science, though they often do not recognise this subject as an independent discipline. Biologists, computer scientists, economists, engineers, mathematicians, sociologists and many others are realising that in many instances one cannot hope to master the systems studied by adopting the classical approaches to science.

The latter are largely based on a bottom-up approach. Individual elements are first studied and mastered. When these are at least partially understood, one proceeds to examine the properties and behaviour of more complex assemblies of these elements. Conceptually at least the methodology may be indefinitely extended to study ever larger and more complex systems as built up in more or less structured fashion from sub-systems and primitive elements.

In practice, however, the mathematical and other descriptive tools used for such studies rapidly become unwieldy. They tend to break down unless a system is effectively homogeneous, as in physics for example, or has a simple, regular structure, as in crystallography. In general, the properties of a system observed as an entity are not readily discernible, nor do they follow easily from one's knowledge and understanding of the attributes and behaviour of its components. Despite the fact that one may be concerned with global characteristics, these cannot be directly or totally related to or inferred from elemental behavioural patterns.

Thus progress in revealing the nature of the physical world has necessitated the development of newer methodologies that do not rely on the study of individual phenomena in exquisite detail. A system, a process, a phenomenon may be viewed in the first place from the outside, observing, clarifying, measuring and modelling identifiable attributes, patterns and trends. From such activities one obtains increasing knowledge and understanding based on the behaviour of both the system and its sub-systems, the process and its sub-processes. Following through developing insight in structured fashion, this top-down, outside-in approach in due course leads to an understanding of, and an ability to control, the individual phenomena but in the context of their total environments.

The global, systems, viewpoint has been fruitfully applied, directly or indirectly, in many areas of the natural sciences. Epidemology, Genetics, Information Theory, and Thermodynamics are typical areas of applied science where the approach has yielded important results of both theoretical and of practical significance.

In recent years, however, interest has had to increasingly focus on, in Simon's words,[(1)] the "sciences of the artificial", on the behaviour of systems created by man. And people invariably constitute sub-systems or elements of the artificial systems, if only by virtue of the fact that they design, build and use them.

Naturally one has sought to develop a theory of and for these systems in terms of the concepts, the techniques, the languages, the mathematical tools of the epidemologist, the geneticist, the information theorist, or the thermodynamicist, for example. Further, after abstractions that remove dependencies on the specifics of the original system studied, one could expect to deduce observations or hypotheses about systems in general. Thus a discipline of Systems Science is developing and with it the systems scientist. His ultimate objective will be to isolate and reveal those attributes of system behaviour that arise by virtue of their being a system or some specific system.

2. SYSTEMS

But what are systems? Many alternative definitions have been proposed.[2] For our purpose a system may be viewed as a structure of interacting, intercommunicating components that, as a group, act or operate individually and jointly to achieve a common goal through the concerted activities of the individual parts. We note that, in general, the partitioning of a system into components is by no means unique. Selected components may be viewed as elemental or atomic but under further scrutiny each will, in general, be seen to be itself structured and to contain components that together satisfy the definition of a system in their own right. The component will constitute a sub-system of the original system. Equally any system will, in general, be a sub-system of some other system or systems; of its environment.

The Concorde is clearly a system in the sense of the above definition. So are the total grouping of people and equipment that co-operatively ensure the safe transportation of the craft and its passengers, or of successive loads of passengers, between points of departure and destinations. A computer installation with its equipment and its staff represents another class of systems; one in which the perceived attributes and characteristics will be as much dependent on the attitudes, skills and managerial decisions of the staff as on the technical capabilities of the equipment. Equally each of its sub-systems, the central computer for example, is a system in its own right. Its sub-systems, in turn, include storage, arithmetic, control and peripheral access units.

3. SYSTEM MEASUREMENT AND SYSTEM MODELS

Attempts to measure the performance of such installations have not been very successful so far. Any measures must not refer directly to detailed workload or machine characteristics. The sheer variety and quantity of available data would quickly make the data and its analysis unmanageable. Moreover, measures are often desired to facilitate the comparison of alternative machines or workload environments, or the installation at different times and when some components will have inevitably changed. Inclusion of detailed characteristics of the installations and/or environments to be compared in the measures would, in practice, invalidate the measure. Differences in measure will in fact often be due to differences in characteristics and/or to changes in the system that cannot be uniquely or meaningfully quantised or ordered.

Thus we require to define system measures that are global in nature. For example, we might determine the average turn around time (interval between submission of a job request and receipt by the requestor of the result of its execution) for all jobs or for jobs of a particular class or type. This measure would depend both on machine capability and on the performance of installation personnel. Given some further measures that relate system performance to job stream and environmental characteristics, one might then produce a model of the installation of system behaviour at some point in time. And given several such models one might arrive at a generalised installation model and theory.

4. THE EMERGENCE OF A PROBLEM

There is, however, at least one snag in this objective. Both the environment that generates the workload and the installation that executes it include people. Their behaviour pattern (in relation to each other and to the system) depends on observation, and interpretation of system behaviour; and also on less tangible factors. The more complete their understanding of apparent system behaviour becomes the more will they modify the system or adjust their behaviour in relating with and to the system, so as to obtain the closest possible approximation to what is considered optimum performance.

In general, as knowledge and understanding of an artificial, man changeable, system increases we attempt individually and collectively to modify the behaviour of that system. We attempt to make it behave in some other, preferred, fashion. The modification is obtained by changes to the system, to usage protocol and so on. The resultant configuration is and must be treated as a different system which requires a new model to represent it. Thus artificial systems and their models appear to be essentially transitory, continuously evolving. Does this constitute an intrinsic barrier to complete understanding and mastery of the system?

5. FIRST ATTEMPTED SOLUTION AND ITS CRITIQUE

One might of course argue that in order to study the system all change proposed in response to observed behaviour be inhibited. In a system model this would be represented as an opening up of the representation of the feed-back path that permits humans to modify the system in response to its behaviour. However, if this is done, one is now studying a different system. And this is not just a superficial difference. This same feed-back loop underlies both system improvement and its adjustment to an ever-changing environment. And adaptability, if it exists, is perhaps one of the most important properties of a system.

System adaptation and evolution reflects and results from the mutual reinforcement of the system and its environment. In non-biological systems human observation, thought, action and reaction play a major role in forcing and guiding system development and evolution. Thus, for example, a computer installation in which procedures, the components or the structure may not be modified in response to observed behaviour, is a different system from one in which initiation of change in response to experience is an essential part of managerial responsibility. An installation in which all change is inhibited is in fact a dead or dying installation.

6. A FURTHER EXAMPLE

A further example may make the basic dilemma even clearer. One of the biggest problems occupying the attention of computer scientists is that of programming methodology. The majority of those studying it, investigate the primary problem of how programmes should be created in the first place.[3-6] Others note the fact that programmes undergo a continuous process of maintenance and change[7,8] driven by the evolutionary pressures mentioned above. The dynamics of the resultant programme evolution can and are being studied[9] for a variety of environments and programmes.

In these studies there have been a number of instances where detailed numeric data has been available relating to the progress of a programming project and the growth of the subject programme system.[9,10] In each case a similar pattern of evolution has been observed, and interpreted as a reflection of programme evolution dynamics.[11] This has permitted the construction of conceptual and statistical models that reflect increased understanding of the characteristics of programme development and maintenance and of the process whereby this is achieved. Consequently it has proved possible to develop techniques based on these models for more precise planning and prediction of the programming process in relation to a specific system.

In parallel with this development the insight gained from these global, system-like studies of the programming process has permitted the development of proposals for improvement of programming methodology in general. But implementation of these proposals immediately invalidates the models of system behaviour since the system, the process, has changed, and since it has changed in response to deductions that were made from these very models. Similarly application of the forecasting and planning data derived from a set of models to modify the output of the system invalidates these same models as representations of the programming organisation, its tools and its activities that together constitute the system. If the output of the models are accepted as essentially correct, activities are reorientated and adjusted to conform to model-based forecasts. The outputs from the models become a self-fulfilling prophecy of the behaviour of a new system. The latter includes the old system together with some new elements, namely the models, the model implementors, its interpreters and so on. But paradoxically the models are based on data generated from a system that excludes the models.

At the other extreme, the output of the models may incite activity to prove them wrong. This also leads to change in the system and negation of the models. In practice the stratagem adopted falls between these two extreme responses. But the consequence is the same. The mere act of studying the system leads to a change in the system which paradoxically invalidates the earlier models.

7. INTRODUCTION TO THE AREA OF IGNORANCE

One may, of course, create a new model of the new system, a model that includes the system model as an element of itself. But, at best, this can only be achieved by applying an iterative procedure. Will such a procedure converge? Can it? Must it? Therein lies the problem, a problem that will force an ultimate admission of ignorance. And we shall see that this ignorance is not just due to insufficient knowledge, understanding or wisdom. We have here an area of uncertainty and indeterminacy that has its roots in the freedom of thought, of interpretation, of choice and of action of mankind, individually and collectively. As such it appears to be absolute and unbreachable.

8. THE RELEVANCE OF GÖDEL'S THEOREM

An initial attempt to resolve the issue of convergence can be based on Gödel's theorem. In simple terms the latter states that one cannot prove the consistency and completeness of an axiomatic system using only the axioms and the rules of inference of that system. Informally one can state that an assertion about a system (and a model of a system represents an assertion about the system) cannot be shown to be absolutely true from within the system; by using only known facts about the system.

Suppose now that we assume that there exists an absolute theory for some artificial system. The latter could be represented by an axiomatic model in which each part and activity of the system reflects either an axiom or a theorem. From Gödel's theorem it follows that the correctness of the model cannot be demonstrated from within itself. That is, the behaviour of the system cannot be predicted absolutely from within the system. But demonstration of the validity of the model, of comparing the predictions of the model with the behaviour of the system, is an essential ingredient of artificial-system behaviour. By its very nature such activity is a part of the behaviour of the system as represented in the model. Hence there does not, in this respect exist an "outside" to the system. Thus one cannot obtain an absolute theory, a demonstrably correct model. Attempts to include the model in itself must lead to non-convergence. Ignorance about the total behaviour of artificial systems is intrinsic to their very existence.

9. UNCERTAINTY

In the above discussion we have gone one step beyond Gödel. We have asserted that the very activity of proving its model correct is itself part of artificial-system activity. Thus there exists no environment external to the system from which its total and absolute behaviour may be observed. Hence exact system science is not knowable, is meaningless, does not exist.

Over and above this theoretical limit there exists an even more significant barrier to total knowledge of the state and behaviour of an artificial system. There is an essential uncertainty which at all times implies some degree of ignorance about the system. The situation can be formulated in an Uncertainty Principle, analogous to that of Quantum Physics. By measuring and modelling an artificial system we increase the extent and precision of our knowledge and understanding of its mode of operation. In general this causes the system, the environment and/or the interactions between them to be changed. Thus the more accurately we measure the artificial system the less we know about its future states, if we believe, as I believe, that in the final analysis the humans or groups of humans that change the system are unpredictable.

The parallel is of course not complete. Heisenberg's principle arises from the fact that the mere act of observation must affect the position and momentum of the object being observed. Measurement, knowledge and change of state come together, they are inextricably bound. In our case, on the other hand, knowledge appears to come first. The system change occurs subsequently, after the information gained has been digested and applied to "improve" the system.

But the difference is a delusion. People are components of the system. Complex artificial systems cannot be designed absolutely correctly *ab initio*. Observation and modification are, in practice, intrinsic ingredients of system operation. Therefore the very act of observation that gives the humans in the system additional information about the system changes the state of the system.

In the uncertainties of the exact sciences we observe paired indeterminacies. Thus conjugate parameters such as, for example, the position and momentum of a particle cannot, by the very nature of things, be simultaneously determined with absolute precision. But the product of the indeterminacies can be bounded precisely. We know just how close to exact knowledge we can get.

In the case of the artificial system on the other hand one cannot (at the present time) predict with precision, what will be observed, how it will be interpreted, what action will be taken as a result. One cannot even determine the state of the total system with any certainty by applying probabilistic judgements as in Game Theory.[(12)] After all each situation, each sequence of events, will occur only once. Thus for our case the uncertainty is of a different order of magnitude. To talk of absolutely conjugate variables as measures is not meaningful. One cannot even identify with absolute certainty, the totality of system measures that are involved in any given measurement activity. Far less can one determine an absolute and general limit to the accuracy with which even the known measures can be observed.

Are these uncertainties absolute? I think "yes", but I do not *know*.

10. CONTINUOUS EVOLUTION

It has already been observed that a system can in general be considered as a sub-system of some other system. Amongst its several implications the discussion of the last section suggests that every artificial system with its environments forms such a family of super-systems in which the system boundary cannot be clearly defined or confined.

Observation of system state, behaviour and performance by humans in the environment leads to changes in the system, in environmental-system interactions or in system usage. All of these represent changes in the system itself, changes that result from the unpredictable interpretation of the human observer.†

In fact as humans in the system and its environment observe system behaviour they adjust their own responses, behaviour and objectives to optimise utility in some sense. This represents a change in system state. It also leads inevitably to imposed changes in system mechanisms, implemented in order to achieve more cost-effective operation in the changed human environment. These change also, provide new opportunities for interface and environmental changes, that, for example, modify system behaviour and/or usage patterns. These in turn suggest and ultimately cause renewed system state changes and so on. Thus we observe the "mutual reinforcement of system and environment", previously referred to. It is the major cause of continuous system evolution.

It may be true that there exists an optimum strategy for system improvement, as we find with the minimax strategy of Von Neumann and Morgernstern.[(12)] But such a strategy merely maximises the minimum benefit. It can in no way lead to an exact system theory, totally predictable system behaviour, a foreseeable evolutionary path to a final state.

11. THE ROLE OF HUMAN THOUGHT AND ACTION

Thus we arrive at a recognition of the fundamental factor that appears to make an inexact science of the science of systems, created, used and controlled by men. The evolution of such

† Note the intrinsic circularity here (as elsewhere in the paper). Until an exact model is known, observed behaviour must be interpreted and the system is strictly unpredictable. But if the system is unpredictable, no exact model can be found. Hence no exact model can be found by the Scientific Method.

systems depends on people's observations, thoughts, interpretations and actions. Man's ingenuity to invent new rules, new twists, new interpretations, new objectives is as unbounded as the language in which he expresses his thoughts.[13] Total knowledge, the final state, can never be reached. Ignorance must always be present.

This observation penetrates the core of the problem and exposes an ultimate area of ignorance. If mankind's ability and expression of ability were bounded, then perhaps there could be an exact Science of the Artificial, despite some difficult questions that remain unanswered and are perhaps unanswerable. But if, as I believe (but cannot know), such a bound does not exist,[8] then the theory of man-made systems, of systems including people or operating in an environment that involves people, can never be complete or precise. How good can it get? That too I do not know.

REFERENCES

1. H. Simon, *The Sciences of the Artificial*, MIT Press, 1969.
2. M. J. Alexander, *Information System Analysis*, SRA, 1974; p. 4 quotes ten definitions which, being slanted to information systems, is by no means an exhaustive listing.
3. E. W. Dijkstra, "The Humble Programmer", 1972 ACM Turing Award Lecture, *Comm. ACD*, 15, 10, October 1972, pp. 859-66.
4. W. Turski, "Software engineering – some principles and problems", Mathematical Structures – Computational Mathematics – Mathematical Modelling, *Paper dedicated to Professor L. Iliev's 60th Anniversary*, Sofia, 1975, pp. 485-91.
5. N. Wirth, "Program development by stepwise refinement." *Comm. ACM*, 14, 4, April 1971, pp. 221-7.
6. D. L. Parnas, "On the criteria to be used in decomposing systems into modules." *Comm. ACM*, 15, 12, December 1972, pp. 1053-8.
7. L. A. Belady and M. M. Lehman, "Programming system dynamics or the metadynamics of systems in maintenance and growth", IBM Research Report RC3546, T. J. Watson Research Center, Yorktown Heights, N.Y., September 1971.
8. M. M. Lehman, "Programs, cities, students – limits to growth?" *Imperial College of Science and Technology Inaugural Lecture Series*, Vol. 9, 1970-4, pp. 211-29.
9. L. A. Belady and M. M. Lehman, "A model of large program development", *IBM Systems Journal* Vol. 15, No. 3, 1976, pp. 225-52.
10. D. H. Hooton, "A case study in evolution dynamics". M.Sc. Thesis, Imperial College of Science and Technology, September 1975.
11. M. M. Lehman, "Notes on the evolution dynamics of large programs", MML/138, February 1976, privately circulated.
12. J. Von Newmann and O. Morgernstern, *Theory of Games and Economic Behaviour.* Princeton University Press, 1953.
13. A. Koestler, *The Act of Creation*, Pan Books, Chapter 7, final paragraph, beginning the bottom of page 176, Pan Books, ed., 1971.

Michael V. Tracey

Chief, CSIRO Division of Food Research in Australia.

President of the Australian Biochemical Society 1970-2 and titular member of the International Union of Pure and Applied Chemistry 1973-7. Foundation Fellow of the Australian Academy of Technological Sciences 1975.

Took part in early developmental work on leaf protein as a food for man, has carried out research into the enzyme breakdown of cellulose and chitin by soil organisms and led a research unit working on wheat proteins.

HUMAN NUTRITION

In the past man has never been as vitally concerned with the quality and nature of the air he breathes and the water he drinks (provided only it be neither brackish nor salt) as he has been with the quality and nature of his third requirement for life – food. Fresh air and fresh water are essentially invariant in composition, the former available in vast excess of his needs over the whole surface of the earth, the latter a condition of habitable areas, for without it none of the living things he eats can exist. The questions man can ask about the suitability of air and water for his needs are few and easily answered. He suffers no lack of air, so it is of little interest to determine his need for it except should he wish to immure himself in a submarine or a space capsule, both recent preoccupations of a very few. Within limits it makes no discernible difference to his health if he drinks more or less water than his neighbour – and there is no alternative to water to have its merits weighed. Food, however, is of many kinds, and each varies in quality. If he eats too little he wastes – too much he may become corpulent; if the quality is inadequate he may suffer from many diseases curable by a change of diet. Air is free, water commonly so, but food in most inhabited areas of the world has a price in labour or money, and that cost varies markedly from one food to another. So man can exercise a choice among foods and has always been aware in some degree that the choice exercised may have results on his well-being. It is not surprising, then, that nutrition has been a lively area for inquiry, a fruitful one for *ex cathedra* statement, an important component of religious practice and a paradise for the enthusiast, the quack and the charlatan.

The questions we ask of those who profess a scientific knowledge of food are on the face of it simple, though framed in a manner conditioned by the advances made in experimental work over the last 200 years on the nutrition of animals. We ask quite simply what are man's requirements for food in terms of usable energy and particular compounds we cannot ourselves synthesize, and what are the ways in which these requirements can be met without ill consequence from available or potentially available sources. Authoritative answers abound from individual experts, from nutritional bodies and from supranational bodies. Broadly the answers agree, but in detail they are inconsistent and none applies to an individual, for they relate to a series of statistical abstractions (the very active man 18-35 years old weighing 65 kg; the woman in most occupations, 18-55 years old weighing 55 kg, for example) and in general there is remarkably little evidence of their correctness. This sweeping statement is so contrary to the impression gained from an examination of textbooks on nutrition and the authoritative statements of the more widely publicized expert bodies that they require some support. Four distinguished British nutritionists, including the doyen of infant nutrition, J. C. Waterlow, wrote a letter[1] to *Nature* in 1972 entitled "How much food does man require?" Its opening sentence is "We believe that the energy requirements of man and his balance of intake and expenditure are not known". They go on to refer to the widely held opinion that 70 per cent of the world's population is undernourished and suggest that on the contrary it is possible that 30 per cent of the world's population are really eating too much and that an unknown proportion of the rest are not undernourished. In 1974 Professor Mark Hegsted of the Harvard School of Public Health reviewed[2] the protein portion of the FAO/WHO Report on Energy and Protein Requirements of 1973. He said in relation to this report on protein requirements: "By the time people" (on a representative committee) "who disagree find an issue they can agree upon, that issue represents nobody's opinion and is likely to be misleading or nearly useless. In my opinion (the report) has some very poor sections because of compromise. When old and accepted concepts become inadequate, but adequate replacements are not yet available, the situation is obviously difficult. I believe that this is the situation now with regard to protein requirements and protein quality or protein utilization." In addition to energy (carbohydrates and fats) and protein, man needs vitamins. At an international conference on Requirements for Vitamins in Old Age, Sylvia Darke of the Department of Health and Social Security said:[3] "Any conclusion can only be one of ignorance . . . Perhaps the crux of the matter is that for most of the vitamins we do not really know with certainty the requirement for younger adults."

It is clear that, contrary to the general impression, the emperor of human nutrition has no clothes but at least he knows what clothes he should have, for his subjects have by now produced an almost complete list of the raiment, haberdashery and accessories necessary – but they cannot be worn with conviction while there is continuing dispute concerning the size and cut of every item in the list.

It is perplexing at first that the science of human nutrition should be in such disarray although it is the science closest to the individual's welfare and often vital to the survival of those in power in a society. A population not getting enough to meet its nutritional needs becomes discontented, its soldiers perform less well than is hoped of them, and discontent may reach the stage of rebellion. As a consequence, pragmatic knowledge of the requirements of man for food existed for thousands of years arising not only from the experience of quartermasters but from the experience of slave owners and slave dealers over the centuries in feeding slaves to ensure maximum work output at minimal cost in food, or bringing an indifferent lot "up to condition" before inspection and sale. The knowledge acquired was, however, never codified, though that acquired of the nutritional needs of male adults in armies formed a useful basis for energy requirements. The word ration, meaning a fixed allowance or share of provisions, came into use in the first years of the eighteenth century, and by the end of the nineteenth century meant literally in army usage 1 lb of bread and ¾ lb of meat (about 90 g protein and 2000 kcal). By the time of the Second World War pragmatic knowledge was enough to ensure an adequate diet in the United Kingdom – perhaps more adequate than in previous years, judging by its success. Whether it was still above requirements in important respects cannot yet be decided.

There are two fundamental reasons for the present undeveloped state of our knowledge of human requirements for food. These are that people are born unequal in their nutritional needs and that they are reluctant to be the subjects of the kind of experiment on which our detailed knowledge of the requirements of domesticated food animals is based. Since in the more illuminating experiments that might be planned on man the lifetime of the experiments is likely to be similar to that of the subjects, and the active life of the experimenter shorter, the multigeneration nutritional experiments common with rats are out of the question – one fairly recent experiment on the effect of polyunsaturated margarine on rats ran for 75 generations and still continues. While human volunteers are obtainable for nutritional experiments, these inevitably occupy a very small portion of the life span and are normally confined to adults. The children who were observed on different controlled diets in Duisberg in 1947-48 were involved in the experiment for a maximum of 2 per cent of their expected life span. Observations on a few individual adults voluntarily restricting themselves to known diets have been made for up to 10 per cent of the expected life span, but strictly controlled experiments under laboratory conditions have been much more limited. The most thorough and perhaps the longest of these took place some 10 years ago; it involved a total of fifteen healthy men for 19 weeks who were placed on a sterile water-soluble diet whose constituents were precisely known and pure. The experiment was designed to evaluate a formulation for use by man in space, and it was funded by the National Aeronautics and Space Administration of the United States. No deficiency in the diet was identified but the experiment occupied only 0.5 per cent of the life span. It is clear, then, that while long-term nutritional effects are readily susceptible to experimental observation in short-lived laboratory animals, it is unrealistic to expect that carefully controlled experiments in man will ever exceed 1 per cent of his normal life span or a very few per cent of his generation time. It is for this reason that human nutrition is and must be primarily an observational rather than an experimental science.

The inability of the human nutritionist to carry out long-term well-controlled experiments on human subjects means that many questions cast in a scientific form are in fact unanswerable with the clarity or precision that the public and its rulers have come to expect from science. That they have such an expectation is a consequence of their awareness of the remarkable successes of the physical and chemical sciences in the last century. Unfortunately the sciences of living organisms are far less able to provide answers not hedged about with qualifications arising from the intrinsic variability between individuals, and in an individual, with time and circumstance. In fact, much of human nutrition falls under the classification of trans-science rather than science. The word "trans-science" was coined in 1972 by Alvin M. Weinberg, Director of Oak Ridge National Laboratory in the United States. He says:[4]

> Many of the issues which arise in the course of the interaction between science or technology and society – e.g. the deleterious side effects of technology, or the attempts to deal with social problems through the procedures of science – hang on the answers to questions which can be asked of science and yet *which cannot be answered by science*. I propose the term *trans-scientific* for these questions, since, though they are, epistemologically speaking, questions of fact and can be stated in the language of science, they are unanswerable by science; they transcend science.

and goes on to give an example of a trans-scientific question which is close enough to the field of nutrition to quote directly – it concerns the biological effects of low-level radiation insults –

> Let us consider the biological effects of low-level radiation insults to the environment, in particular the genetic effects of low levels of radiation on mice. Experiments performed at high radiation levels show that the dose required to double the spontaneous mutation rate in mice is 30 rontgens of X-rays. Thus, if the genetic response to X-radiation is linear, then a dose of 150 millirems would increase the spontaneous mutation rate in mice by ½ per cent. This is a matter of importance to public policy since the various standard-setting bodies had decided that a yearly dose of about 150 millirems (actually 170 millirems) to a suitably chosen segment of the population was acceptable. Now, to determine at the 95 per cent confidence level by a direct experiment whether 150 millirems will increase the mutation rate by ½ per cent requires about 8,000,000,000 mice! Of course this number falls if one reduces the confidence level; at 60 per cent confidence level, the number is 195,000,000. Nevertheless, the number is so staggeringly large that, as a practical matter, the question is unanswerable by direct scientific investigation.

To emphasize the point still further, it must be realized that results on mice cannot be extrapolated to man for many reasons, including the enormous difference in life span, and that a population of 8 billion is considerably in excess of the total number of mankind alive today, and even in excess of all the men alive now and those who have ever lived!

The question of the nutrition of the individual, which is one in which every individual has an interest, is one that is in the realm of "trans-science". We are all born unequal – otherwise we would not be individuals – and each of us differs from every other in our genetic endowment (this is true even of identical twins, if we include extra-chromosomal endowment), in the fine detail of our biochemistry, in our physiology, in our stature and in our habits, and as a consequence of all this we each differ in our nutritional needs and how best they may be satisfied for each of us. An excellent example of this intrinsic variability in our apparent nutritional needs is supplied by the work of Widdowson.[5] She used over 900 subjects in this study, of whom at least 20 boys and 20 girls were in each yearly age group from 1 to 18 years. Among the boys in every age group there was one who consumed twice as much energy as another, and one 16-year-old boy was consuming less energy than one 1-year-old. Much the same was true of the girls. The subjects were accepted as a representative sample of healthy British children, most of them from the middle classes. There was no evidence that a difference in composition of the diets was correlated with the amount eaten, so it can be assumed that the twofold variation in intake of energy applied also to the vitamins, proteins, trace metals and other components of what was eaten. Later Widdowson,[6] in commenting on the results of the study, said: "We must accept the general principle that people do not eat, and do not all require, the same amount of food. The reason they do not has still never been satisfactorily explained." That statement remains true today.

Much of the implicit emphasis so far has been on the needs of man for food – how much he needs and what its nature must be to avoid deleterious consequences. The obverse of this is much more recent – a concern relating to the effects of the intake of an excess over needs. Gluttony has been regarded as a sin by the Christians and by many other religions, more for ethical reasons than from any effect it might have on human life and health. The classic results of excessive intake have been regarded as obesity and gout, but the opportunity to acquire these conditions has until very recently in history been restricted to small sections of the population. Now, however, many societies provide a majority of their members with the opportunity for excess – and in some instances encourage it in relation to some components of the diet, as likely to increase both health and longevity. It has been suggested that a remarkable number of afflictions can now be added to gout and obesity as consequences of a diet supposed to be unbalanced in one way or another but almost always characterized by an intake in excess of supposed needs. It is only necessary to refer the reader to those authors who regard the increased proportion of refined cane sugar in Western diets as a root cause of many ills, to those who emphasize the importance not of total fibre but of cereal fibre, to those who believe that high intake of saturated fat is a health hazard unless balanced by fat in a polyunsaturated form, and those who believe that even a normal varied diet is deficient in ascorbic acid, lecithin, or some other component.

Instead of discussing the insecure foundation of beliefs such as these, a much more strongly rooted – and widely accepted – belief can be examined. We all know, or think we know, that over-eating leads to obesity. In fact a report by Nutrition Canada, [7] which was produced by a headquarters staff, eight expert committees, five regional directors and a considerable field staff, can still make the demonstrably false statement: "When calorie intake is in excess of need, the surplus calories are stored as adipose tissue in the body." Anyone by thinking of the eating habits of his relatives and friends in relation to their weight is likely to get a subjective impression that there must be something wrong with this flat, unqualified statement, and that it must be at least a vast over-simplification of the situation. For more objective evidence we can look at figures for the United Kingdom in 1965. Childless members of the highest social class (determined by income) ate 124 per cent of the recommended energy allowance. The same group in the lowest income classification ate only 114 per cent of the allowance. If the allowances were correct and excess energy intake is laid down as fat, the childless member of the highest social class would have put on an average of half a hundredweight a year, and those of the lowest, 33 lb. It happens that obesity is less common in the higher social classes than the lower, and neither class increases in weight by several stone annually.[8] Fortunately, there are nutritionists who realize the existence of this problem, and it is not too much to hope that it will be investigated experimentally. The sooner this is done the better, because the existence of a means of disposing of surplus energy metabolically implies the possibility that knowledge of its details will lead to an understanding of obesity, and, it is to be hoped, to its control other than by restriction of food intake – a method notorious for its inefficacy in practice.

I hope that I have made it plain that in human nutrition there are many areas of ignorance in respect of the needs for many dietary components for the statistical abstractions of the population groups segregated by age, weight, sex and activity. These areas include energy needs, protein needs and vitamin needs. Furthermore, we know too little about the need for essential trace metals in man's nutrients to be yet aware of the dimensions of our ignorance in this area, for while we have indications of the magnitude of needs for individual metals, we are only now beginning to realize that they may be interrelated rather than independent. Of the needs of the individual in normal health we know only that they will almost certainly lie between 30 per cent less than that of the mean, and 30 per cent more than the mean for his statistical subdivision – subject, of course, to the all too little confidence we can at present place on those means. Much progress in our knowledge of human needs for foods and its components will come, but the trans-scientific nature of the questions the concerned individual wants to ask must be accepted and we must accept further that it is unreasonable to berate the professionals for their inability to answer them in a concrete and satisfactory manner.

REFERENCES

1. J. V. G. A. Durnin, O. G. Edholm, D. S. Miller and J. C. Waterlow, *Nature (Lond.)* **242**, 418 (1972).
2. D. M. Hegsted. in *Nutrients in Processed Foods*, Amer. Med. Ass. 5 (1974).
3. S. J. Darke in *Nutrition in Old Age*, Symposia of the Swedish Nutrition Foundation, X, 107 (1972).
4. A. M. Weinberg, *Minerva*, **10**, 209 (1972).
5. E. M. Widdowson. A Study of Individual Children's Diets. Med. Res. Counc. Spec. Rep. Ser. 257, H.M.S.O., 1947.
6. E. M. Widdowson, *Proc. Nutr. Soc.* **21**, 122 (1962).
7. *Nutrition: A National Priority. A report by Nutrition Canada*, Information Canada, Ottawa, 1973.
8. D. S. Miller, *New Sci.* **44**, 116 (1968).

Patrick D. Wall

Director of the Cerebral Functions Group and Professor of Anatomy at University College, London.

WHY DO WE NOT UNDERSTAND PAIN?

Pain is. My pain, as it grows, is an imperative, an obsession, a compulsion, a dominating, engulfing reality. Your pain is a different matter. I observe you and listen to you. I sympathise with you by guessed analogy. If your state is beyond my ken, a woman in childbirth, a man passing a kidney stone, I am in particular difficulty. Even if I have experienced your particular situation, if I see you hit your thumb with a hammer, I not only remember my own pain but I recall my behaviour and now I am in the dangerous situation of assessing the appropriateness of your behaviour. Doctors and patients become extremely angry with each other when there is a mismatch between disease and the amount of pain which the doctor expects, especially if the patient has the impertinence to fail to respond to accepted therapy. At this point, the doctor begins to question if the pain is real or in the mind. What is this curious question asked by the observer but never by the person in pain? Before we launch into a search for mechanisms, we must first define the class of phenomena whose mechanism we wish to discover.

Is Pain a Sensation?

This question is simply not asked in standard works which begin to discuss the subject of pain having covered the other classes of less dramatic events which may be detected such as touch, pressure, hot and cold. These are presented as the primary modalities of cutaneous sensation, that is to say the categories of touch, pressure, hot, cold and pain within which the events on the skin are classified by the reporting sensory nervous system. It will immediately be seen that there is an error in this series of subdivisions. Pressure and temperature are objective events to be measured either by the detectors of one's sensory nervous system or by any number of other instruments available to oneself and others. Pain as we have said is absolutely not in this class. Injury can be objectively detected. However, it is classically and intuitively assumed that injury means pain and pain implies injury. Injury and pain are believed to be so tightly linked that the words can be interchanged. Here begins the trouble and the confusion. As we shall see the two are by no means interlocked to the point of identity. There is no doubt that most tissues of the body contain nerve fibres which begin to signal the existence of injury or changes in tissue which show that serious damage is impending. In the embryo, cells close to the spinal cord send out paired nerve fibres; one branch grows out into tissue and establishes an end station where it will detect events; a second branch grows into the spinal cord where it establishes contact with a relay cell. These cells will transmit information from the peripheral nerve toward the brain. Some of the endings in the tissue generate nerve impulses only if the tissue is damaged. They detect temperatures above 45°C or very heavy pressure or the presence of chemicals released by damaged tissue. Some of the medicine against pain works by blocking the appearance of these pain-producing compounds. The smallest of the nerve fibres with diameters of 0.5-2.5 microns are the ones connected to the injury detectors. A false injury signal can be generated by electrical stimulation of nerve fibres. Similarly, some diseases of peripheral nerves will generate impulses in these small nerve fibres and send false signals into the central nervous system where the signals are treated as though there really was an injury. Recordings can be made of the traffic of nerve impulses along these fibres both in man and in animals. If impulses are detected in this type of fibre, a specific statement can be made that injury has occurred in the specific location of the fibres' terminals. The intensity of the injury is mirrored in the frequency of the nerve impulses. Having achieved the satisfactory proof of the normal cause of the origin of the observed impulses, it is natural to perform a logical volte-face and to predict the sensory consequences of the existence of these nerve impulses, i.e. pain. In the classical way of thinking, an objective linking phenomenon has been observed between injury and pain, i.e. the presence of nerve impulses which can be called alternatively injury or pain nerve impulses in pain or injury detecting nerve fibres.

Is there a Sensation of Pure Pain?

One of the implications of the word sensation is that it represents an intermediate stage in the brain's analysis of events. It is the truest report of the actual nature of the disturbance which can be detected by the peripheral receptors and transmitted as a coded message in units of nerve impulses to be presented in the brain as some faithful representation of the sensory world. The existence of this stage is accepted without question in much philosophical and scientific thinking. It is the given truth on which the mind can operate. This sensory stage leads to a fundamentally different set of mental processes: perception, cognition, recognition, emotion, memory and perhaps learning and reaction. Now, it is quite easy for me to imagine that my mind is working in this fashion if I sit in a psychological testing laboratory and a small red light is turned on in front of me. In a very detached and neutral way, I can examine this stimulus and dissect its components. I know that if this identical red light came on when I was piloting a plane, I might add many things; meaning, "fire in the starboard engine"; memory, "the emergency drill"; emotion, "anxiety", etc., but this would not interfere with my contention that under simpler circumstances, I could simply sense a pure red light with the simplest unique perception and cognition to follow. I wish to submit that I have never sensed a pure pain in a similar fashion. If I sense a pain, it comes in a packet with such changes as fear, loathing, anxiety, dislike, urgency, etc. There is one unique and special exception to this statement. That situation is in an experimental laboratory where I have volunteered to take part in a pain test where the circumstances are very peculiar: confidence in the experimenter that he will not injure me; confidence in the machinery that it will not break down; repeated experience of the exact type of stimulus. Here I can play a special type of intellectual game with the gadgets and the experimenter. I guess at the intensity of the stimulus on the basis of my feelings. Much of the literature on pain assumes that it is only in these laboratory circumstances with trained subjects that the sensation of pain can be studied. I believe the situation to be so bizarre as to represent a misleading cartoon of the essentials of pain existing in the real world of people.

There is a second peculiarity of pain which differentiates it from most of our sensory experience. Pain is given. Most of our sensory experience is taken. Very rarely we may be sitting passively when an event occurs in our world and impinges on our sensorium. Even then the next move is to explore. We alert, orient, look, listen and search. In this second phase, we are taking an active role in the sensory experience, acquiring more information which allows us to make a discrimination. If we are limited to a passive role of simply receiving imposed stimuli, our discriminatory abilities are extremely poor unless we have been given prior training about the exact nature of the impending event. Normally we search and explore, questioning the world about us and, in a way, creating our sensory world by our own active movements. Pain, however, hits us as passive receivers in the first instance. It may trigger investigation but this rarely modifies the feeling. There are those strange members of our society who actively seek pain in sex, sport, initiation ceremonies and warfare but this apparently paradoxical active pain is so obviously associated with real or imagined reward that we need not wonder about its relation to the basis of pain. Just as the trained experimental subject has learnt to report on the epiphenomena of pain, the masochist has learnt to concentrate on the consequences.

Are there Circumstances where Injury Occurs Without Pain?

These are surprisingly common and are obviously crucial to test the depth of understanding of pain. Some of the situations have quite trivial explanations. The only way in which the brain can "know" about the injury of tissue is by way of the injury detecting nerve fibres. If these have failed to develop in the embryo or have been cut by injury or destroyed by disease or blocked by the injection of a local anaesthetic, no pain is felt when injury occurs. This allows for an important dissection of the link between injury and pain. A patient, with local anaesthetic injected around his lower spinal cord, may visually observe a surgical operation on himself. He may become upset, anxious, nauseated, panicked but he feels no pain. He is completely intellectually aware that his body is suffering repeated severe injury but the situation is as though he is witnessing an operation on a close friend. This would seem to support the classical view that pain is triggered by the injury impulses and not be any other signal. However, there are much more mysterious but common circumstances of uncoupling. A few people are born without the ability to feel pain although all other feelings are normal. The most detailed examination of their peripheral and central nervous systems fails to reveal any defect which has so far been detected. Their mental development can be quite normal in spite of the fact that painful punishment has no meaning to them. They do, however, demonstrate the biological usefulness of the injury-detection mechanism. Burns occur because hot things are not noticed. Fractures may not attract attention and go untreated. These people quickly learn to use other cues and to avoid dangerous situations but this is not enough. The commonest sign in these patients is severe joint disease originating from their failure to guard a limb after the minor sprains and insults to which we subject ourselves in the most careful living.

The real challenge comes when we look at war or sports injuries. Over 70 per cent of soldiers who had been hit with bullets or metal fragments reported their first feelings with neutral words "thud, blow, thump, bang, etc.", and on direct questioning said they did not feel pain. Some casualties may cry out or scream but on investigation and treatment, it often turns out that this is their expression of their understandable and fully justified terror, anxiety and anguish rather than a reaction to pain. Casualties may be fully aware of the extent and consequences of their injury without feeling pain. A girl with a leg blown off in an explosion said "Who is going to marry me now?" but felt no pain. A man with a foot amputated in an industrial accident said "There goes my holiday" and "People will think I am a fool to have let this happen" but was not in pain. One should not imagine that such injuries always occur under conditions of great action and drama or where the injury may provide some reward, a medal, evacuation out of battle or a pension. Painless reactions to injury can occur in calm unexciting dull circumstances where the injury is a disadvantage. Nor should one brush off the phenomena by supposing these are brave silent sufferers since, by the next day these casualties are always in pain. This delay in the onset of pain for hours after full awareness of the injury makes nonsense of those definitions of pain which stress the immediacy of the feeling. All evidence agrees that these pain-free states are associated with massive discharges of impulses in the injury-detecting fibres.

These naturally occurring failures of pain to be associated with injury lead us to a consideration of induced states, the placebo reactions. Some 30 per cent of cancer patients whose disease and pain is judged so severe that they are given narcotics, experience a complete relief of their pain if they are given a blank tablet which the patient and the nurse believe to contain a pain reliever. The tuning of this response to social expectation is shown by the fact that 40 per cent of such patients are relieved by an intramuscular injection and 50 per cent by an intravenous injection. In a trial of a new therapy for treatment of slipped discs by direct injection of an enzyme into the disc which was seen to be out of place on X-ray examination, 70 per cent of patients responded to the injection of an inactive compound. Here, in a double blind trial, the full paraphenalia of modern surgery had engulfed the patient, admission to a famous hospital, elaborate X-ray investigation, operating room, anaesthesia, recovery room, recovery. In spite of their repeated observation of the placebo response, thoughtful and humane doctors still cannot believe their eyes and tend to fall back on the assumption that their initial diagnosis must have been wrong. If a patient responds to an inactive compound, then the pain could not have been "real" and must have been psychogenic. Such a statement is a completely unjustified conclusion which attempts to prop up an outmoded dualism. It is true that there is a rare type of patient who hallucinates the existence of injury and pain often for symbolic reasons and might be called a psychogenic pain patient but it is equally true that such patients are quite resistant to placebos and often all other forms of therapy.

The best known pain-free states associated with injury are those which a therapist claims to produce by way of hypnosis or acupuncture. The most important fact about these two manipulations is that their successful use is extremely rare. The Chinese succeed in operating on less than 10 per cent of their surgical patients with acupuncture and this usually includes a back up of sedative, narcotic and local anaesthetic drugs. Hypnosis for surgical anaesthesia succeeds in a much smaller group of people. Success involves a close personal interaction between patient and therapist. The Hilgards' book *Hypnosis in the Relief of Pain* shows that hypnosis involves a high-level intellectual verbal change. On the other hand, the rare successful cases of acupuncture are best explained by a combination of anxiety relief, suggestion and distraction, all three of which are known to decrease the expression of pain.

Does Pain Occur Without Injury?

Again the answer is, yes. There are a number of situations where severe pains occur where the apparently damaged tissue is in fact quite healthy. Some of these have relatively trivial explanations and occur because false signals are being generated along the injury-signal pathway. Subtle damage or disease of nerve fibres may lead to their generating a continuous barrage of impulses in the absence of a stimulus. If cells in the spinal cord or brain lose their nerve fibres which normally drive them they become spontaneously active. In the case of amputations, some patients experience pain apparently originating from the absent limb, phantom pain. The great majority of such amputees are reacting to false signals originating from damaged nerves in the stump of their limbs. However, there is a rare type of phantom limb pain which originates from within the central nervous system and which is a complete mystery. A very common example of pain without injury is "referred pain". Here there is damage existing but the patient is quite convinced that the injury exists in some distant place. Inadequate blood flow in the heart produces a feeling of pain on the surface of the chest and in one arm, an impacted wisdom tooth produces an ear ache, shoulder joint problems produce neck and head pain, etc.

If there are complete mismatches between injury and pain, it is not surprising that there are always quantitative variations between the amount of damage and the amount of pain. Here a mass of ethnic, educational, personal and situational factors come into play. Obviously it is first necessary to dissect private suffering from public display since these change independently.

These huge variations of threshold, strength assessment and tolerance have been sufficiently carefully studied for one to be quite certain of their existence and to be quite puzzled by their meaning and origin (Zborowski, 1969; Melzack, 1973).

Is Pain in a Class of Feelings which Include Hunger, Thirst and Suffocation?

I wish now to make a strong case for this view. A great deal of the body's machinery is involved in maintaining a stable internal environment. If this shifts from its normal range, a series of more and more complex and elaborate reactions are triggered. These states of awareness have many properties in common. One is urgency. As awareness grows, not only is attention riveted but activity is more and more monopolised in attempts to abolish the state. So great is the urge to do something that these states have been associated each with its drive and we would add to the commonly accepted drives to eat and drink, a drive to preserve the body intact. Unlike what is ordinarily called sensation where the nature of the stimulus is paramount, these states are at least as much concerned with the nature of impending or actual behaviour. Hunger may or may not be triggered by an empty stomach, but it certainly implies a growing monopolisation of behaviour towards eating. Imperative behaviour is not a monopoly of these states. A fire alarm bell may be equally imperative but this is a learnt response to a stimulus which under other circumstances could be quite neutral. This neutrality does not exist for hunger, thirst or pain. We see then why false signals cannot be treated as hallucinations and why habituation does not occur. Attention is normally initiated and directed by preconscious mechanisms which direct mental processes as to the subject matter for mental consideration. In the case of pain, not only has a preconscious selection of subject been initiated by the time of the first onset of felt pain but appropriate action has also been initiated. Like attention, only one form of behaviour is engaged at any one time. An overall decision of the most appropriate behaviour is selected from the available repertoire of behaviour. This can provide us with a clue to explain the paradoxical uncoupling of injury and pain. Where some alternative behaviour is deemed appropriate, the mental world of the individual does not contain pain. The hungry, thirsty, injured football player about to score a goal is completely unaware of his internal body state. As the game finishes, awareness appears. Of course the details of the behaviour are not given, it is the intention of the behaviour which is apparent. I will not discuss the nature of the aware entity, the "me" who is now in pain. My intention is only to shift the discussion of the phenomenon of pain away from the stimulus to a realm of reactions which are designed to preserve the internal milieu. The existence of injury may or may not trigger reaction depending on the circumstances. Pain is associated with the turning on of the drive for reaction. The nature of the reaction will depend on a different set of circumstances. This shifts pain away from the stage of event detection towards a stage of reaction decision.

The Mind-body Problem. The Identity Solution

A common philosophical solution to the mind-body problem is to propose that for each mental state there is a state of activity in brain structures and that the two are complementary ways of describing the identical phenomenon. Superficially this sounds like a straightforward and verifiable statement.

What is observed in the brain?

There is considerable evidence that rapid events in brain and behaviour are completely describable in terms of nerve impulses running along observable structures to mix and interact by observable chemical and electrical means at the junction points where nerve fibres make contact with subsequent cells. At these junction points, the synapses, decisions are taken depending on the convergent messages. Unfortunately, the present state of technique does not allow the activity of all of the obviously involved structures to be observed. The structures are small and the observing probes are destructively large. Even on the single cell–single impulse level, we are in a state of indeterminacy because to observe with our present tools is to change.

What is known is that injury leads to the generation of nerve impulses in certain types of small diameter fibres. It is on the edge of practicability to observe all of the nerve impulses in all of the peripheral nerve fibres in the presence of a localised defined injury and to describe the complete nature of the message. All of the indications are that this message would have identical properties from individual to individual. A test of this belief is that an artificial generation of the coded message when injected into the nervous system convinces the person that he is injured and he reacts accordingly. As soon as we examine the first cells to receive this injury message, the situation complicates. The cells are not simple messenger boys. They receive convergent information from other sources. Some of these nerve impulses arrive from the periphery and tend to exaggerate or diminish the effects of the injury signal. Furthermore, all of the cells are under powerful descending control from the brain. This means that the transmission of the injury message through this *gate-control* region is contingent not only on the arrival of the message but on the presence of other events in the periphery and also on the set of the brain which has decided whether to permit the invasion of the brain by the message. The transmission and the strength of the signals are under control. So much for the concept of a "pure sensation" reporting the nature of peripheral events. It is evident that central decision processes have determined if such evidence is admissible and in the event of a negative answer they suppress evidence of the event close to its source.

Since information transmitted toward the brain is contingent on the activity of the source of the original message plus the activity of the peripheral and central modulating mechanisms, one would need to make simultaneous observations of the state of activity of several structures to understand the origin of the actual transmitted message. Here we come to a second technical source of indeterminacy. As the number of independent observing probes multiplies, the destruction they produce increases. It is theoretically possible to provide a particular solution to a particular "N" body problem but not if the means of observation and manipulation forbid the collection of the data needed to establish a causal chain.

Even though our means of study of even the first stages of transmission are crude and inadequate, they provide useful clues as to the type of mechanism involved and to possible therapies. The discovery of the existence of peripheral modulating fibres has led to the development of electrical stimulation of those nerve fibres which decrease the gain on the transmitting system. The presence of descending control systems allows the artificial implantation of electrodes to activate the turn off signals. It now seems highly likely that the analgesic effect of narcotics such as morphine comes from the fact that these drugs artificially fire a descending pathway. Furthermore, a naturally occurring substance in the brain, enkephaline, has narcotic properties and it seems likely that the nervous system normally activates this descending control by releasing its own internal narcotic.

We know that the injury signal is transmitted normally to the brain by way of nerve fibres running in a particular quadrant of the spinal cord. Pain may be abolished on one side of the body by surgically cutting these fibres, an anterolateral cordotomy. This operation is fully justified because of the relief it provides for those about to die of inoperable cancer but, if the patient survives for many months, the pain returns often with additional unpleasant qualities. This distressing phenomenon has led to a search for other regions in the brain which may be safely destroyed to abolish the sensation. This search has been a miserable failure at best leading to temporary relief and often producing unacceptable limitations of thought and behaviour. Similarly, centrally acting pain-controlling drugs are uniformly disliked by patients who must accept progressively more serious intervention in their mental and physical life in return for the abolition of their pain.

What Would One Expect to Observe in the Brain?

Evidently the search for a "pure pain" centre and its input has been unsuccessful. We have given reasons to suppose that the search was based on a misinterpretation of the nature of the phenomenon. Those who continue the classical search fall back on the essentially mystical concept that there must be repeated redundant alternative pathways and centres to the considerable disadvantage of the patient who is subject to more and more extensive lesions. Let us consider the consequences of our alternative suggestion that the search should be directed at those parts of the brain where decisions on relevant behaviour are made. We know next to nothing about these regions and there is no satisfactory way of thinking about them. We have here a vast area of factual and conceptual ignorance. We might try the *gedanken* experiment of recording all nerve impulses in all cells in man in pain to see if we could by-pass the theoretical problem by cracking the code in the classical fashion by noticing a correlation of nerve impulse pattern and the declaration of pain. We would have to exclude the arriving messages and the leaving motor orders to try to focus on the source, nature and location of the decision-making process. But supposing, as seems likely, that the process depends not only on injury but on existing and past circumstances, then the pattern becomes individual to the subject under observation. Furthermore, if each stimulus changes the pattern of activity by adding memory and recall, the code is a shift code which is uncrackable. We have a problem.

REFERENCES

Hilgard, E. R. and Hilgard, J. R. (1975) *Hypnosis in the Relief of Pain*, Kauffmann, Los Altos.
Melzack, R. (1973) *The Puzzle of Pain*, Penguin Education, Harmondsworth.
Zborowski, M. (1969) *People in Pain*, Jossey-Bass, San Francisco.

J. H. P. Willis

Consultant Psychiatrist at Guy's Hospital and King's College Hospital, London.

Fellow of the Royal College of Physicians of Edinburgh and the Royal College of Psychiatrists.

Concerned with the treatment and investigation of heroin addicts and the study of drug abuse and dependence.

DRUG ADDICTION

Sir Francis Bacon said that "Ignorance begets confidence more surely than there is knowledge". This about sums up the present state of uncertain certainty that continues to surround the study of drug and alcohol addiction or to give it its more fashionable name, dependence. Drug addiction is a complex phenomenon which has generated a good deal of heat and invited much speculation but which continues to defy understanding and analysis as fast as it grows. The literature has expanded far beyond the limits of information saturation at a rate comparable to the growth rate of the literature on a topic such as schizophrenia where research, publication and clear understanding lie in inverse proportion.

We may start with the fact that we are indecisive about the real meaning of the term addiction – a word that has only been used comparatively recently and which has led many to prefer the word "dependence", possibly because it is a longer word but probably because it appears to have lost the moral overtones of a word which when applied to someone can make that person appear somehow less acceptable to society to someone who is not an addict – providing, of course, that the addiction/dependence concerned is one that society does not condone. Condoned addiction or dependence involving substances such as nicotine and ethanol is often tolerated and encouraged by people promoting the sale of these products, however, non-condoned addiction, e.g. to opiates, amphetamines and halluconogenic drugs, is not tolerated and is discouraged even to the point of legal sanctions which often astonish us by their indiscriminate and inappropriate savagery. And it is this observation about addiction that is perhaps fundamental, namely that it is a topic which is surrounded by inconsistencies which no doubt reflect our ignorance of what exactly is the essence of addiction at all. Cannabis is regarded as a "bad" drug yet most pharmacological studies to date suggest that it is probably one of the safer mind-altering drugs in existence. This does not mean to say that this is a drug that has no dangerous short- or long-term affects, far from it. And it might be added that cannabis has perhaps a special position in that it is not in any way legally permissible in Western society and yet at the same time the Western population endures considerable morbidity and mortality from prescribed mind-altering drugs and others bought across the counter. Common examples of this would include the hundreds of thousands of overdoses from tranquillisers and sleeping pills prescribed by doctors, the overdoses and deaths caused by antidepressant medications and the high incidence of gastric bleeding and severe overdoses from Aspirin which can be bought across the counter of any chemist shop. These examples are merely intended to reinforce the notion that addiction and drug abuse are areas where inconsistencies in attitude abound.

Addiction or dependence then are words that are used to encompass states in which a person becomes apparently unable to stop taking a drug either because it gives pleasure or because the person experiences withdrawal symptoms if the drug is discontinued.

The World Health Organisation has defined some of these terms relatively clearly; for example, a drug is defined as "any substance which when taken into the living organism may modify one or more of its functions". In medical practice this covers a wide range of therapeutic substances; antibiotics for infections, antimetabolites in leukaemia and in cancer, insulin in diabetes, sedatives, hypnotics, tranquillisers, etc. – in fact a vast range of highly active and desirable chemicals that are used by doctors in the treatment of a wide range of illnesses. In so far as addiction is concerned, the drugs involved are those which are broadly categorised as "mind altering" – in other words, they can affect a person's thinking, feeling and perception or behaviour or indeed any combination of these. Drug dependence is defined as a state which may be either physically or psychologically determined or both in which a person's continued taking of the drug leads to a state where the person is unable to stop either because of withdrawal symptoms or because of the person's continued involvement with the subjective pleasures caused by the drug. Although there is general agreement that there are such phenomena as drug abuse and drug dependence, there is considerable disagreement about the essence of these conditions. We may start by noting that the severity of various types of addiction is increased because of straightforward technicological advances. First, it is now possible to produce pure and more powerful types of mind-altering drugs; and secondly, because there are more effective ways of delivering these drugs into the body thus producing more intense and severe types of addiction. However, it is really not good enough to talk about addiction as an established fact without trying to understand some of the basic mechanisms involved and here we enter an area of real ignorance. Very often it appears that speculation and non-verifiable assertion have been considerably mixed together when various authorities have tried to explain the nature of addiction. A good example of this is found in the "psychosocial" approach to addiction. In this there appears to be a general hypothesis that people can become addicted to drugs as a result of a confluence of factors in which social and psychological forces operate forcefully within the individual, as it were, driving him inexorably towards being a drug addict. This had led to an enormous amount of speculation which has found perhaps its most absurd expression in the notion put forward by a number of psychologists and psychiatrists that there is a special type of personality which is, as it were, addiction prone. Unfortunately, however, no one has ever been able to substantiate this claim in any credible way.

Social inquiries into the causation of addiction have on the whole been equally unhelpful. American studies, for instance, tended to relate severe forms of addiction to poverty, educational and job deprivation and membership of particular ethnic groups. At least this was the case until drug addiction began to spread more widely through the American population. Thereafter emphasis on the social factors, though still stated, became less emphasised. In our own culture many would undoubtedly agree that socially disadvantaged people are probably more at risk to drug and alcohol abuse and addiction than the average person but nobody could really definitely go further than this and pick out clearly a precise way in which the social factors might operate. These references to psychosocial factors in addiction are mentioned first of all since there has been a recent tendency to emphasise the psychosocial factors of addiction as being of prime importance. It is almost as if it has become unfashionable to be concerned about the basic pharmacological aspects of addiction, i.e. the truly drug-related aspects. The reasons for this are comparatively simple to spot. In the first instance pharmacological studies are time consuming, often unrewarding, usually expensive and unhappily seem to feel failed to meet the needs of society that requires instant answers to difficult problems preferably yesterday. But whichever way we look at it, ultimately it is the pharmacological aspects of drug dependence that must lead to a proper understanding of the condition. An important topic here is the phenomenon of tolerance in which the drug taker requires bigger and bigger doses of the drug whether it be pain relieving, centrally stimulating or sedative in type in order to produce a required effect. This fundamental pharmacological problem accounts for the fact that pharmocologists to date have been unable to synthesise any really effective pain-relieving drug that does not induce tolerance. Current interest in pharmacology focuses considerably on the existence of drug receptors within the central nervous system. It should be realised that the action of a drug is a complex process by which a drug brings about a change in some pre-existing physiological function or biochemical process within an organism. The majority of drugs therefore act at a cellular level and for this reason the concept of drug receptors is at the moment one of central importance in pharmacology. It is postulated that drugs combine reversibly with receptors at a macromolecular level and bind to them by means of ionic bonds, van der Waals' forces and hydrogen bonds. It is suggested that the formation of these different types of bonds produces a stable drug receptor complex which leads to the pharmacological effects and these interactions are reversible. However, it should be realised that the concept of drug receptors is essentially hypothetical although it has been given considerable impetus comparatively recently by the apparent discovery of enkephalin which appears to be a specific opiate receptor within the brain, i.e. a protein binding molecule in the central nervous system at the level of central nervous system transmission. Presumably the binding affect of such a receptor reduces the effectiveness of the drug concerned and therefore increases the need for higher doses. If the work to date is valid and it appears to be, this will represent a most vital breakthrough in the global ignorance which we have about the essence of opiate addiction. Drugs of addiction in fact, are usually classified in terms of their dominant activity into four groups: first of all the drugs with predominantly depressant action, this includes sedatives and hypnotics and major tranquillisers: secondly, central stimulant drugs of the amphetamine type; thirdly, the narcotic analgesics, e.g. opiates and opiate-like drugs; and fourthly, the hallucinogenic drugs such as LSD and cannabis.

However, even though it may appear that the basic problem such as that of the existence of specific drug receptors may have been at least partially solved, we are left with many unanswered questions though we can classify addictive drugs in various ways as previously described, and although we know that they all have psychoactive effects we remain remarkably ignorant about their metabolic pathways. For instance, the pathways by which LSD is metabolised within the brain must be extremely complex but the precise nature is as yet undetermined. It has to be admitted that our understanding of central transmission within the brain itself is still very far from clear although it is a better organised area of knowledge than even 10 years ago. We are left ultimately with a number of vital but unanswered questions, some of which may appear comparatively simple but which do call for answers. These include:

1. Of all the mind-altering addictive drugs which are those that really matter as health hazards and why?
2. Is any type of treatment of drug addiction that is used of greater value than any other, and why?
3. What are the best criteria by which we may judge recovery?
4. How much do we really know about the long-term effects of mind-altering drugs in the physical and psychological sense?

This last question is very important since for many years there has been a tendency to assume that "drugs do not alter the personality". This is really a speculative assertion which has never been properly tested under any degrees of rigour and yet it is a vital one central to any understanding of whatever the particular psychological factors may be in determining whether or not someone is more vulnerable to addiction, and if they are whether the drug concerned is going to produce any long-lasting psychological effects. Unhappily, therefore, we have to concede at the end of it all that outside the laboratory, where precise and measurable work can proceed with animals and human volunteers, it is a fact that much of the research that goes on in the field of addiction can very rarely be subjected to any degree of rigorous control at all and so, for this reason, we are left with the unhappy suspicion that our ignorance of the essence of addiction is likely to persist for some time to come.

Wilse B. Webb

Graduate Research Professor of Psychology at the University of Florida and a former Fellow of Cambridge University.

A world authority on sleep and a founding member of the Association for the Psychophysiological Study of Sleep and the European Sleep Society.

SLEEP

Writing in the *Idler* of Saturday, 25 November, 1758, Samuel Johnson wrote remarkably and profoundly on sleep:

> Among the innumerable mortifications that way-lay human arrogance on every side may well be reckoned our ignorance of the most common objects and effects, a defect of which we become more sensible by every attempt to supply it. Vulgar and inactive minds confound familiarity with knowledge, and conceive themselves informed of the whole nature of things when they are shewn their form or told their use; but the Speculatist, who is not content with superficial views, harasses himself with fruitless curiosity, and still as he enquires more perceives only that he knows less.
>
> Sleep is a natural state in which a great part of every life is passed . . . . Yet of this change so frequent, so great, so general, and so necessary, no searcher has yet found either the efficient or final cause; or can tell by what power the mind and body are thus chained down in irresistible stupefacation; or what benefits the animal receives from this alternate suspension of its active powers.

If by "what powers the mind and body are thus chained down" can be taken to mean proximal conditions in the Humean sense we are making progress. But regarding "the final and efficient cause" little certainty has emerged in the more than 200 years.

To some extent we may invoke a familiar excuse relative to the life sciences. Sleep research, until very recently, has been a limited and spasmodic endeavor over these years. Even within the lagging life sciences – physiology, biology and the late blossoming psychology – sleep has not been a central interest. There was an occasional curious wandering researcher whose attention strayed from the more primary concerns of his discipline. A crumb of data or speculation was dropped here and there from the main tables. Mostly the sleeping giant of behavior slept.

The modern era of sleep research emerged from a convergence of disparate events. The moment to moment measurement of the sleep process became possible. This occurred when it was noted that "brain waves", the constantly changing electrical potentials of the brain that were measurable by the electroencephalogram (EEG), showed a distinctive change at the onset of sleep and a complex, dynamic and unique patterning throughout sleep. This was in the late 1930s. In the 1950s the periods of dreaming during sleep were pin-pointed by the combined use of the EEG and other physiological measures. At about the same time the central nervous system began to be explored intensively and sleep was included as a "bench mark" of massive central nervous system changes. And, most importantly, during the 1950s and 1960s there was an abundance of research money.

In the 1960s and 1970s sleep has received the attention of a wide range of special interests: biologists, biochemists, endocrinologists, psychiatrists and psychologists. There are now some 600 papers each year related to the phenomena of sleep. I would note, however, that these still represent marginal interests. In my discipline of psychology, for example, only about 1 paper in 100 are sleep related.

It is reasonable to argue, however, that our lack of knowledge about the "efficient and final cause" has been due to our concentration on the "form and use" of sleep. Almost the entirety of our efforts have been concerned with the description of sleep and the search for immediate or proximal determinants of sleep.

We have indeed learned a great deal about the form of sleep. Consider the wonder of a few selected facts about this dark kingdom:

– From more than seventy species of animals we know that a wide range of sleep patterns exist. Armadillos and opposums, for example, sleep 18 or more hours per day. On the other hand, ungulants, such as goats and sheep, sleep less than 4 hours a day as does the elephant. A particular type of shrew may not sleep at all. Some animals may sleep in stretches of long hours like the gorilla while others will sleep in bursts of minutes like the cow or small rodents. Some will sleep at night, some during the day, and some like many bats or desert animals, will be crepuscular or awake at dawn and twilight.

– Within a species there is typically a range of individual differences. For example, in the human, the average amount of sleep is between 7 and 8 hours, but between individuals the natural range is between 5 and 10 hours of sleep.

– During sleep there is a complex and highly organized pattern of sleep stages. In the human there are five stages of sleep and each occupies different time periods and proportions of sleep. The stage associated with dreaming occupies between about 20 to 25 per cent of each person's sleep and occurs in 90-minute intervals. Deep sleep amounts to about 18 per cent of sleep and occurs in the first third of sleep.

– Both the internal structure and patterning of sleep show common patterns of unfolding within species with development. In humans, for example, sleep changes systematically from the infant's sleep of about 15 hours in amount occurring in intermittent bursts across the 24 hours to a pattern of a long nocturnal period of about 8 hours in the adult. Like walking or weaning, these changes are in a natural and orderly sequence.

– When sleep is displaced in time – such as on shift work or during jet intercontinental travel – sleep patterns are disrupted but will become reorganized in its original pattern in time.

As for "function" of sleep – and here I believe Johnson meant the "mechanics" – we know yet more. We know about the activity of single nerve cells during sleep, we know the various structural parts of the central nervous system involved, and we are beginning to know about the biochemical changes associated with sleep.

An almost universally held answer to "why" we sleep was stated clearly by Shakespeare: "... sleep that knits up the ravelled sleve of care." Ask almost anyone why they sleep and, in terms varying from precise to vague, they will most likely say that we do so in order to "rest". Pushed further as to why we "rest" (which is just another way of describing sleep) they are likely to say we rest to "recover" or "restore" ourselves.

I have become dissatisfied with that answer. There are a number of reasons for this dissatisfaction. The most prominent of these is that our intensive search within the "mechanics" of sleep has failed to find what, if anything, is being restored. Simply, nothing has been found to happen during awakening which is reversed during sleep. To date there is no substance in the blood or the brain that builds or restores during sleep. If there is a busy little house maid scurrying around setting our daily disordered house into good order she is most elusive and most lazy.

Moreover, the range of individual differences both within and across species does not fit a restorative pattern. Since sleep occurs within a 24-hour "box" the longer an organism is awake, the shorter the amount of sleep. This means that elephants must "recover" from 20 to 22 hours of wakefulness in 2 to 4 hours while an oppossum requires 18 hours to recover from 6 hours of wakefulness. A human who typically is awake for 18 hours must recover in 6 while one who is awake for 14 recovers in 10 hours.

The main problem for me is that a "restorative" theory, while sounding satisfactory, fails to provide heuristic predictions about the phenomena of sleep.

And I am a "Speculatist" who must "harrass himself" with what I hope is not "fruitless curiosity". As such, I worry about the "efficient and final causes" of sleep. In my worrying I have followed the path of Tinbergen and the ethologists. Tinbergen specifically urged that we pay attention to "the *effects* of a behavior; of the ways in which behavior influences the survival of a species (in relationship to its particular environment) . . . to understand the state of adaptedness and the process of evolutionary and adaptation".

As a result of this approach I have been championing the notion that sleep is an adaptive process that evolved out of a species-environment interaction. This proposition specifically states that sleep has been shaped by and functions in relationship to the optimum survival of a species and, in particular, in relationship to its food gathering. A few examples may give the idea. The elephant sleeps for so few hours because it must forage the rest of the time to avoid starvation. Only those elephants that could sleep limited periods of time survived and reproduced short-sleeping elephants. Grazing animals must forage widely and as a herd since they have no safe sleeping place nor are they dangerous. Only short and intermittent sleepers can survive and reproduce in such circumstances. Man functions and survives poorly at night. Only those humanoids who could remain inactive (sleep) during the approximately 7 to 8 hours of darkness survived to reproduce our species that sleep as it does.

If I am correct in this notion that sleep is an innate, evolved and inherent process that emerged to fit us to our environment of eons ago, then sleep in our modern day faces severe strains. Modern times have brought the Edison Age of electric lights and fill that time with continuous entertainment and arousal. Industry has placed man around the tireless and arhythmic machines across the 24 hours. The jet aircraft tosses sleep across multiple time zones. Drugs hold out promises of bending sleep to our momentary demands. Pervasively, we raise our strident cries and push our self-centered demands that sleep be subservient to our whimsy, bend to our needs, pressures and terrors. We ominously move toward viewing our failures to sleep as "illnesses" to be "cured".

Again Tinbergen has put it well. ". . . the developments of physics and chemistry have shown (that) knowledge of causes provides us with the power to manipulate events and 'bully them into subservience. . . .' Man, particularly Urban Man, is inclined even in his biological studies to ape physics, and so to contribute to the satisfaction of his urge to conquer nature . . . we have changed our environment . . . (and) as a result our behavioral organization is no longer faced with the environment in which the behavior was molded, and as a consequence, misfires."

I believe that we must remember our inherent nature. In a reasonably natural and stable environment sleep will serve its adaptive functions well as a silent and well-trained servant. It is our "misbehaviors" in relation to sleep, goaded by a changed environment and a thoroughly anthropomorphic arrogance about nature, which "fails" sleep as it is pushed beyond its natural limits. From my perspective, anchored in my adaptive theory of the "efficient and final" cause of sleep, we must, rather than learn the proximal causes of sleep, learn the laws of sleep. In turn we must act in accord with those laws. I agree with Sir Francis Bacon of 500 years ago: "Nature cannot be commanded except by being obeyed."

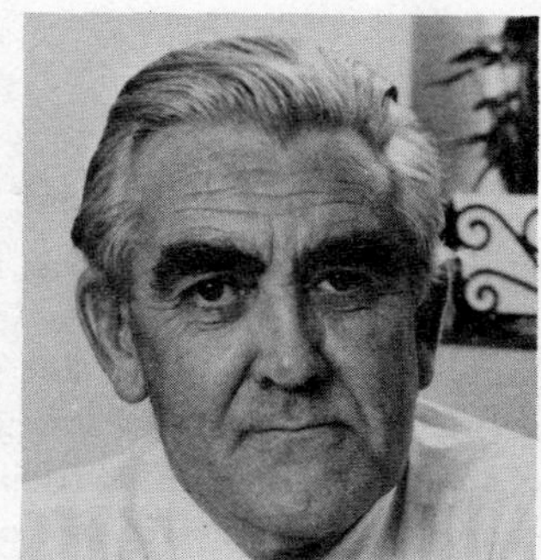

Ewan Cameron

Ewan Cameron

Senior Consultant Surgeon of the Argyll and Clyde Health Board.

Fellow of the Royal College of Surgeons, Glasgow; the Royal College of Surgeons of Edinburgh and the Linus Pauling Institute of Science and Medicine.

Linus Pauling, F.R.S.

Professor of Chemistry at Stanford University.

Awarded the Nobel Prize for Chemistry in 1954 and the Nobel Peace Prize in 1962.

ASCORBIC ACID AND THE GLYCOSAMINOGLYCANS: AN ORTHOMOLECULAR APPROACH TO CANCER AND OTHER DISEASES

INTRODUCTION

Orthomolecular medicine is the preservation of good health and the treatment of disease by varying the concentrations in the human body of substances that are normally present in the body and are required for health.[22, 23, 25] Of these substances, the vitamins are especially important, and ascorbic acid, in particular, may have much greater value than has been generally ascribed to it. Irwin Stone[31–33] has advanced arguments to support the concept that the optimum rate of intake of ascorbic acid is about 3 g per day under ordinary conditions, and larger, up to 40 g per day, for a person under stress. An argument based on the fact that only a few animals species require an exogenous source of ascorbic acid and on the amounts of ascorbic acid contained in a diet of raw natural plant food has led to the conclusion that the optimum daily intake of ascorbic acid for an adult human being is about 2.3 g or more.[24] Ascorbic acid is a substance with extremely low toxicity; many people have ingested 10-20 g per day for long periods without serious side effects, and ingestion of as much as 150 g within 24 hr without serious side effects has been reported.[16] In this respect ascorbic acid may be considered an ideal substance for orthomolecular prophylaxis and therapy.

GROUND SUBSTANCE

Cells in the tissues of the body are embedded in ground substance. This ubiquitous material pervades every interspace and isolates every stationary cell from its neighbours. It must be traversed by every molecule entering or leaving the cell. There is evidence that the interface between a cell membrane and the immediate extracellular environment is the crucial factor in the whole proliferative process. Variations in the composition of the extracellular environment exert a profound influence on cell behavior, and in turn the cells possess a powerful means of modifying their immediate environment. This interdependence is involved in all forms of cell proliferation and is particularly important in cancer. A proliferating cell and its immediate environment constitute a balanced system in which each component influences the other. Recognition of this relationship and an understanding of the means of controlling it could lead to rational methods of treating cancer and other cell-proliferative diseases.

Until recently cancer research has tended to concentrate almost exclusively upon the cell, and to ignore the other half of the proliferation equation. The intercellular substance is a complex gel, containing water, electrolytes, metabolites, dissolved gases, trace elements, vitamins, enzymes, carbohydrates, fats, and proteins. The solution is rendered highly viscous by an abundance of certain long-chain acid mucopolysaccharide polymers, the glycosaminoglycans and the related proteoglycans, reinforced at the microscopic level by a three-dimensional network of collagen fibrils. The principal glycosaminoglycans so far identified are hyaluronic acid, a long-chain polymer with high molecular weight (200,000-500,000) and simple chemical structure (alternating residues of N-acetylglucosamine and glucuronic acid), and varieties of chrondroitin (alternating residues of N-acetylgalactosamine and glucuronic acid) and its sulfate esters. Other glycosaminoglycans may also be present. The chemistry of the ground substance and the intercellular environment has been reviewed recently by Balazs.[1]

An important property of the intercellular substance is its very high viscosity and cohesiveness. This property is dependent upon the chemical integrity of the large molecules.

The viscosity can be reduced and the structural integrity destroyed by the depolymerizing (hydrolyzing) action of certain related enzymes (the endohexosaminidases, the β-N-acetylglucosaminidases, the β-N-acetylgalactosaminidases, and β-glucuronidase), known by the generic name "hyaluronidase". It is probable that most cells in the body are able to produce hyaluronidase.[3, 10] The interlacing molecular network of the intercellular ground substance is in a constant state of slow dynamic change, with synthesis of glycosaminoglycans (polymerization) balanced by their breakdown (depolymerization by hydrolysis) through the catalytic action of hyaluronidase, and subsequent excretion. It is within this slowly changing environment, called the "milieu interieur" by Claude Bernard, that all cellular activity takes place. The normal cell and the cancer cell both thrive and die within this environment.

HYALURONIDASE AND CELL PROLIFERATION

Some years ago the hypothesis was advanced that all forms of cell proliferation depend upon one fundamental interaction between the cell and its immediate environment.[5] The hypothesis may be stated as follows: all cells in the body are embedded in a highly viscous environment of ground substance that physically restrains their inherent tendency to proliferate; proliferation is initiated by release of hyaluronidase from the cells, which catalyses the hydrolysis of the glycosaminoglycans in the immediate environment and allows the cells freedom to divide and to migrate within the limits of the alteration; proliferation continues as long as hyaluronidase is being released, and stops when the production of hyaluronidase stops or when the hyaluronidase is inhibited, and the environment is allowed to revert to its normal restraining state.

In normal healthy tissues cell division is taking place at a constant slow rate, corresponding to normal cell replacement. This normal "background" rate of cell division results in a slow metabolic turnover of ground substance, with liberation into the blood stream and then escape in the urine of the partially depolymerized fractions of the intercellular glycosaminoglycans, produced in the immediate vicinity of the dividing cells. These degradation products of ground substance depolymerization can be recognized and measured by various biochemical methods. Depending upon the analytical procedure used different fractions have been given different names. For our purpose they may be grouped together under the general term "serum polysaccharide". In health the serum polysaccharide concentration remains within a relatively narrow "normal" range.[6, 35] The process is kept in check by the presence in the tissues and the blood of a substance called "physiological hyaluronidase inhibitor" (PHI). In health the serum PHI concentration lies within a well-defined "normal" range.[9, 19]

In conditions in which excess cell proliferation is occurring, such as inflammation, tissue repair, and cancer, depolymerization of ground substance can be demonstrated histochemically in the immediate vicinity of the proliferating cells,[34] and there is also a significant increase in concentration of both the serum polysaccharide[6, 35] and the serum PHI.[9, 19]

NEOPLASTIC CELL PROLIFERATION

It follows from this hypothesis that cancer may be no more than the permanent exhibition by some cells of a fundamental biological property possessed by all cells. We suggest that the characteristic feature of neoplastic cell proliferation is that these cells in becoming malignant have acquired, and are able to bequeath to their descendants *in perpetuo*, the ability to produce hyaluronidase continuously. Wherever they travel, these cells will always prosper, multiply, and invade within the protective independence of their own self-created depolymerized environment. These renegade cells are autonomous only because they possess this specific ability, the ability to isolate themselves permanently from "contact" and all the usual "controls" governing tissue organization and growth restraint.

By endowing a clone of cells with this single property of continuous hyaluronidase release it is possible to provide a reasonable explanation for many of the morphological features of malignant invasive growth.[5] The methods whereby cells might acquire this property in response to a wide variety of carcinogenic stimuli have also been outlined, together with the experimental evidence in support of the concept.[5]

THERAPEUTIC CONTROL OF CELL PROLIFERATION

Assuming that cell proliferation depends upon depolymerization of the ground substance by cellular hyaluronidase, we see that there are two methods of exerting some therapeutic control of cancer and of other disease states in which excessive cellular proliferation is a harmful feature. We may attempt to increase the resistance of the ground substance to enzymatic depolymerization, that is, to strengthen the ground substance, or to directly neutralize the cellular hyaluronidase by decreasing its production or inhibiting its action.

TREATMENT BY STRENGTHENING THE GROUND SUBSTANCE

The resistance of ground substance to the action of hyaluronidase can be increased in several ways, some of which are already established as useful methods for retarding cell proliferation.

Radiotherapy, irradiation with X-rays, is an example in which the result of the treatment is that some of the amorphous ground substance has been replaced by a dense deposit of collagen.[15] The direct cytotoxic effect of radiotherapy is thus reinforced by a permanent reduction in the susceptibility of the ground substance in the treated region to the action of hyaluronidase, with a consequent long-lasting diminution in proliferative activity.

Hormone therapy is effective because the physical-chemical state of the ground substance is profoundly influenced by many endocrine factors; the experimental evidence has been reviewed elsewhere.[5] Resistance to the action of hyaluronidase can be increased by the administration of cortico-steroids, estrogens, androgens, and thyroxine, and these effects are enhanced after adrenalectomy and hypophysectomy. These hormones, although differing widely in their special effects on particular target cells, all exert to a greater or lesser extent, and roughly in the order stated, the same effect on the intercellular field, namely, the absorption of amorphous ground substance and its replacement by a more resistant fibrous substance. Without wishing to enter into the current debate about the mode of action of endocrine therapy in malignant disease,[30] we are content to note that these hormones are the ones used with some success in the palliation of various forms of human cancer.

Other agents may also be effective in altering the intercellular environment and indirectly exerting some controlling influence on the behaviour of cells. It has been pointed out[5] that an explanation is provided of the "Haddow paradox", that substances which are locally carcinogenic (by creating a local intensely impermeable carcinogenic environment) have also some anti-proliferative value when administered systemically in experimental cancer (by bringing about similar generalized changes in the resistance of the ground substance to hyaluronidase and thus decreasing cell proliferation).

Because of the complexity of the intercellular ground substance and its responsiveness to external influences, many of the innumerable "cancer treatments" that have been hopefully advocated year after year might have some element of truth behind them. It is also true, however, that no form of cancer treatment based on the antineoplastic effect of modification of ground substance can ever be more than palliative, because to render the ground substance totally resistant to hyaluronidase would create a situation incompatible with life itself.

TREATMENT BY INHIBITION OF HYALURONIDASE

Although the indirect methods of retarding cell proliferation, described above, are of great interest and in special circumstances of undoubted value, the direct inhibition of cellular hyaluronidase offers more spectacular therapeutic possibilities.

Hyaluronidase may be inhibited by drugs and by immunological methods, but the approach most likely to succeed appears to be that of utilization of the naturally occurring inhibitor, PHI.

Spontaneous regression of advanced cancer has been well documented in a number of fortunate patients as a direct consequence of massive inter-current infection with hyaluronidase-producing bacteria.[8] A possible explanation for this remarkable phenomenon is that the depolymerizing action evoked an upsurge in the serum concentration of PHI of sufficient magnitude to inhibit totally the malignant capability of the neoplastic cells.[5] It is known that such infections are always associated with an increase in the serum PHI concentration.[9, 19] It has been independently demonstrated in experimental cancer that the injection of "Shear's polysaccharide" induces not only carcinolysis[12] but also a sharp and significant rise in PHI concentration.[11] The problem is how to employ this suppressive mechanism in practical therapeutics.

ASCORBIC ACID AND HYALURONIDASE INHIBITOR

There is strong evidence to indicate that ascorbic acid is involved in some way in the synthesis of physiological hyaluronidase inhibitor. A strong suggestion to this effect is provided by the manifestations of scurvy, resulting from a deficiency of ascorbic acid. If ascorbic acid were required for the synthesis of PHI, a deficiency of ascorbic acid would cause the serum PHI concentration to decrease toward zero. In the absence of such control of hyaluronidase by PHI, background cellular proliferation and release of hyaluronidase would produce a steady and progressive enzymatic depolymerization of the ground substance, with disruption and disintegration of the collagen fibrils, intraepithelial cements, basement membranes, perivascular sheaths, and all the other organized cohesive structures of the tissues, producing in time the generalized pathological state of scurvy. These generalized changes, tissue disruption, ulceration, and hemorrhage, are identical to the local changes that occur in the immediate vicinity of invading neoplastic cells. This concept of scurvy, as involving uncontrolled depolymerization of ground substance, explains why scurvy is always associated with a very high level of serum polysaccharide.[26] It also explains why very small amounts of ascorbic acid have profound effects in the treatment of scurvy. The total body content of ascorbic acid is estimated to be around 5 g.[7] and yet this small amount controls the health of the whole body content of intercellular material, which must amount to many kilograms of substance. The symptoms of scurvy are relieved by the ingestion of a few tenths of a gram of ascorbic acid. It seems clear that ascorbic acid is not an important constituent unit of the intercellular ground substance, as had been suggested; instead, it may well be involved in the synthesis of PHI, the circulating factor that controls intercellular homeostasis. Ascorbic acid is, of course, required for the conversion of proline to hydroxyproline, and is accordingly essential for the synthesis of collagen. It may well serve in several ways in determining the nature of tissues and the state of health of human beings.

A preparation of PHI has been reported to have molecular weight about 100,000 and to consist of 94 per cent protein and 6 per cent polysaccharide.[20] It is our opinion that it is the polysaccharide that has the power of combining with the active region of the enzyme and inhibiting its activity, and the following discussion is based on that opinion. PHI has a general chemical similarity to the glycosaminoglycan polymers of the ground substance.[9, 19] The PHI serum concentration rises significantly in all conditions in which excessive cel proliferation is a feature,[9] but PHI is known not to be a simple breakdown product of ground-substance glycosaminoglycan. Its precise chemical composition is still unknown. It has recently been suggested[4] that it is a soluble glycosaminoglycan residue in which some or all of the glucuronic acid units are replaced by the somewhat similar molecules of ascorbic acid. The general theory of enzyme activity and the action of inhibitors[21] involves the idea that the active region of the enzyme is complementary to the intermediate complex, with the structure corresponding to the maximum of the energy curve (at the saddle-point configuration, intermediate in structure between the reactants and the products) that determines the rate of reaction. This theory requires that inhibitors of the enzyme resemble the activated complex, rather than either the reactant molecules or the product molecules. Accordingly it is unlikely that PHI would be a fragment of hyaluronic acid or a fragment of any other glycosaminoglycan. It would instead involve at least one residue of a related but different substance. It is possible that a residue of ascorbic acid resembles the activated complex, and that incorporation of such a residue would produce an altered glycosaminoglycan that could function as an inhibitor of hyaluronidase. It is also possible, however, that the chemical activity of ascorbic acid, such as its reducing power or its power to cause hydroxylation reactions to take place, could cause conversion of an oligoglycosaminoglycan into PHI. Whatever the mode of action of ascorbic acid in synthesis of PHI, whether it involves incorporation of an ascorbic acid residue or some other reaction, the therapeutic implications of the concept that ascorbic acid is involved in the synthesis of PHI are considerable.

The hypothesis that ascorbic acid is required for the synthesis of PHI and is itself destroyed in the course of the synthesis explains why in such conditions as inflammation, wound repair, and cancer the individual always appears to be deficient in ascorbic acid, on the basis of measurement of its concentration in the serum, measurement of urinary excretion, and saturation tests.[2, 7, 18] It is clear that the total body requirement of ascorbic acid has become abnormally high, as would result from an increased synthesis of PHI with incorporation or destruction of ascorbic acid.

POSSIBLE USES OF HYALURONIDASE INHIBITOR AND ASCORBIC ACID

If the basic concept of cellular proliferation is correct, PHI might be a valuable therapeutic agent in directly controlling all forms of excessive proliferation, including cancer. It is a naturally occurring substance found in the serum of all mammals,[20] and should be safe and free from dangerous side effects. Determination of the chemical structure of PHI and its synthesis should not present insuperable difficulties. However, it may not be necessary to synthesize the substance. It is possible that, given enough ascorbic acid, the body could synthesize a proper quantity of PHI.

The concentration of ascorbic acid in blood plasma is about 15 mg/l when the rate of intake is about 200 mg per day. With larger rates of intake the concentration in plasma increases only slowly, because of urinary excretion, reaching about 30 mg/l for an intake of 10 g per day. Hume[13] has reported that an oral intake of around 6 g per day is required to correct the measured leucocyte ascorbic acid deficiency encountered during the course of the common cold. Leucocytes are especially rich in ascorbic acid, and one of their functions might be to act as a mobile circulating reservoir of ascorbic acid ready to be used for local production of the protective substance PHI at any site where excessive "inflammatory" depolymerization is taking place. The finding of Hume *et al.*[14] gives some support for this view. In their studies of myocardial infarction they found that, in addition to the usual leucocytosis, the ascorbic acid content of the circulating leucocytes undergoes a very sharp depletion accompanied by an increase in the concentration of ascorbic acid in the tissues at the site of the infarction. In the therapeutic situations here envisaged, with ascorbic acid being prescribed to control excessive cell proliferation, a daily intake of 10-50 g, or even more, and with the bulk of that administered intravenously at first, might be necessary to achieve the desired effect. McCormick[17, 18] and Klenner[16] have been advocating and using this form of treatment for many years. Their combined clinical experience indicates that very large doses of ascorbic acid, in the range mentioned above, can be given intravenously with perfect safety and with apparent benefit in a wide variety of disease states.

The hypothesis that ascorbic acid is required for synthesis of PHI and can thus control harmful depolymerization of glycosaminoglycans explains why the vitamin is effective in curing scurvy and in improving the healing of wounds. The potential therapeutic uses of this relatively simple substance may, however, be much greater.

It has been postulated for years that the administration of ascorbic acid would increase tissue resistance to bacterial and viral infections by improving the integrity of the tissues. We are now in a position to suggest that, through the action of PHI, the administration of ascorbic acid in sufficiently high dosage may provide us with a broad-spectrum agent effective against all those pathogenic bacteria, and perhaps viruses, that rely upon the release of hyaluronidase to establish and spread themselves throughout the tissues. The dramatic clinical successes reported independently by McCormick[17, 18] and by Klenner[16] in a wide variety of infective states support this contention.

The effectiveness of the water-soluble anti-oxidant ascorbic acid on ground substance and especially on cell membranes may be increased by the simultaneous administration of the fat-soluble anti-oxidant vitamin E. Shamberger[28, 29] has reported observations on the effectiveness of ascorbic acid and also of vitamin E, together with selenium, in inhibiting the growth in mice of tumors initiated by 7,12-dimethylbenz(α)anthracene and promoted by croton oil.

The hypothesis also indicates a safe and elegant method of control in many inflammatory and auto-immune diseases where, although the individual causes are still unknown, the essential harmful feature is always excessive cell proliferation and ground-substance depolymerization. A trial of orthomolecular doses of ascorbic acid seems justifiable and preferable to the use of corticosteroids, irradiation, and all other indirect methods currently employed.

Most important of all, we are led to the conclusion that the administration of this harmless substance, ascorbic acid, might provide us with an effective means of permanently suppressing neoplastic cellular proliferation and invasiveness, in other words, an effective means of controlling cancer. Ascorbic acid in adequate doses might prove to be the ideal cytostatic agent. Regressions might be induced in a few patients with rapidly growing tumors with precarious blood supplies, but in the great majority the effect of the treatment is expected to be to "disarm" rather than to "kill" the malignant cells. "Tumours" would remain palpable and visible on X-ray examination, but all further progressive malignant growth might be stopped. Hopefully malignant ulcers would heal, and pain, hemorrhage, cachexia, and all the other secondary distressing features of neoplasia would be brought under control. This desirable outcome might be termed carcinostasis, with what had been neoplastic cells now rendered harmless and re-embedded in intact ground substance, subject again to normal tissue restraints, and persisting in the body in heterotopic situations as "paleoplastic" collections of essentially normal cells. A suggestion of the possibilities of the use of ascorbic acid in the control of cancer has been provided by the report by Schlegel *et al.*[27] of its effectiveness against cancer of the bladder. It is our hope that a thorough trial will be given to this safe substance, ascorbic acid, which may turn out to be the most valuable of all substances in the armamentarium of orthomolecular medicine.

ACKNOWLEDGEMENT

This article originally appeared in *Oncology* 27, 181-192, Karger, Basel, and appears here in slightly modified form.

REFERENCES

1. E. A. Balazs, *Chemistry and Molecular Biology of the Intercellular Matrix*, Academic Press, New York, 1970.
2. O. Bodansky, "Concentrations of ascorbic acid in plasma and white cells of patients with cancer and non-cancerous chronic disease". *Cancer Res.* **11**, 238 (1951).
3. A. J. Bollet, W. M. Bonner and J. L. Nance, "The presence of hyaluronidase in various mammalian tissues". *J. Biol. Chem.* **238**, 3522-7 (1963).
4. E. Cameron and D. Rotman, "Ascorbic acid, cell proliferation, and cancer". *Lancet* i, 542 (1972).
5. E. Cameron, *Hyaluronidase and Cancer*, Pergamon Press, New York, 1966.
6. E. Cameron, A. Campbell and W. Plenderleith, "Seromucoid in the diagnosis of cancer". *Scot. Med. J.* **6**, 301-7 (1961).
7. S. Davidson and R. Passmore, *Human Nutrition and Dietetics*, William & Wilkins, Baltimore, 1969.
8. T. C. Everson and W. H. Cole, "Spontaneous regression of cancer: preliminary report". *Ann. Surg.* **144**, 366-83 (1956).
9. D. Glick, "Hyaluronidase inhibitor of human blood serum in health and disease". *J. Mt Sinai Hosp.* **17**, 207-28 (1950).
10. H. Grossfield, "Studies on production of hyaluronic acid in tissue culture. The presence of hyaluronidase in embryo extract". *Exp. Cell Res.* **14**, 213-16 (1958).

11. Z. Hadidian, I. R. Mahler and M. M. Murphy, "Properidine system and nonspecific inhibitor of hyaluronidase". *Proc. Soc. Exp. Biol. Med.* **95**, 202-3 (1957).
12. H. F. Havas and A. J. Donnelly, "Mixed bacterial toxins in the treatment of tumors. III. Effect of tumor removal on the toxicity and mortality rate in mice". *Cancer Res.* **21**, 17-25 (1961).
13. R. Hume, Personal communication (1972).
14. R. Hume, E. Weyers, T. Rowan, D. S. Reid and W. S. Hillis, "Leucocyte ascorbic acid levels after acute myocardial infarction". *Brit. Heart J.* **34**, 238-43 (1972).
15. B. Jolles and P. C. Koller, "The role of connective tissue in the radiation reaction of tumours". *Brit. J. Cancer* **4**, 77-89 (1950).
16. F. R. Klenner, "Observations on the dose and administration of ascorbic acid when employed beyond the range of a vitamin in human pathology". *J. Appl. Nutr.* **23**, 61-88 (1971).
17. W. J. McCormick, "Ascorbic acid as a chemotherapeutic agent". *Arch. Pediat., N.Y.* **69**, 151-5 (1952).
18. W. J. McCormick, "Cancer: a collagen disease, secondary to a nutritional deficiency"? *Arch. Pediat., N.Y.* **76**, 166-71 (1959).
19. M. B. Mathews and A. Dorfman, "Inhibition of hyaluronidase". *Physiol. Rev.* **35**, 381-402 (1955).
20. J. K. Newman, G. S. Berenson, M. B. Mathews, E. Goldwasser and A. Dorfman, "The isolation of the non-specific hyaluronidase inhibitor of human blood". *J. Biol. Chem.* **217**, 31-41 (1955).
21. Linus Pauling, "Chemical achievement and hope for the future". *Am. Scient.* **36**, 51-58 (1948).
22. Linus Pauling, "Orthomolecular somatic and psychiatric medicine". *Z. Vitalstoffe-Zivilisationskr.* **12**, 3-5 (1967).
23. Linus Pauling, "Orthomolecular psychiatry." *Science*, **160**, 265-71 (1968).
24. Linus Pauling, "Evolution and the need for ascorbic acid". *Proc. Nat. Acad. Sci., Wash.* **67**, 1643-8 (1970).
25. Linus Pauling, "The significance of the evidence about ascorbic acid and the common cold". *Proc. Nat. Acad. Sci., Wash.* **68**, 2678-81 (1971).
26. C. L. Pirani and H. R. Catchpole, "Serum glycoproteins in experimental scurvy". *Arch. Path.* **51**, 597-601 (1951).
27. J. U. Schlegel, G. E. Pipkin, R. Mishimura and G. M. Schultz, "The role of ascorbic acid in the prevention of bladder tumor formation". *J. Urol., Baltimore*, **103**, 155-9 (1970).
28. R. J. Shamberger, "Relationship of selenium to cancer. I. Inhibitory effect of selenium on carcinogenesis". *J. Nat. Cancer Inst.* **44**, 931-6 (1970).
29. R. J. Shamberger, "Increase of peroxidation in carcinogenesis". *J. Nat. Cancer Inst.* **48**, 1491-7 (1972).
30. B. Stoll, *Endocrine Therapy in Malignant Disease*, Saunders, London, 1972.
31. I. Stone, "Hypoascorbemia, the genetic disease causing the human requirement for exogenous ascorbic acid". *Perspect. Biol. Med.* **10**, 133-4 (1966).
32. I. Stone. "On the genetic etiology of scurvy". *Acta Genet. Med. Gemellol.* **15**, 345-50 (1966).
33. I. Stone, "The genetic disease hypoascorbemia". *Acta Genet. Med. Gemellol.* **16**, 52-62 (1967).
34. J. M. Vasilief, "The role of connective tissue proliferation in invasive growth of normal and malignant tissues: a review". *Brit. J. Cancer* **12**, 524-36 (1958).
35. R. J. Winzler and J. C. Bekesi, "Glycoproteins in relation to cancer". In Busch, *Methods in Cancer Research*, 2, 159-202 (Academic Press, New York, 1967).

Photograph by Sergei Alexyevich Kurnikov of Moscow, USSR

Preston Cloud

Research Geologist U.S. Geological Survey and Professor Emeritus of Biogeology and Environmental Studies at the University of California, Santa Barbara. Was for 10 years Chief of Paleontology and Stratigraphy Branch of the U.S. Geological Survey.

Member U.S. National Academy of Sciences, American Philosophical Society, American Academy of Arts and Sciences and holder of the Penrose Medal of the Geological Society of America.

THE VEILS OF GAIA

Knowledge advances like the concentric ripples that spread outward from a pebble tossed into a mill pond. Its expanding front is in contact with an ever-widening periphery of ignorance as growing comprehension generates new and more subtle questions. Gaia (or Gaea), Mycenaean goddess of the earth, like a veiled dancer, reveals her secrets only to the skilled and persistent explorer. Geologists, geophysicists, oceanographers and other members of her priesthood must learn how to ask the right questions, how to identify and probe the critical areas.

As the growing probability of continental drift sent Earth scientists in search of mechanisms, so plate tectonics (growth and destruction of Earth's crust along the seismically active margins of giant mobile slabs or plates) has provoked searching questions about deep subcrustal circulation and the tectonics of plate interiors. As advances in geochemistry provide new geothermometers, geobarometers and geochronometers, so new questions arise from the need for reconciliation of previously unrecognized anomalies. As advances in biogeology and experimental biogenesis bring fresh insights into the origin and early development of life, so they excite deeper inquiry into interactions with other evolutionary processes. As lunar science and planetary geology have evolved, so likewise have they challenged former ideas about the early evolution of the earth and demanded new models of planetary accretion and differentiation. As the needs of man have grown, once mainly scientific pursuits such as the study of geothermal energy have entered the arena of practicality and thrust upon us new systems to be analysed and interpreted. The growth of energy and materials shortages, the imminence of earthquake prediction, and the growing importance of geologic hazards in an overcrowded world work to create unforeseen problems along the boundary between geology and the social sciences.

A favorite game of scientists is trying to identify and articulate the big problems that lie beyond the outer rim of knowledge. Any imaginative scientist can think of dozens of such problems without much effort. The possibilities are illustrated by some questions chosen at random, to which we would like to have answers but to which we cannot now respond with very high levels of confidence.

How did the earth form and its concentric structure develop? What was the age and nature of the oldest solid crust? How thick was it and what was the rate of flow of heat from Earth's interior through it? What went on before there was a solid crust? When did the earth become cool enough for a hydrosphere to condense from water previously present only as steam? When was it cool enough to support simple forms of life and how soon thereafter did life first appear? When did the advanced types of cell-division called mitosis and meiosis arise, setting the stage for higher forms of life, and what were the steps leading up to this? Can we use the earth as a paradigm for the study of planetary evolution generally? And so on.

As a result of the findings of lunar and planetary science, geologists are turning back to the old idea of an initially molten Earth, but with a difference. We now think it started some 4.6×10^9 years ago as a falling together of solid particles within the contracting solar nebula but that the conversion of gravitational energy to heat led to rising temperatures and, at some size, to a molten proto-earth. It would probably have taken some hundreds of millions of years after final aggregation for such an earth to cool off enough to generate a solid crust, considering additions to the early heat budget from the decay of radioactive potassium to argon and calcium and the decay of uranium-235 to lead. About 400 to 500 million years is a common guess for the time required. However, we have as yet no direct evidence from which to reconstruct any concrete event within the first 800 million years of Earth history.

Following that first 800 million years or so, primary and metamorphic mineral assemblages tell us something about the pressure-temperature (P-T) regimes under which formed, but attempts to convert such data to estimates of crustal thickness and thermal gradient have suggested different numbers to different people and to the same people at different times. Not so long ago many geologists believed the early crust was thin, hot, and basaltic. Now a contrary view is gaining ground. New laboratory and field investigations into the P-T conditions of formation and natural habitats of minerals, mineral associations, and geochemistry of particular rock types are converging on these problems but with no clear indication as yet as to what the final answers will be.

Until recently it was generally accepted that the great depletion in noble gases of the terrestrial atmosphere as compared with cosmic abundances was clear evidence of a secondary origin for the atmospheric gases. But this too is now questioned by some, and the whole problem of the modes and time or times of terrestrial outgassing and atmospheric evolution is undergoing reexamination. Atmospheric oxygen, although certainly secondary and doubtless in large part of photosynthetic origin, has in some part also originated by non-biogenic photolysis of water vapor. How much we do not know, because we do not know where *all* of the oxygen now is, or how much of the carbon dioxide that has passed through Earth's atmosphere was originally carbon monoxide, or what the flux rates of hydrogen have been, or whether there was much or little methane and ammonia in the initial atmosphere.

As for life, there is indirect evidence that it was already present when the oldest yet known sediments were being deposited about 3.8×10^9 or more years ago. But we do not see structurally preserved organisms until somewhat less than 2.5×10^9 years ago when the oldest demonstrable microflora is recognized. Where is the older record we think should exist? Is it comprised of the simple spheroids and nondescript linear features that some accept as an older fossil record, or is it yet to be found? If we were to find representatives of the initial microorganism, would we be able to give them a high objective confidence rating as once-living structures? Altogether we now have only a few dozen authentic records of life for the first 4×10^9 years of Earth history, many of them still undescribed. With so limited a sample what level of confidence can we assign to generalizations about the time and mode of origin of major biologic innovations before the first appearance of recognizable animal life about 680 to 700×10^6 years ago.

Geologists are now riding a wave of euphoria based on the well-founded expectation that the modern theory of plate tectonics will provide the matrix out of which will grow new and more fundamental understandings of many sorts. It must not be overlooked, however, that plate tectonics itself reveals new areas of ignorance; evokes questions we had not thought of before. If fold mountains characteristically form compressionally at converging plate margins, how may we explain the formation of the folded and thrust-faulted Rocky Mountains *within* the North American plate? If rising "hot spots" explain isolated uplifted intracontinental massifs, what makes the hot spots come and go? And why do not long-lasting hot spots always leave traces, like the Hawaiian chain of islands, as moving plates drift over them? If the porphyry copper deposits that compromise most of the world's copper reserves are related to subduction zones where oceanic rocks are carried deep beneath overriding continental margins, does this mean that some subduction zones dipped steeply beneath the continents, as under the Andes, while others, as beneath the southwestern U.S., were flat or multiple? And, finally, when did plate tectonic mechanisms begin; how, when and where were they turned off and on; and what identifying criteria might we use to delineate old plate margins as a way of defining prospective new metallogenic provinces and episodes?

It is safe to say that future geological research will pursue even the most esoteric of these questions and others no matter how pressed civilization may become for practical solutions to daily problems. For pure science is among the most relevant of all human activities. We will always need to know in a deep sense how the earth in all its aspects evolved to its present state – both to satisfy mankind's thirst for the poetry of planetary evolution and because we must understand the fruits of knowledge before we can turn them to our own ends.

Paradoxically, the extent of our present ignorance is a telling reflection of progress. For a high level of prior understanding is, after all, essential to the formulation of the kinds of questions geologists are now asking and attempting to answer.

A. G. Cairns-Smith

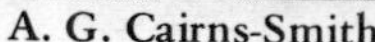

C. J. Davis

A. G. Cairns-Smith

Lecturer in the Department of Chemistry at the University of Glasgow.

Working on an alternative to "chemical evolution" as a key to the explanation of the origin of life on earth.

C. J. Davis

Worked with Dr. Cairns-Smith on practical and theoretical aspects of the design of an alternative molecular biology.

THE DESIGN OF NOVEL REPLICATING POLYMERS

Polymer chemistry has reached the stage at which the design and synthesis of novel replicating molecules has become a feasible research aim. It is hard to exaggerate the importance that success in this direction could have: it could provide routes into new man-made molecular biologies, to the manufacture in quantity of co-polymers with specific and complex sequences – to new kinds of information-containing macromolecules which, like proteins and RNAs, could have intricately specific tertiary structures. There exists in principle a field of engineering at the colloidal level accessible in principle to the technique of repeated selection of replicable and mutable polymers.(1) Here we will try to indicate the variety of possible approaches to the first most difficult step in such an enterprise – achieving a faithfully replicating synthetic polymer.

STAGE-WISE REPLICATING CYCLES

Let us start by thinking about molecules that are like DNA in that they are linear co-polymers and ladders of some sort. Defining our use of terms we may say that there are four main *processes* in the replication of such a molecule: chain separation, monomer location, alignment of located monomers and on-chain polymerisation. Figure 1 illustrates a scheme in which these processes take place through a number of distinct *steps* (a, b_1, b_2, etc.). In this way the problem of replication is divided into a number of separate problems that can be tackled one at a time: the molecules can be driven round a *replicating cycle* through a specifically contrived sequence of external conditions. This is in line with the usual strategy of organic synthesis and we suggest that for early designs of replicating ladder polymers the aim should be such a *stage-wise* replication. This is unlike DNA which can replicate (with an enzyme) under a single set of conditions – i.e. *continuously* – and we should not expect polymers designed to replicate without an enzyme and through a stage-wise cycle to be very like DNA. Let us consider some design problems associated with the four main processes.

CHAIN SEPARATION AND MONOMER LOCATION

Unlike most ladder polymers, a replicable ladder would have to be separable into single strands. We are not, however, restricted to DNA-like hydrogen bond pairing for making the rungs of the ladder. Indeed *covalent bonds*, such as ester linkages, might provide simple and unambiguous locating mechanisms (i.e. mechanisms that both recognise and hold monomer units). The ladder would then be a fully covalent structure, e.g.

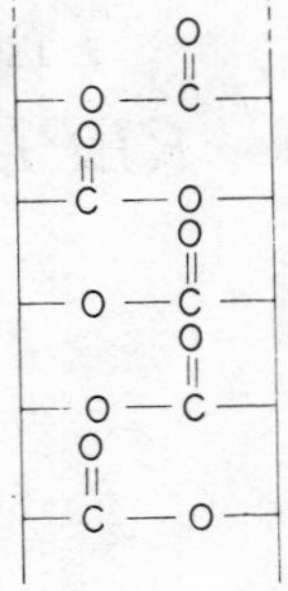

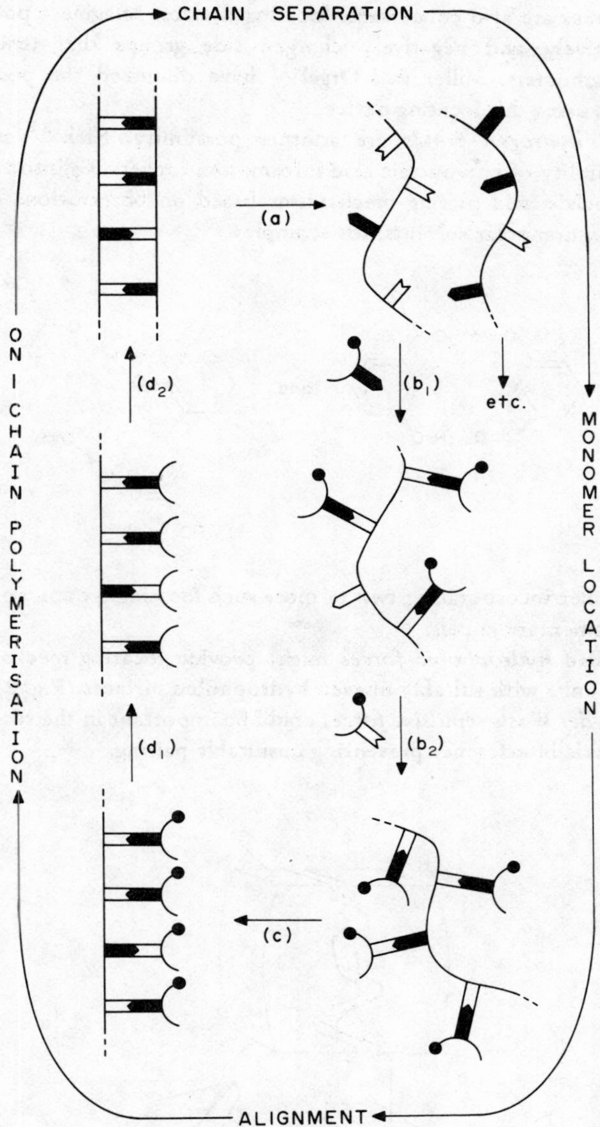

Fig. 1. Example of a stage-wise replicating cycle for a ladder copolymer. The rungs of the ladder are broken (a). Monomers (blocked to prevent polymerisation) are located successively (b_1 and b_2) and aligned (c) before deblocking (d_1) and main bond formation (d_2) gives two new ladders (one for each strand produced in (a)).

Here the rungs would be hydrolysed under conditions (a) in Fig. 1, the hydroxyl- and carboxyl-containing monomers then being successively located under a more or less complex set of conditions (b).

A major advantage of using covalent locating bonds for a replicating cycle can be seen from Fig. 1: the locating bonds must not be significantly reversible under conditions (c) and (d) or broken daughter chains would be produced.

Electrostatic forces are also conceivable locating devices. Imagine a polymer with a specific sequence of positively and negatively charged side groups that tend to associate with complementary monomers. Miller and Orgel[2] have discussed the possibility of primitive replicating proteins using this locating device.

Unconventional *hydrogen bonds* are another possibility. Rich[3] and Eglinton[4] have discussed the possibility of non-nucleic acid information carriers. Eglinton considered a number of possible non-nucleic acid pairing mechanisms based on observations of specific hydrogen bond associations in non-polar solvents, for example:

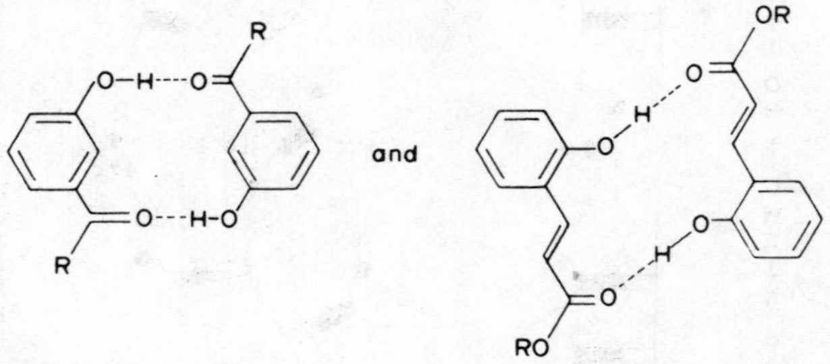

A replicating polymer incorporating two or more such locating groups would produce identical rather than complementary copies.

Even the so-called *hydrophobic forces* might provide locating mechanisms for a polymer consisting of large units with suitably uneven hydrophobic surfaces (Fig. 2).

In addition *van der Waals repulsive forces* could be important in the recognition element of a location device, steric interference preventing unsuitable pairing.

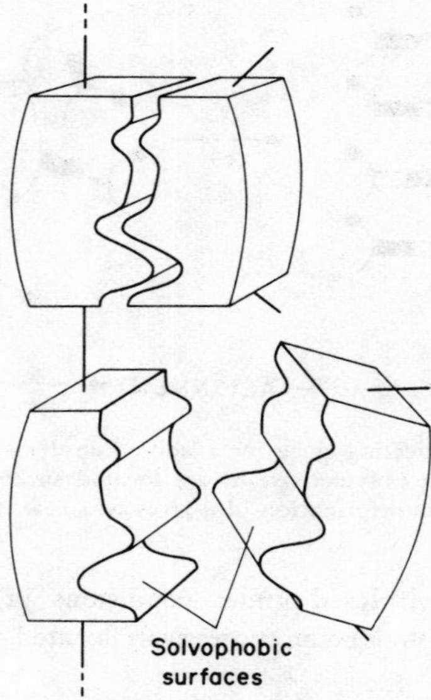

Fig. 2. Formalised view of a locating mechanism that depends on two kinds of units each of which has one solvophobic surface which makes a complementary fit with the other.

ALIGNMENT

Monomer units might be properly located on a parent polymer chain, in the sense that the monomer sequence was correctly specified, and yet fail to polymerise in that sequence by not being properly aligned. Fig. 3 illustrates some possible consequences of faulty alignment. These

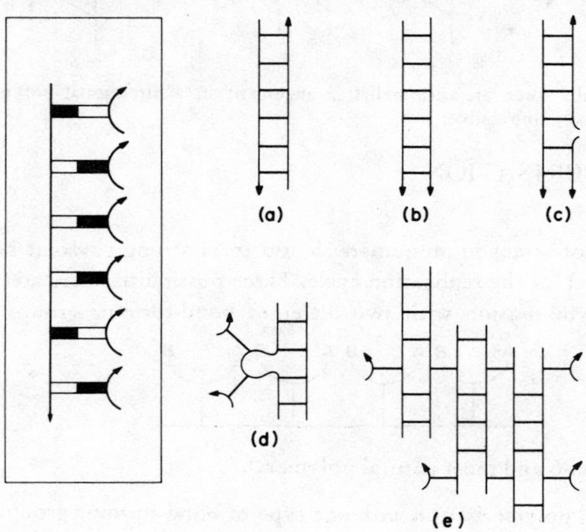

Fig. 3. Polymer with monomer units attached (boxed). If the alignment is maintained as shown this should give (a) after polymerisation of the monomer units but, if the rungs of the ladder can rotate on their own axes, (b) and (c) are possible outcomes. If the main chain is flexible, errors due to exclusion of one or more located monomers from the newly forming chain are possible (d). Crosslinking between chains would be a still more serious error that might arise through polymer flexibility or through rotation of some of the rungs on polymer axes during polymerisation (e).

problems arise because in general linear polymers are flexible (and those that are not tend to be very insoluble) and because pairing bonds may be more or less rotatable. Three possible approaches might be:

(i) The use of rigid polymers that contain bulky, possibly charged, side groups to improve solubility.

(ii) The use of flexible polymers that can be held on some kind of support during polymerisation – for example, on a grooved crystal surface.

(iii) By designing polymers that become rigid, before the polymerisation of attached monomers, through secondary associations between the attached monomers. For example, the located units might tend to stack, as in the nucleic acids, or there might be inter-unit hydrogen bonds as in the Hermans conformation for cellulose.[5] We will consider later a specific possibility (Fig. 10).

Even if the monomer units are held so that subsequent polymerisation will join them together in the correct sequence there may still remain the problem of controlling the direction of that polymerisation (Fig. 3 (a, b and c)). In principle this problem can be solved by stereochemical restraints: for example, the normal conformation for the ladder might be twisted, as in DNA, or slanted as in Fig. 4.

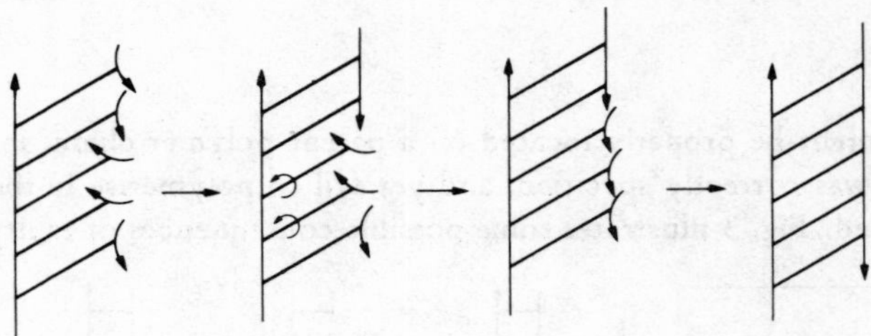

Fig. 4. Slanted rungs could force an antiparallel arrangement in a directional polymer through parallel polymerisation being sterically impossible.

ON-CHAIN POLYMERISATION

The polymerisation of attached monomers should form strong covalent bonds stable under all the conditions needed for the replication cycle. Three possibilities here are:

(i) Directional polymerisation with two different bond-forming groups:

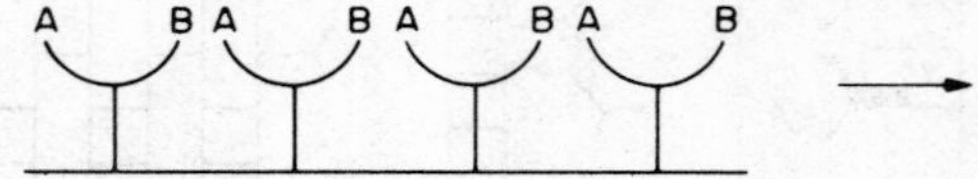

(compare nylon-6 and most natural polymers).

(ii) Non-directional polymerisation with one type of bond-forming group:

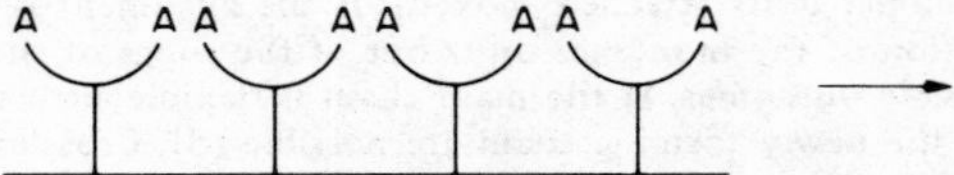

(compare polydisulphides or polydiacetylenes).

(iii) Non-directional polymerisation with two bond-forming groups:

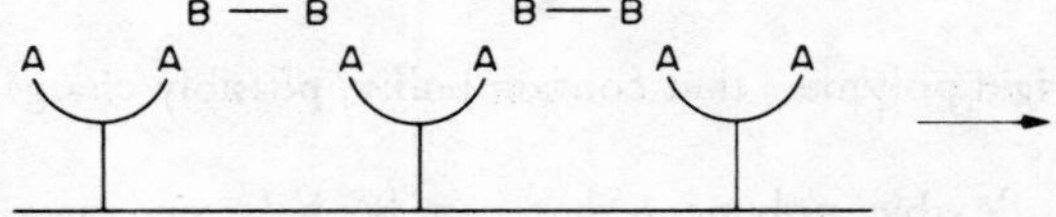

(compare most synthetic condensation polymers, e.g. nylon-66, terylene, etc.). Here the more reactive group might be B: it could be kept out of the system until the polymerisation when a "clip molecule" B-B would be added.

TWO TENTATIVE EXAMPLES

One starting point for thinking about a specific type of replicating polymer is shown in Fig. 5. This is a secondary polyamide ladder with Schiff base (anil) rungs. (Anils are relatively stable, but like Schiff bases generally they can be easily formed and easily hydrolysed.) A CPK

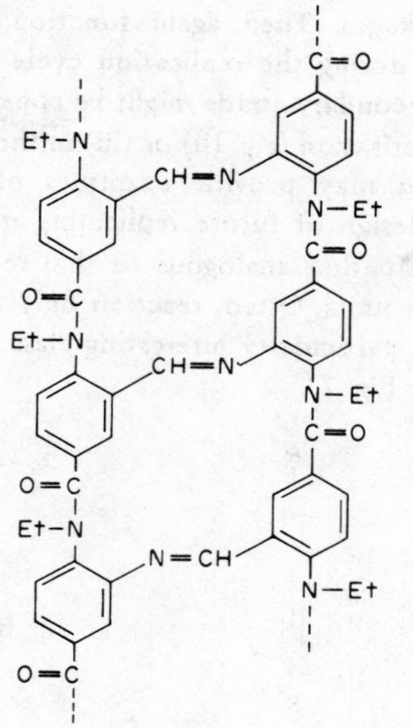

Fig. 5. A hypothetical replicable ladder polymer with Schiff base rungs.

model[6] of this structure can be made in which the main chains are in the same plane. Here there is an angle between the aromatic groups in each of the anil rungs of about 20-30° (α in Fig. 6). This is well below the range (about 50-60°) found in crystalline anils.[7] By twisting the

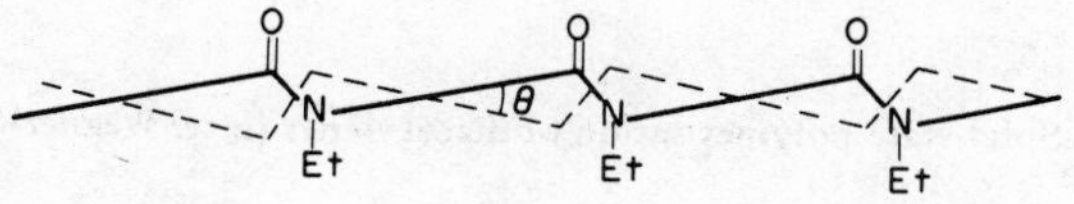

Fig. 6. Simplified side view of a CPK model of the structure shown in Fig. 5. The long sloping lines represent the planes of the aromatic rings, θ is about 20°.

ladder out of the plane into a long pitched double helix, however, the anil angles are easily altered to come within the 50-60° range. The model illustrates that slanted rungs greatly favour an anti-parallel arrangement (compare Fig. 4).

Figure 5 is the simplest of a whole series of similar structures – it is very much a "mark I" design capable of considerable modification in the light of experience. For example, the units:

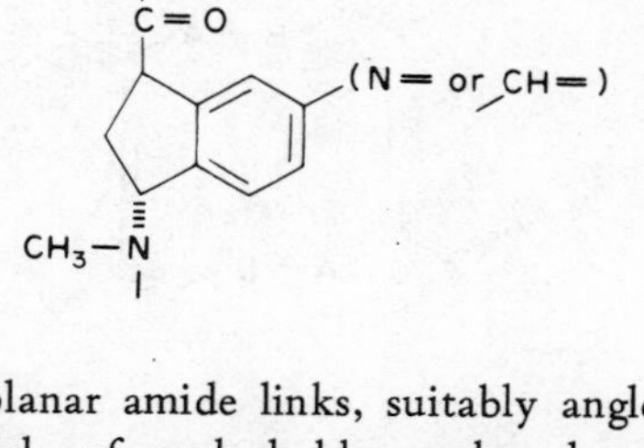

also make a good ladder – with planar amide links, suitably angled rungs and no very bad non-bonding interactions. In models of such ladders, phenol ester rungs can usually be substituted for the anil linkages. Then again functional groups could be blocked in the monomers to be de-blocked during the replication cycle (compare Fig. 1). Still further from "mark I", bonds other than secondary amide might be considered for the main chains and other strategies for on-chain polymerisation (e.g. (ii) or (iii) in the last section).

Solid-state polymerisations may provide examples of aligning and main chain-forming mechanisms for use in the design of future replicating molecules: these reactions take place within the somewhat rigid situation analogous to that required for the controlled zipping of located and aligned monomer units. Often, reaction only takes place to a significant extent in the solid. Diacetylenes are a particularly interesting class that polymerise when appropriately arranged in the crystal[8] (see Fig. 7).

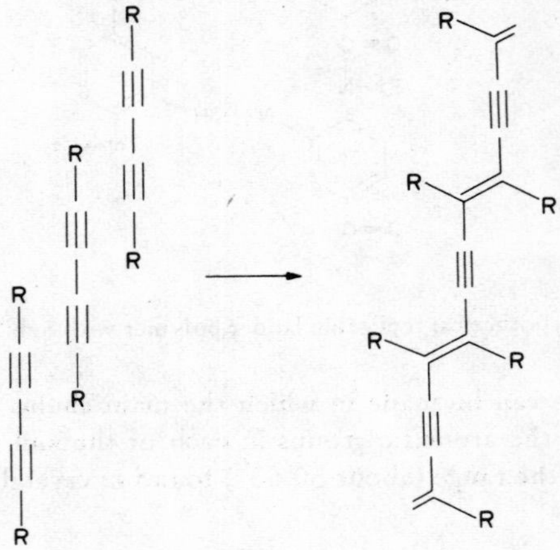

Fig. 7. Solid-state polymerisation of diacetylenes (after Wegner[8]).

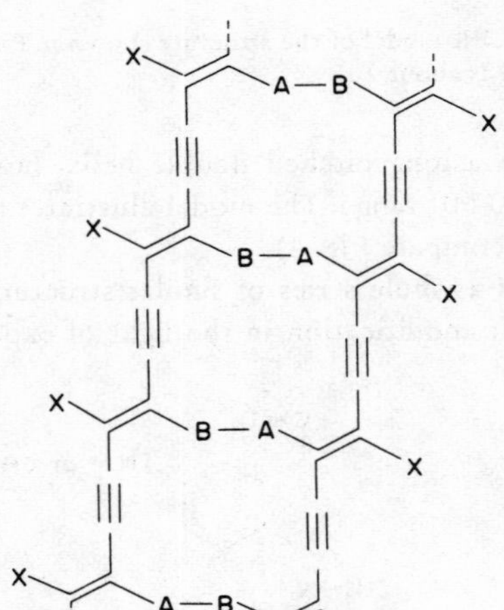

Fig. 8. A hypothetical replicable molecule based on diacetylene polymerisation.

Figure 8 illustrates a hypothetical replicating molecule based on a diacetylene polymerisation. Here two monomers would be required:

$$A - \equiv - \equiv - X \qquad \text{and} \qquad B - \equiv - \equiv - X$$

with A and B as pairing groups and X as an aligning and/or solubilising group. The studies of Wegner and his school on the effects of side chains in diacetylenes on their ability to polymerise in the solid may provide insights into suitable X groups. "It seems that only those diacetylenes are reactive having substituents which favour an arrangement of the monomer molecules within the lattice by which the individual molecules are arranged in a ladder-like structure with the conjugated triple bonds forming the rungs and the substituents forming the side pieces of a ladder (e.g. by involvement of hydrogen bonds)".[8] Such a ladder (which should not be confused with a replicating ladder) is shown in Fig. 9.

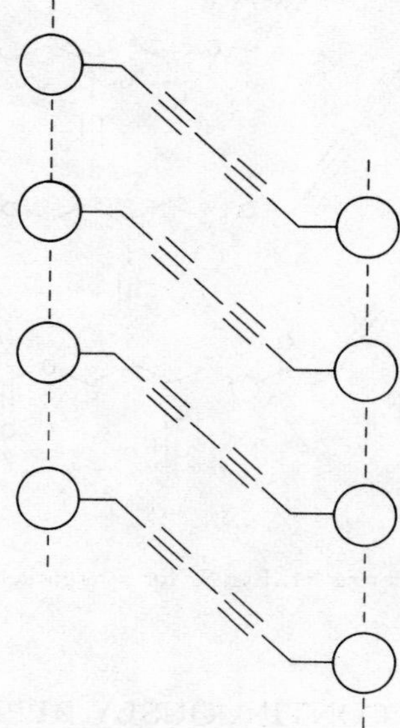

Fig. 9. Ladder-like arrangement of diacetylene monomers proposed by Wegner[8] to be an essential prerequisite for their polymerisation in the solid state. Dashed lines represent secondary aligning forces between side groups.

We have made a CPK model of the structure shown in Fig. 8 in which A-B is an ester linkage:

$$-CH_2-\underset{\underset{\displaystyle O}{\|}}{C}-O-CH_2-$$

and X a urethane that was found by Wegner to be particularly effective as an aligning group:

$$-O-\underset{\underset{\displaystyle O}{\|}}{C}-\overset{H}{N}-C_6H_5$$

We have also made a model of the corresponding single-strand polymer with monomers attached. The attached monomers can readily be aligned in a "Wegner ladder" as shown in Fig. 10 (compare Fig. 9).

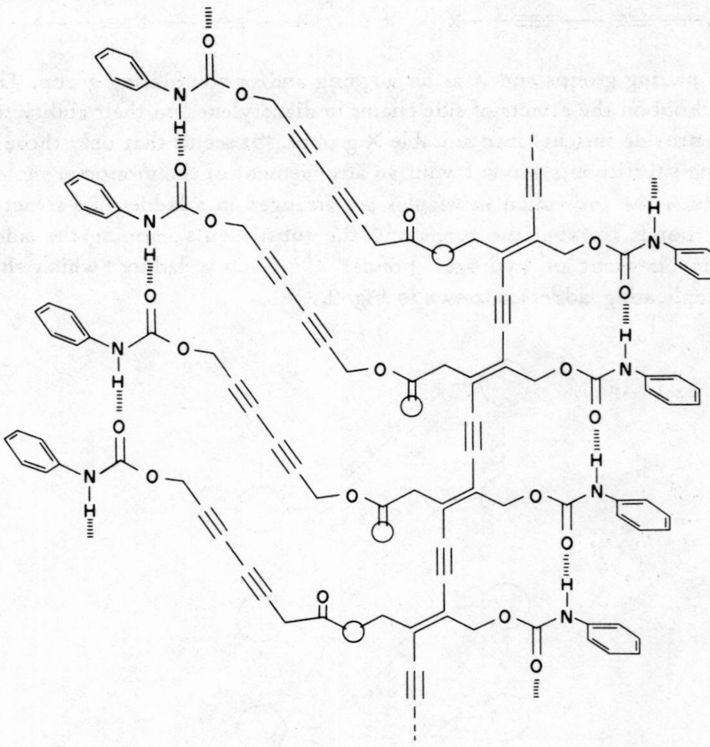

Fig. 10. An example of an alignment device (cf. Fig. 9) for a hypothetical replicating polymer of the type shown in Fig. 8.

TOWARDS THE DESIGN OF CONTINUOUSLY REPLICATING MOLECULES

A replicating cycle is essentially a stage-wise (i.e. non-continuous) process. The number of stages required might indeed be less than the six steps shown in Fig. 1. If, as with diacetylenes, polymerisation required accurate alignment then unblocked monomers might be used. Also alignment might well take place without a change of conditions. And it might be possible to add the monomers together – in large excess perhaps, or under conditions in which the rung bonds were slightly reversible and where alignment forces favoured located monomers as opposed to monomers paired with each other in solution. But it would be very difficult indeed to find conditions under which, in addition, completed ladders completely dissociated. At least one alternation of conditions, corresponding to (a) and (b+c+d) in Fig. 1, would seem to be called for to overcome the essential difficulty that replication of a ladder demands both the breaking of rungs and the making of rungs. For continuous replication a somewhat different strategy would seem to be needed.

A CONTINUOUS PROGRESSIVE MECHANISM

Watson and Crick[9] originally suggested that DNA might replicate through a cycle (chain separation, "crystallisation" of monomers onto the separate chains, and their polymerisation), but we know now that DNA replicates continuously through "replicating forks" moving down the molecules and leading to Y-shaped or more complex forms (Fig. 11, f→h).

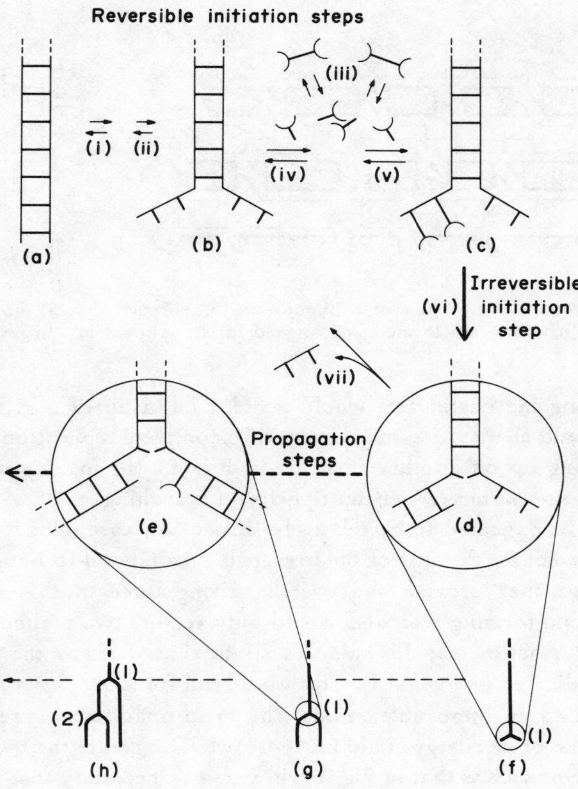

Fig. 11. A proposed scheme for a continuously replicating molecule. The conditions should be such that the rung bonds are sufficiently reversible for ladders occasionally to "fray" at one end ((i) and (ii)), and for there to be a reasonable concentration of unpaired monomers (due to (iii)). Occasionally, through (iv) and (v), a critical structure such as c is formed in which at least two monomers are aligned. This is followed by a rapid, energetically downhill step (vi) which provides the overall energy for the reaction. Ideally part of the energy of (vi) raises the ground state for the next rung-breaking reaction (i.e. d is in some way strained until the adjacent rungs are broken). Propagation processes follow through structures like e (cf. f→g→h). Propagation should be rapid to minimise reactions like (vii) and excessive branching (cf. h), since these could lead to incomplete replication.

A progressive mechanism might be suitable also for non-enzymic replication of a synthetic ladder polymer – for example, the scheme shown in Fig. 11 (a-e). Here the locating bonds are reversible so that rung making and breaking can proceed under the same conditions – while problems of incomplete replication that would then arise for a cycle are largely avoided. A nice balance of reaction rates would be required, however. The overall rate of initiation (up to structure d in Fig. 11) should be much slower than the subsequent propagation of the

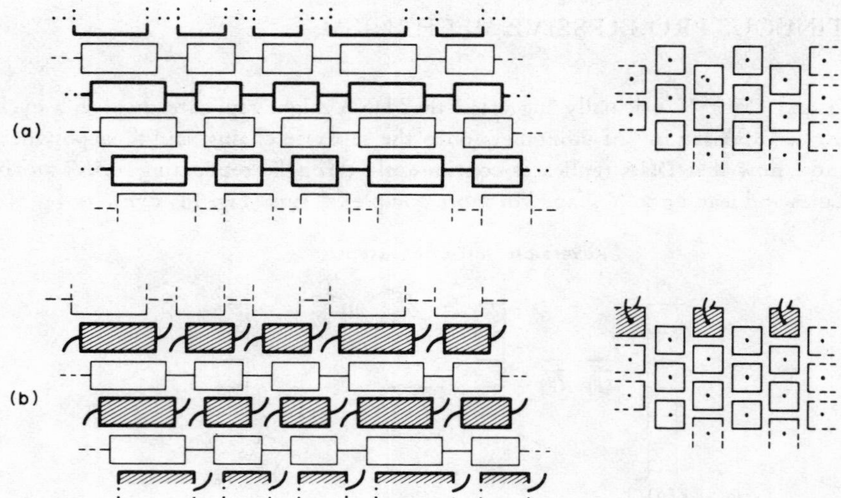

Fig. 12. (a) Top and end views of part of a hypothetical copolymer crystal showing surface grooves. (b) Unpolymerised units (hatched) are located and aligned in the grooves in a manner similar to that in the polymer.

replicating fork along the chain: this would avoid a build up of d and so minimise the side reaction (vii). It would also reduce the chances of incomplete replication arising from one fork in a molecule making up on another (e.g. (2) in Fig. 11 (h) making up on (1)) – because relatively slow initiation would reduce multiplicity of branching.

Initiation might be expected to be relatively slow in any case since it involves a sequence of four reversible steps before the first of the irreversible main bond-forming reactions. (It is these irreversible reactions that provide the overall driving force in this scheme.) Each of the subsequent main bond-forming reactions would only require two preliminary steps. Ideally the main bond-forming reaction should induce a strain that weakens the adjacent rungs in the parent ladder (as indicated in Fig. 11d). This would call for some clever molecular engineering, but it is by no means an impossible device: the main bond-forming reaction is energetically downhill so that some of its energy could be "borrowed" to create the required strain.

Finally, for a scheme such as that in Fig. 11, it would be necessary that polymerisation of free monomers was slow. The solid-state reactions that we have referred to indicate that such a state of affairs is quite possible. This requirement would add, however, to an already daunting list of specifications. It is no wonder that DNA uses an enzyme: perhaps such a continuous mechanism is only possible either with an enzyme or with very large and complicated monomer units.

CRYSTAL REPLICATION OF ORGANIC POLYMERS

One of the attractions of the idea that the most primitive genetic materials might have been crystals (see "Synthetic Life", p. 408) is that locating and aligning processes are inherent in the three-dimensional growth of crystals. In connection with the origin of life, inorganic crystals seem the most promising, but a similar idea might be applicable to the design of organic polymers that could replicate without ladder structures being involved.

Figure 12(a) illustrates part of the surface of a hypothetical crystallite formed from a collection of co-polymer molecules with an identical sequence of long and short units. Surface grooves, arising from the packing of these chains, locate and align the units (Fig. 12(b)) which then polymerise *in situ* to give a new surface with another set of grooves – and so on. The polymer crystal thus grows from units in the surrounding solution with copying of the original sequence. The replication cycle is eventually completed by mechanical break up of the crystal so formed into smaller crystallites. Thus, in the cycle, there are neither single strands nor discrete ladders.

Large information units (corresponding to the blocks in Fig. 12) would be more likely to adsorb strongly in the grooves. These units might best be themselves oligomers of different types that have a similar cross-sectional area. Such oligomers could be expected to bind well along most of their lengths on the corresponding parts of the crystal grooves. One would expect some misfit at the ends of the oligomers, however, since it would be here that they would differ structurally from the polymer out of which the crystals were made. Such misfitting would be relieved as a result of the polymerisation reaction: thus, rather as the relief of local misfitting is thought to be an important part of the mechanism of lysozyme action[(10)], polymerisation of the adsorbed monomers might be specifically catalysed by the grooves.

CONCLUSION

Synthetic replicating molecules are an important and feasible goal for organic polymer chemistry. Of three general modes of replication considered here: (i) stage-wise ladder replication, (ii) continuous ladder replication and (iii) crystal replication; the first can be most easily planned in terms of the known techniques of organic chemistry; the second, although most like DNA replication, is the least promising for early designs; the third might provide the simplest practical system using relatively unsophisticated molecules, but it depends on crystal-packing considerations which are difficult to predict. Polymer crystallographers are perhaps the best placed to design or come across suitable systems here.

REFERENCES

1. The power of this technique is discussed in *The Life Puzzle* by A. G. Cairns-Smith (Oliver & Boyd, Edinburgh, 1971), pp. 89-105. Compare also the technique discussed for evolving catalysts in "Synthetic Life" (this volume, p. 409) and that for evolving RNA molecules (D. R. Mills, F. R. Kramer and Spiegelman, "Complete nucleotide sequence of a replicating RNA molecule", *Science,* **180**, 916 (1973), and references therein).
2. S. L. Miller and L. E. Orgel, 1974, *The Origins of Life on the Earth*, Prentice Hall Inc., New Jersey, p. 155.
3. A. Rich, *Origins of Life*, edited by L. Margulis, pp. 274-81, Gordon & Breach, New York, 1970.
4. G. Eglinton, *ibid.*, pp. 281-92.
5. D. A. Rees, "Structure, conformation and mechanism in the formation of polysaccharide gels and networks". *Adv. Carbohydrate Chem. Biochem.* **24**, 267 (1969).
6. Corey-Pauling-Koltum space-filling models: designed for the study of non-bonding interactions, side chain packing, etc., in biological molecules.
7. H. B. Buergi and J. D. Dunitz, *Helv. Chim. Acta,* **53**, 1747-64 (1970).

8. G. Wegner, "Topochemical polymerisation of monomers with conjugated triple bonds". *Macromol. Chem.* **154**, 35-48 (1972).
9. J. D. Watson and F. H. C. Crick, *Nature*, **171**, 964 (1953).
10. D. C. Phillips, "The three-dimensional structure of an enzyme molecule". *Scient. Am.*, **215**, No.5, 78 (1966).

A. G. Cairns-Smith

Lecturer in the Department of Chemistry at the University of Glasgow.

Working on an alternative to "chemical evolution" as a key to the explanation of the origin of life on earth.

SYNTHETIC LIFE FOR INDUSTRY

I think that it should be technically fairly easy to set up conditions under which a set of microscopic objects would form and start to evolve towards objects that many, perhaps most people, would say were alive. As I see it the problem of making life in this sense is at present not so much practical as theoretical: I do not think that the conditions required will be particularly complicated or difficult to contrive when we know what these conditions should be.

In rather general terms we know what the initial objects should be like. They should be able to reproduce. More particularly they should be "hereditary machines": they should consist of or contain structures, analogous to the genes of modern organisms, that can hold and replicate some kind of pattern which can affect the reproductive efficiency of the objects. The genetic patterns must be stable and accurately printable so that effective ones can be retained through many generations; but there must also be occasional minor errors so that improved patterns can arise by chance. Such chance events will be comparatively rare, but when they do happen the structures to which they give rise will then tend to become common types increasing the chances that further modifications will yield still further improved types – and so on. In principle such an evolution could continue for as long as there were ways in which increasingly particular genetic patterns could, by some means or other, increase reproduction rates.

This is easily said, but we seem to be talking about some set of microscopic machines that would be far from easy to make. Yet if you believe that life arose spontaneously on the earth, and not as some freak event but as a matter of course, then I think that you must also believe that there is at least one class of indefinitely evolvable objects members of which should not be very difficult to make – that class from which life on earth first arose.

Biological nature can produce wonderfully ingenious structures that are far beyond our present manufacturing powers. But invariably the intricate engineering feats of biology have been the result of prolonged evolutionary processes operating on fully reproducing systems. Physical nature is never very ingenious, and it was by definition physical nature that made our first ancestors. This is the basis for believing that a simple kind of synthetic life is a feasible research project: because we can already put together molecular arrangements that are far more specific than anything that physical nature could have contrived. For example, we can doggedly string together amino acids in almost any sequence that we choose up to lengths of 100 or so. When we do this kind of thing, when we perform any very long organic synthesis, we are making a structure of gigantic improbability, a structure that unguided nature would not have hit on in a billion years even with the known universe to experiment in.[1] If nature unguided by natural selection did hit on our most ultimate ancestors within such a time (and on present evidence life appeared on earth at least within a billion years of the conditions being right – and possibly almost at once) then we should be able to mimic nature's performance within quite a short time – say about a fortnight. That we have not succeeded in what should be technically a relatively simple matter is an indication that we have not been looking in the right place.[2] I suspect that we have been concentrating too much attention on the chemistry of the nucleic acids and proteins – and that this is not the right place.

Natural selection, although not quite blind – it knows a good thing when it sees it – is nevertheless totally unimaginative – it never looks ahead. We are so used to the idea that biological structures are appropriate that it is easy to come to think that evolution can somehow foresee what is going to be appropriate in the future. Nucleic acids and proteins are clearly such splendid molecules that it may seem that it would only have been sensible for life to have brought them in as soon as possible. But would they have been brought in at the start? Would they, then and there, have been the most appropriate structures? Or are nucleic acids and proteins essentially sophisticated components of an essentially sophisticated system?

That nucleic acids are now used by all organisms on earth for holding and printing genetic messages is no argument that this is the only possible genetic material, nor does it argue that nucleic acids or anything like them must have been used by our most remote ancestors. Virtually the whole of our central biochemical machinery is common to all life on earth now, but it is hardly plausible to suggest that this whole machine was there from the start – that it arose as some initial physico-chemical accident. As far as one can see only natural selection could have refined the multitude of ingenuities inherent in it. Natural selection no longer operates significantly on the common central biochemical machinery – that is why it has remained common. But there must have been an earlier phase, before some last common ancestor, when this machinery was evolving. Of this phase we have no direct knowledge, but it is not safe to suppose that earlier versions of the now common machine were merely simplifications of it. In particular there is no safety in the assumption that a process which ended with nucleic acids and proteins started with either.

Consider a much later part of the evolutionary story. It might be argued that our ape-like ancestors were unsophisticated versions of ourselves. But what of the reptiles, amphibia and fish in our ancestral line, or for that matter our still more remote single-celled forefathers: are these to be seen as simplified human beings? As we think backwards such an anthropromorphic view becomes increasingly absurd. So, I think, if we could trace backwards the still earlier, now long finished, phase of central biochemical evolution our biochemical prejudices would be seen to be increasingly misleading. Evolution has been by no means simply an increase in complexity and sophistication of some set of mechanisms that were there from the start. Typically, in that part of the evolutionary story for which we have evidence, quite new mechanisms emerged at levels at which they became possible and useful, and these often displaced earlier structures. There is no reason to suppose that the now invisible biochemical phase of evolution was any different in this respect.

In looking for an evolutionary process that could be initiated in the laboratory from simple starting conditions it is, I think, earlier genetic structures that we should in particular try to imagine – structures which like nucleic acids can hold and print stable patterns, but which unlike nucleic acids do not require specially contrived conditions for their operation. We seek truly primitive genes of the kind that could have initiated the evolution of life on earth.

There is no need to suppose that the primitive genetic material was chemically at all similar to ours, nor that it evolved into ours through a succession of intermediate types. It may be true that evolution generally takes place through a succession of small design modifications, but it is not always true that a structure carrying out a particular function now originated through the transformation of some earlier structure performing the same function. The lungs did not evolve through a gradual transformation of gills. Lungs evolved from accessory respiratory bladders in certain fish which, for subsequently evolving land animals, was to take over the whole function of breathing. This final takeover followed a long period during which lungs and gills were both present.

We can imagine a similar takeover mechanism operating in the evolution of the genetic machinery of organisms. It might have happened like this.[3] A first genetic material forms the basis for the evolution of some alien (i.e. non-nucleic acid directed) life system. This creates, eventually, supplies of molecules not previously cleanly available. New kinds of accessory genetic materials become possible within the now quite highly evolved primitive organisms. One of these accessory genetic materials in one species somewhere gradually evolves the ability to direct the synthesis of protein. When the protein synthesising machinery is sufficiently accurate, enzymes are discovered through which the new gene-catalyst system gradually takes over the control of the synthesis of all the components necessary for its independent operation. The process is gradual, but there is no direct continuity between the old and new genetic materials: rather there is a takeover by the new material whose qualifications did not include the need to be at all chemically similar to the material that it was displacing.

It might be argued that even if nucleic acids and proteins were comparatively late inventions some of the small molecular components of present-day life, such as amino acids, sugars, etc., must have been incorporated in the first organisms since such molecules have been shown to form under conditions simulating the primitive earth. It is true that the small molecular components of modern organisms are often remarkably easy to make. They turn up under a variety of uncontrolled conditions – not all of them relevant to the primitive earth. But several other molecules are even easier to make. It is difficult to say which easily made molecules might have been favoured by the first evolving systems. It seems hardly likely that the exact set of all the present-day amino acids, sugars, etc., were required. At least some were surely selected during early evolution whatever view you take. How do we know that any were truly primitive? If you abandon the idea that nucleic acids and/or proteins were necessary initial components there remains no particular reason. Whatever happened during early evolution we should not be surprised at the outcome: a machine using a limited set of easily made parts each with several, often diverse, functions. This is good engineering. But how and when was this set decided on?

I am inclined to think that the most promising initial hereditary and control machinery was not organic at all; that a microcrystalline mineral of some sort would best fit the specifications for a primitive genetic material.[(4)] (I will return to this later.) Let us summarise these specifications. Any genetic material has to be able to *hold* information – that is some kind of particular pattern out of a large number of possible patterns, something like a sequence or a two-dimensional array. Then the information must be *replicable*. Finally the information must *control* the immediate environment, and for a primitive genetic material this control must be very direct.

Considering first the control function, we should not insist that this should be initially catalytic. What matters is that the genetic patterns should influence the environment somehow so as to increase the chances that they, the patterns, are replicated. This might be by helping the structures that hold them to stay in the place where the surrounding solutions contain the units required for replicative synthesis – perhaps by helping the structures to stick to the sides of a stream. Eventually genetic patterns would have to appear that did have some direct catalytic effects, but these could be at first quite unspecific, converting some more or less complex mixture of organic molecules in the environment to some other more or less complex mixture that is better at promoting gene propagation – perhaps it buffers the acidity more appropriately, or binds interfering cations, or is an anchor or ultraviolet screen. There are many useful things that even only vaguely controlled mixtures of organic compounds can do. (After all, most organic materials that we use – plastics, paints, fuels, foods, etc. – are mixtures.)

Now the trouble with trying to make a primitive genetic material out of an organic polymer is that that polymer must be made from rather specific and complicated molecular units. Particular complex molecules are hard to make and have a way of indulging in very complicated reactions unless tightly controlled. Indeed organic reactions without either organic chemists or enzymes tend to yield either rather simple substances like carbon dioxide or graphite, or exceedingly complicated mixtures like raw petroleum. On the other hand, inorganic crystallisation processes are more limited. These can take place from simple units, such as dissolved silica and alumina, available through rock weathering in natural environments. A mineral primitive genetic material would have the great advantage of not generally reacting chemically with organic molecules while being able to control such molecules through adsorptive and catalytic effects. This leads to a view of primitive organisms as having been made from two sets of units. First the (inorganic) genetic units performing the limited but precise function of pattern replication through some kind of crystal-growth process, and second a (largely organic) collection of units making up a surrounding "jam" whose semi-chaotic composition is under loose control by the inorganic genetic particles embedded in it. The genes make the jam which helps the genes. (By being a buffer, anchor, shield or whatever.)

What kind of continuously crystallising mineral might fit the specifications for a primitive genetic material? Clays are a possibility. They can form continuously in an environment such as that provided by water percolating through a porous bed. They can, in principle, hold a great deal of information in the form of patterns of substitutions of aluminium atoms for silicon atoms in their structures. Also they adsorb organic molecules profusely and can catalyse many organic reactions. Furthermore, the aluminium substitutions, because each creates a local negative charge, can affect interactions with organic molecules as well as the way in which the microcrystalline clay particles interact with each other. The substitutions are located in the very thin (about one nanometer) platelets, stacks of which, together with sandwiched cations, make up the individual crystallites. The stacks can readily come apart exposing the substitution patterns. The platelets themselves are quite flexible so that one can imagine structures such as tubes and grooved surfaces being specified through particular modes of self-folding, crumpling, or interlocking caused by particular charge patterns.

There remains a crucial question. Do charge patterns in clays ever replicate? Can a clay platelet holding one particular "picture" be made to print that "picture" into a new clay platelet forming on top of it? Something of the sort does indeed seem to happen under suitable conditions as Armin Weiss has recently shown.[5] Under suitable conditions of crystal growth, characteristic charge densities are inherited by newly forming material as are catalytic and adsorptive properties that depend on charge arrangements.

Perhaps, then, the indefinitely evolvable microscopic objects that I referred to at the beginning have already been found. There remains the important question of whether there is here a sufficiently accurate pattern-replicating mechanism operating; but at the very least Weiss's experiments indicate that inorganic crystal genes are a reasonable experimental goal. If really effective ones can be found that can print detailed information accurately it should not, I think, take very long to bring about the evolution of highly physicochemically improbable structures – by repeatedly breeding from crystallites with some chosen property that can depend in a detailed way on charge arrangements. The ability to catalyse a particular reaction could be such a property. At first one would not be looking for very specific catalysis. It would be a long time before specificity comparable to that of modern enzymes could be expected, but as remarked earlier many valuable industrial products are complex mixtures of organic molecules and replicable catalysts that had been bred suitably to alter the compositions of such mixtures could be of value too. Consider the formose reaction. This is the conversion of formaldehyde into complex mixtures of (mainly) sugars. A number of catalysts, including some clays, can bring this reaction about. Unfortunately the product (formose) is inedible for higher animals and even micro-organisms seem to find it rather indigestible. There are just too many kinds of molecules there, most of which – the L-sugars among them – are not components of our biochemical system. But suppose that you could breed catalysts for the formose reaction from single seed particles. You would expect to produce much more specific mixtures resulting from the particular catalytic effects of particular grooves and niches that happened to be characteristic of the ancestral seed. One might expect batches derived from different initial seeds to have different activities with respect to D- and L-sugars, an effect that could be improved, probably, by further repeated selection. Then perhaps an edible formose might come onto the market.

A primitive genetic material, then, a material through which catalysts can be bred towards desired activities and specificities is, I suggest, a modern philosopher's stone that should be attainable – and stone is quite possibly the right kind of material. Key fields are the processes of inorganic crystal growth and the origin and early evolution of life. Indeed the origin of life, which has only comparatively recently emerged as a respectable scientific quest, may lead us to one of the most useful imaginable inventions of science. It may lead us to an alien synthetic life for industry.

NOTES

1. A given 100-long amino acid sequence, for example, is one out of 20^{100} possibilities which is far more than the number of electrons in our galaxy (less than 10^{70}).
2. Spiegelman has found conditions under which RNA molecules evolve, but these conditions are not simple, they include among other things the presence of an enzyme.
3. I have discussed this recently in greater detail in *Proc. Roy. Soc. (Lond.)* B, **189**, 249-74 (1975).
4. *J. Theor. Biol.* **10**, 53-88 (1966).
5. Paper presented at the 4th International Conference on the Origin of Life at Barcelona (1973).

Martin Holdgate

J. W. L. Beament, F.R.S.

Martin Holdgate

Director of the Natural Environment Research Council Institute of Terrestrial Ecology at Cambridge.

Former Director of the Central Unit on Environmental Pollution in the U.K. Department of the Environment and Deputy Director of the Nature Conservancy.

J. W. L. Beament, F.R.S.

Professor of Agriculture and Head of the Department of Applied Biology in the University of Cambridge. Fellow of Queens' College, Cambridge.

Formerly Reader in Insect Physiology in the University of Cambridge.

Member of the Natural Environment Research Council. Awarded the Scientific Medal of the Zoological Society in 1962.

THE ECOLOGICAL DILEMMA

Ecology is the study of the relationships between living things and their environment. The frontier between the two components in the relationship is not a fixed one. For a single organism, whether a man, a bacterium or an oak tree, the "environment" is the whole outer world – other people, other species, or the physical components of atmosphere, soil or ocean. But it is also possible to speak of the ecology of a species, or of a population of that species, or of a community of different kinds of plants and animals living together in a defined area and collectively interacting with the physical environment and with external biological influences. Ecology encompasses all scientific approaches to the organism:habitat interactive system, and is concerned with many kinds of interrelationship, at many levels.

Experimental biology in the laboratories of the middle decades of the present century was preoccupied with classical experimentation in which the system under study was reduced to the simplest attainable level. The aim was to examine one variable at a time, holding all the rest constant, and to compare the responses of the plant or animal or physiological system under examination with an identical "control" given precisely the same treatment except that there was no experimental modification of the variable under test. Ecology as a science grew up during this period of classical experimentation and was influenced by it. It is not surprising that there have been attempts at defining as precisely as possible the "organisms" and "habitat" in the ecological interaction.

The individual organisms with which all biologists, including ecologists, deal, are grouped by taxonomists into categories we call species, and ecologists commonly use species as one element in the interactive system they study. We call such an analysis of the responses of single species to environment "autoecology". But the species concept is no more than a convenient abstraction, representing a general statement of the average characteristics, and range of variation at a particular time in a series of continually developing genetic lines. These genetic groups are continually changing, shaped by processes of natural selection which depend on just those interactions between species and environment that are the business of the ecologist. There is obviously something of a paradox here. The ecologist studies the interactions that are the very cause of evolution: in doing so he describes his system in terms of units which an understanding of evolution demonstrates are at best transitory. "Species" are statements of the average condition of a genetic line: natural selection works on the within-species variation many ecologists ignore! Ecological science clearly needs to think four-dimensionally about the organisms it is considering and develop a descriptive system which takes fuller account of the contemporary and continuing variation in the genetic lines that are its basic biological groupings.

For any species (or genetic group) it is possible to elaborate a "medicine": that is, to describe its responses at biochemical, physiological, or population level to the variables in the environment. These responses form a series of "levels" rather like those defined by the pioneer physiologist, Hughlings Jackson. A variation in the conditions in the habitat, or in the composition of an animal's food may, if it is slight, cause no more than a shift in the rate or scale of a few biochemical reactions, and the organism as a whole may adjust to this perturbation so that there is no detectable influence on its behaviour or physiology. A more severe disturbance may, however, bring about obvious changes in the behaviour, metabolism, or reproductive success of the whole organism. Even this, however, may not affect the capacity of that species to maintain a viable population in the area subject to the habitat variation – for the inherent variability of plants and animals is likely to mean that the population has some that are more "robust" than others in the face of the variation, and the capacity of all life forms to produce more young than are needed to replace the parents means that the reproductive failure or the death of the most sensitive members of the population can be made good. At biochemical, physiological and population level, organisms have a "homeostatic" capacity – a capacity to respond and counteract external disturbances. The "medical science" it is theoretically possible to compile for each species would need to document this capacity, quantitatively, for all foreseeable perturbations, singly or in combination.

To do this for any organism is obviously almost impossibly complex (as is well exemplified by the limitations to our medical understanding of man!). To do it for the many million species in nature is impossibly daunting: we can predict that it will not be done. Abstraction is necessary: we require some general laws. Physical scientists indeed criticise biologists for this lack of rigorous general descriptions of the behaviour of organisms. But it is easy to see why they have not been forthcoming. The cause is the manifest variability in biology. Within any genetic line at any time there is such substantial variation in so many characters that the simplest generalisations of taxonomy can be questioned as a foundation for exploring the evolutionary dynamics. This contrasts with the physical environment where fundamental particles, stars, or physico-chemical processes have a more restricted variation. It makes the type of rigorous law we would like to have in ecology much more difficult of attainment.

Faced with the impossibility of deriving by classical experiment the responses of all organisms to all variables – a task that would take longer, and cost much more than the writing of the works of Shakespeare by Gullivers' monkeys – ecologists have been concerned to analyse how far we can make valid general statements about the behaviour of a higher order of abstraction – the community. Have such complexes of individuals drawn from many taxa, some kind of identity? The concept of the "ecosystem", comprising assemblages of plants and animals and the environment with which they interact, has wide acceptance as an abstraction and implies that such systems should have collective responses which can be described by general laws. Some ecologists consider that there is a relationship in some systems, where there is a high order of competition and a long history of co-adaptation, between diversity (i.e. number of components) and stability. In other systems, where the habitat is subject to more continuous disturbance, the diversity is less and the biological stability may be quite different. But concepts like diversity and stability may not be wholly appropriate for ecological systems if based upon counts of the taxonomic units present rather than the quantity of living matter and apportionment between genetic types or biochemical systems. Modern ecological studies, for example in the International Biological Programme and its successor enterprises the Man and Biosphere Programme of UNESCO and the programme of SCOPE, the Scientific Committee on Problems of the Environment, have responded to this problem of meaningful abstraction from the intense variability of nature, and the quest for meaningful general statements about the behaviour of these complex systems of interaction between so many variables by new developments in mathematical modelling. Initially, again perhaps because of the traditions of experimental biology, some ecologists sought after deterministic models. These are models in which substitution of measured terms in equations would provide precise predictions of the responses of the systems, as an engineer predicts the response of a structure using known laws of physics and accurate measurements of the strengths of materials and the distribution of stresses. Such models are less easy to apply to biological systems because we do not have the measurements we need about all the components of the system, and how they interact, and we seem unlikely to get them. Consequently modern interest is in models based on probabilities (so-called Markov chains), stating the likelihood of certain responses to certain complexes of factors. Progress is being made, although we are still a long way at least from quantitative laws in ecology. While our abstractions of these highly multivariate systems and our models are alike imperfect, there is now a prospect of advance: much more will be needed if ecology is ever to become a rigorous science.

Even if ecologists number among their ranks altruistic theoreticians, a main reason for supporting the science is its potential practical use. Man himself has an ecology. He depends on the natural world for food, for breathable air, and for the disposal of wastes. He is still unable to control the great systems of nature. Recent research has emphasised how greatly the present atmosphere and the chemical cycles of elements like oxygen, carbon, nitrogen, sulphur and phosphorus are the product of life. It was living organisms that gave us our present atmosphere dominated by nitrogen and oxygen: neither of these would exist naturally in elemental form in a lifeless world. Those forms of life that work without oxygen – so-called anaerobic organisms, mostly bacteria, in waterlogged swamps and oozes – emit methane which in turn puts a brake on the oxygen levels in the air rising to the point where lightning would fire the forests. Ecology is concerned with these great interactions, and with predicting man's impact upon them, as well as with the more circumscribed impacts of agriculture or industry.

Man operates both by modifying the habitat like any other very dominant large mammal, and by altering chemical factors through what we call pollution. We are very ignorant about how best to describe man in ecology. In theory, we should be able to treat him as a component in the ecological system and devise general laws which embrace human behaviour, biomass and activities. Some models already do this – though the probabilities of social response are not easy to assess! But we remain ignorant about many of the precise effects, especially in the long term, of the chemical substances man has emitted into the environment. Obviously, pollutants act like any other chemical variables, progressively perturbing organisms at the biochemical, physiological and population levels. Some individuals, at some stages of the life cycle, are less resilient than others. But rigorous "criteria", defined as the relationship between the exposure of an organism to a pollutant and its response, exist for few species and few pollutants, and we are far from understanding the interactions between the substances we emit and those naturally present in the environment. It is possible that some substances may perturb the biochemical machinery of living things sufficiently not to cause direct harm themselves but to release genetically determined latent defects, and thus produce diseases which are not manifest under normal circumstances. Similarly, the stress of pollution can be expected to alter the balance in ecological systems, but just how such complex entities (if they are entities) respond is far from clear.

Agriculture is a form of applied ecology and indeed one in which some basic principles of ecology have been abstracted by the normal scientific device of restricting and controlling variables. Man has only ever attempted to domesticate a tiny selection of the total range of species available, and has developed the domestication of but a fraction of these. The species chosen are readily digestible, easily cultivatable; they can be harvested and stored traditionally. Our thinking about domestication has been almost entirely conditioned by the primitive agricultural practices of man in Asia Minor and the plant and animal types domesticated there. It has been developed towards adapting these few species to an unreal range of ecological conditions, and to the limits of their efficiency, rather than applying modern principles of productivity, nutrition and mechanisation to a new range of species. Other species, could if developed by new manipulative processes, offer high productivity, and better balance with the habitat of areas where traditional agriculture produces small yields, or no yield at all. It might be better not to try to make the desert blossom as the rose, but to cultivate the prickly pear. Genetic engineering offers an area where the performance of organisms might deliberately be developed to make them more efficient in particular habitat situations, and in interaction with other species.

The amount of the Sun's energy falling on British hills in a day is equivalent to the U.K. use of energy in a human generation; the amount which is taken into the bodies of sheep, which is our main crop, is infinitesimally small. We are cropping organisms which play a relatively small part in the ecosystem. Soil animals are frequently a higher component of secondary production, and it might be logical to extract and crop earthworms, as it is to use biochemical extraction of leaf protein and other food materials from a very wide range of species. One of the edges of ecological knowledge is in this area of man's interaction with living systems, his optimal cropping of them to sustain our own numbers, and the utilisation of the productivity of vast areas which could nevertheless retain their aesthetic and amenity value.

Another massive area of ignorance is in the ecology of man. In practical terms this is perhaps the most critical of all. Man has escaped over the past two centuries from the kind of influence (such as disease, competition, food shortage and shortage of physical space) that regulate the numbers of other animal species. Such regulatory mechanisms lie in wait for us if we do not relate our numbers to the "carrying capacity" of the environment. There has been much speculation about when such a limit might begin to apply, and whether the brake would come from the stifling effects of pollution, shortage of food and energy, or sheer conflict for space. Many uncritical extrapolations, using over-simplified models and inadequate data, are self-evident nonsense: even a computer cannot make bad data into good. But it remains a fact that the one thing we can be sure of about curves that show a population or its resource consumption mounting at an ever-increasing rate is that sooner or later the growth rate will slow and come to a ceiling. For human numbers the question is how and when, and at what size and standard of living. There are human societies – some we castigate as "primitive" – that have already attained such a balance by behavioural means including late marriage and taboos that reduce fertility. The ecologists cannot alone provide a solution to the human dilemma of how to raise living standards for the deprived masses of the world yet ensure that this good is not undone by reproductive excess – but ecology has a contribution to make to the statement of the options. This contribution must come from rigorous research both on the biology of man (for example, how far does crowding really impose stress: what are basic nutritional needs) and sounder models of the man:environment interaction. It must encompass work on the potential productivity of soils, plants and animals we do not now crop. It will need to take account of a fact commonly forgotten by those who construct over-simplified world models: that the world is a diverse place environmentally, filled with diverse cultures with widely ranging standards of living. The "carrying capacity" and life style sustainable in tropical desert lands and northern temperate countries will always differ. If disaster strikes, it will not strike everywhere at once. No one solution will fit everywhere. Local models, leading to local solutions, are likely to be more useful than global syntheses. After all, the human community is not administered globally, and it is predictable that the predicaments of relating man to environment will have to be tackled at regional and national level: ecological analysis and models will be most useful if worked out at that level rather than generalised to a global statement that may be intellectually stimulating but is likely to be less useful and less meaningful. From such models we may be able to help plan human communities that, under various environmental and social circumstances, maximise human longevity and happiness.

We are beginning to have some time series, and we are beginning to realise what might be measured to create time series which indicate longer-term trends and cycles. A science becomes adolescent when it can predict, and of age when those predictions can be explained in fundamental terms. We may therefore be able to look forward to an increasing extent of ecological prediction and, if it is successful, to the development of underlying principles.

It may be that ecology will never be a useful science because of the sheer complexities of the systems that it is attempting to deal with. We may have to classify living material in more manageable units if we are to understand it at all and some kinds of holistic ecology may be delusions and dangerous ones at that. It may be better, for example, to look at the ecology of man in isolation, thereby simplifying the relationships, and only consider other organisms in so far as they are relevant to humanity on a significant scale. We may be well advised similarly to exclude from ecological considerations a very large part of the total network of interactions and thereby to provide ourselves with simplified models. There is a dangerous tendency in the human mind to strive for the comprehensive and it may be that the lesson of ecology is that that way lies utter confusion. The greatest area of ignorance of all in this field is how to make meaningful abstractions that are not themselves dangerous because they exclude important components.

N. L. Falcon, F.R.S.

Chief Geologist of British Petroleum from 1955 to 1965 and an Honorary Member of the American Society of Petroleum Geologists.

Served on the Natural Environment Research Council from 1966 to 1969 and has been a Vice-President and Murchison Medallist of the Geological Society of London.

IGNORANCE BELOW OUR FEET

The petroleum age in which man is living for a time was initiated in 1859, in Pennsylvania, U.S.A., by a bore hole drilled in search of brine close to an oil seepage. The hole accidently produced petroleum at 69 feet, thereby setting in train deliberate oil exploration, leading to many of the conveniences and complexities of modern living as well as accentuating the problems of humanity's future as far as these can be foreseen. Since 1859 millions of bore holes, some of them over 5 miles deep, have been drilled throughout the world wheresoever the geological circumstances of the sedimentary rocks have suggested hydrocarbons may have accumulated in them. Very many but a lesser number of bore holes have been drilled for other utilitarian purposes such as mineral exploration, coal evaluation and water finding. Deep bore holes drilled for purely scientific purposes have until recently been extremely rare, mostly for detailed information on sedimentary rocks with, of course, notable exceptions such as the 1950 hole on Eniwetok atoll in the Pacific that reached a basaltic basement at 4208 feet to prove finally, after some insufficient penetration by other holes, that Darwin's explanation of atolls visualised in 1837 (subsidence commensurate with coral growth) was valid.

It is remarkable that since oil exploration advanced into the marine areas of the world (to the great benefit of geological knowledge) there has been scientific exploration of the oceans by drilling bore holes in the ocean floor from a ship specially constructed, an activity very much stimulated by the expectancy that something useful might be discovered there. The Deep Sea Drilling Project of the U.S.A. has carried out three phases of a research programme called JOIDES (Joint Oceanographic Investigations of Deep Earth Sampling) during which (up to July 1975) 381 holes have been drilled at critical places in the world's oceans, the data obtained being available to all. JOIDES has been so successful in producing scientific results that it is now to be followed by another programme called IPOD (International Phase of Ocean Drilling) which differs from the first phases only in the increase of the international share of the funding. The main stimulus for these ocean drillings has come from ocean surveys by geophysicists, whose indirect methods of investigating the geology of the Earth's crust produce theoretical interpretations which require confirmation before final acceptance. Without bore holes the great expense of marine geophysics would leave unchecked many important scientific speculations about that part of the Earth's crust accessible to man's curiosity below the sea. The Deep Sea Drilling Programme required a great effort in engineering innovation to make it possible to penetrate the deep ocean floor to significant depths.

It is a strange phenomenon in the history of human endeavour that no international programme of deep drilling *on land* has yet materialised and rarely been discussed – the problems are there, the tools available, the cost in effort and resources far less than ocean drilling (or soil sampling on the Moon or other planets) and the possibility of making "accidental" discoveries of useful resources not insignificant.

The Chinese are credited with bore holes (for brine) drilled with bamboos to a depth of over 1000 feet more than 1100 years ago. In the 1850s Western man seldom drilled deeper than 250 feet; by 1918 the world's deepest bore hole was 7386 feet; in 1930 it was about 10,000 feet and in 1939 about 15,000 feet. Now it is 31,441 feet (almost 6 miles). The technical capacity to improve on this depth is certainly available.

The impact of bore holes on man during the last hundred years has been immense. They have contributed largely to his way of life and affluence. Geological knowledge has been greatly advanced by them (as geologists were reminded so eloquently by H. D. Hedberg in 1971 in the 24th William Smith Lecture of the Geological Society of London). Further deep drilling for scientific purposes on land could not only provide great interest to an increasingly informed public but could help to mitigate some of humanity's future problems. Man is accustomed to look to the mountains, skies and stars for spiritual uplift but has tended to leave the bowels of the earth to fears of disasters and hellish punishments. Such directional prejudice is now irrational: potential natural disasters can come from all directions on many different time scales whereas most material benefits have come from mother earth beneath our feet.

On the land areas of the world traditional geological and geophysical investigations continue, with no "experimental" drilling to check the more important deep speculations. A few nationally funded scientific bore holes are drilled but usually these are of restricted local interest to specialists. The cancelled U.S.A. "Mohole" plan to drill a well in the deep ocean, at a location where the geophysically discovered change of seismic velocity called the Mohorovicic discontinuity is at its shallowest, was cancelled, because of expense, and for geological reasons, although the target might have been accessible somewhere on land. Major crustal problems of importance are available for investigation by deep drilling in the mountains, valleys and rifts of the Earth's mobile belts on land; many of these have been known for a long time but earth scientists have accepted their inability to procure factual data without complaint – now that bore holes are being drilled in the oceans they should react more aggressively.

In 1967 the writer was asked by the Geological Society of London to contribute a paper to a symposium on the relationships between surface geological structure and deep earth displacements in mountain belts. He gave his views based on personal knowledge of an area (the Zagros range of Iran) where an unusual amount of surface geological data, sub-surface geophysical data and deep bore-hole information is available. His conclusions, now published (in *Time and Place in Orogeny*, Geological Society of London, 1969) were as follows:

> It is hoped that the few examples mentioned above, from areas where some borehole evidence is available, will serve as a reminder that the relationship between deep displacements and surface structure in the sedimentary veneer overlying the basement is quantitively indeterminable except at relatively shallow depths where the rock succession is known. Real advances in deep structural knowledge, as distinct from subjective hypothesising, necessarily come from deep and expensive boreholes unlikely to be drilled without some economic incentive. The incentive usually begins with geological analogy and must be checked as far as possible by geophysical observations. The interpretation of structure at depth should always be suspect until facts are provided by the drill, and even then multiple hypotheses are usually required to explain the facts obtained which are rarely definitive. In trying to solve these problems qualitatively no alternative is available to the exploration in depth procedures followed by the extractive industries.

The symposium referred to was part of a project to assemble in a systematic way factual data on mountain belts within the continents from as wide a range as possible. The project terminated in 1974 with the publication of an expensive volume containing papers on 47 areas each well known to the 80 authors involved (*Mesozoic-Cenozoic Orogenic Belts*, Geological Society of London, 1974). A perusal of this volume will reveal many important large-scale structural features which could be erroneous interpretations. On the basis of this book alone a long-term world programme of crustal deep drilling could be planned by a suitable panel of experts. Prolonged discussion would be necessary to arrive at an agreed priority of world targets but this has usually been possible in other major international scientific endeavours. The philosophy of any such programme of scientific deep drilling on land should avoid the dissipation of funds by the mere accumulation of detail and should include the potential discovery of useful resources as an important factor in decision making. The potential benefits from "accidental" discoveries are in the mineral resources and energy fields.

Geologists rarely advertise the fact that many of the important economic and scientific discoveries furthered by their researches have been "accidental" or discovered for the wrong reasons. Thus the extraordinary low angle dislocations below the Rocky Mountains of Canada would certainly not have been correctly appreciated without the many deep bore holes drilled in search of hydrocarbons. Nor would the structure at depth in north-western Greece, nor the history of the development of the Red Sea which has to include an explanation of the presence of one of the world's enormous accumulations of rock salt – about 2 miles thick over the greater part of the Red Sea area. It was an unsuccessful search for coal which led to the accidental discovery of rock salt near Northwich in 1670: it was the search for oil in Yorkshire, itself inspired by an oil seepage in a potash mine in Germany, that discovered the valuable Yorkshire potash field; the Sussex gypsum field was discovered a hundred years ago by a sub-Wealden exploration bore hole drilled for scientific curiosity and the first proof of coal measures under Kent was obtained during the same period by a bore hole drilled in the interests of a Channel tunnel.

Experienced economic geologists, responsible for the location of most of the world's deepest bore holes, will be able to recall many examples where one deep bore hole drilled for scientific curiosity in the early days of an exploration campaign would have revolutionised the planning of their project. In the oilfield belt of the Zagros mountains the nature of the original area of interest was misunderstood and geological work concentrated on the accumulation of surface structural detail that was largely irrelevant to the real problem at depth subsequently revealed by bore-hole evidence. In the early years of the exploration of the Niger delta for hydrocarbons many years of scientifically entirely rational work were done in areas subsequently found to be unproductive. Had one deep bore hole been drilled initially in the delta area, close to the coast, for scientific curiosity, as was suggested by one junior at the time, many years and several millions of pounds would have been saved. In the early years of oil exploration in the Middle East completely erroneous ideas were held by respected professional advisers about the geology at depth on the south side of the Gulf in Arabia: one deep bore hole drilled anywhere in that coastal area at that time would have advanced the discovery of Arabian oil by a decade – in this case perhaps not to the health of the world oil industry then in no condition to withstand a sudden increase of supplies.

Geology is a four-dimensional subject and structural geology is three-dimensional as regards the geometry of the rocks. It requires a deeply dissected topography for its successful prosecution to any considerable depth. Small-scale problems can be solved by small-scale topographical dissection but larger, more fundamental, questions require great ridges and valleys to allow access to rocks that otherwise would be miles below our feet. It was in the Swiss Alps that man first found evidence that great slabs or slices of country had been transported laterally over younger rocks for many tens of miles. There has, of course, been a great uplift of the Alps to allow the deep dissection by erosion there to take place, but great problems remain even in that area that can only be solved by deep drilling. Ideas generated by the study of the geology of the Alps have been applied to the explanation of the structure of many other mountain ranges, perhaps erroneously in some cases.

The preferred topography of the geophysical surveyor is the flat plain, and it has been found that the most important seismic surveys are particularly rapid and obtain the best results below water. Geological knowledge below the oceans is so limited, however, so sketchily known from the very widely distributed grab samples and bore holes, that it can only be a satisfying picture where the structural situation is relatively simple, corresponding to the gently inclined sediment filled basins on the continents. It is irrational to expect simplicity at depth in most areas. Having regard to our known degree of ignorance of geology at depth on land, in most of the belts that have suffered greatly from the past earth movements, it would be prudent to keep in mind the improbability that the number of bore holes likely to be funded in the oceans will be able to settle all major outstanding problems unless comparison is possible with evidence from deep bore holes in similar geological circumstances on land where vastly more surface data is known. Land areas are known where major geological problems of oceanic type are available for deep investigation, now uplifted.

One example of this is the Troodos Range of Cyprus, thought on good geological and geophysical evidence to be an uplifted, sliced piece of the oceanic crust of the earlier Mediterranean, comparable to oceanic crust produced by the process termed "sea-floor spreading" associated with submarine mountain ridges such as the mid-Atlantic Ridge and similar features below the ocean elsewhere. Iceland is another area of oceanic ridge conveniently above water level, and was proposed for a deep scientific bore hole in 1974.

Some examples of disputed geology at depth on the continent where a bore hole could settle, or improve on, very old contentions can be mentioned. An area of long continued interest to earth scientists is the Great African Rift Valley, thought to be a very old line of crustal weakness by some geologists, by others to be a not so old line of developing separation in continental crust which could ultimately develop – in some tens of millions of years – into a Red Sea like feature or become part of the Indian Ocean. The two views are not incompatible, but lack of factual knowledge at depth is a great handicap to the interpretation and real understanding of the geophysical data. The Rift Valley is a zone of high heat flow from the Earth's interior, the type of area that has some potential for thermal energy research. Another example is the Musandam Peninsula, the sharp mountainous promontory of northern Oman that dives down to the north below the Strait of Hormuz at the entrance of the Iranian/Arabian Gulf. This promontory is composed of a thick slab of limestones deposited during the time interval 100 to 300 million years ago. Geologists familiar with the region think that this great slab from 1 to 3 miles in thickness, is a great slice of the old edge of the Arabian continental crust that has been pushed or slid in some way towards the west over the continental edge when an earlier ocean lay on the east side of Arabia – a relatively rapid geological event that took place about 100 million years ago, before the Oman mountains had been uplifted, eroded and tilted. No amount of surface geology or geophysical surveying can be expected to prove or disprove this scientific speculation based on analogy with Alpine structures, but one medium depth bore hole (say to a maximum depth of 2 miles) could do so. It would also throw a flood of light on the geological nature of a larger area of interest to earth scientists working in the plate tectonic field and to the mineral and petroleum industry. Another example, nearer home, is in North Devon, the basic underlying structure of which has been discussed in scientific papers since 1891. A recent paper examining the problem – whether the northern coast is or is not underlain by a great structural slide of Alpine or Rocky Mountain type – lists forty-one references to relevant scientific papers, to which several more must be added since the paper was written. This problem, also of some interest to the mineral and petroleum industry, can only be settled by a bore hole. Again a maximum depth of 2 miles would seem adequate, located on the oldest rocks outcropping in the Lynton/Lynmouth area.

Enormous expenditure has been required to bring back a small collection of rocks from the Moon. This interesting and marvellous technical achievement has provided great uplift to the human spirit. It also makes it more surprising that so much ignorance is allowed to remain below our feet on Earth when much could be dissipated by comparatively inexpensive action. This is not due to lack of intellectual interest but to a resigned acceptance of the unavailability of funds for such mundane adventures.

The many problems of world-wide geological interest now being studied by earth scientists in international commissions, such as the Geodynamics Commission of the International Union of Geodesy and Geophysics, and the International Geological Correlation Programme of the International Union of Geological Sciences and UNESCO, are all susceptible to a flood of understanding should their deliberations be accompanied by bore holes. But in the ever-increasing paper work being produced by international earth science, bore holes seem to be taboo. If a text is needed to change this "Ask, and it shall be given you; seek and ye shall find; knock, and it shall be opened unto you" might be appropriate.

The Times for Saturday, 21 June 1975, contained an article by Sir George Porter entitled "Science and the quest for human purpose". Sir George, distinguished Director of The Royal Institution that has given so much to humanity's knowledge and well-being, starts with the following paragraph:

> What is it that we want of ourselves, of man, of our earth, of the universe? What is the purpose of it all? In the past, these questions have been answered by the theologians and the answer, being rather pleasant, was readily accepted. But man's reason does not permit him to think happy thoughts which are irrational and many have had to discard the traditional religions on these grounds. Our great dilemma is that science has not yet helped man to find a new religion which in any way replaces the old ones. There are philosophies of life, such as humanism which provide a *modus vivendi*, but they do little to solve the basic questions answered so confidently by the old religions.

Some 1250 words later, after discussing the interrelations of religion and science and ethics, he concludes as follows:

> There is, then one great purpose for man and for us to day, and that is to try to *discover* man's purpose by every means in our power. That is the ultimate relevance of science, and not only of science but of every branch of learning which can improve our understanding. In the words of Tolstoy, "The highest wisdom has but one science, the science of the whole, the science explaining the Creation and man's place in it".

Many will agree with this view. Many will think it is impossible for man who is part of the animal kingdom to comprehend the whole, for should he do so he would become MAN and equivalent to the Creator. Fortunately curiosity about the marvels of nature can last everybody a lifetime whatever his degree of educational sophistication, for they seem to be inexhaustible. But in exercising our scientific curiosity let us not forget the earth beneath our feet; this 4.5×10^9 year old near spherical cornucopia has had its surface until now merely pricked by our inquisitiveness to a depth of 1/660 of its radius. Great intellectual and material treasures may await discovery by deeper prickings.

Roger W. Sperry

Hixon Professor of Psychobiology at the California Institute of Technology.

Member of the United States National Academy of Sciences, the American Academy of Arts and Sciences, the American Philosophical Society and of the Royal Society of London.

PROBLEMS OUTSTANDING IN THE EVOLUTION OF BRAIN FUNCTION

As indicated in the title, I am more concerned with the functional than with the morphological properties of the brain, and more with remaining unsolved problems than with the solid progress over which we already can beat our chests.

I wish to skip the beginning steps in the evolution of the human brain and pick up the story at about the culmination of the latter half of the age of hydrogen gas. In such a way we can bypass what is by far the most difficult of all the unsolved problems in brain evolution, namely, how, when, and where did the hydrogen age and the whole business start? This problem we can leave to the proponents of the "steady state", the "periodic pulsation", and the "big bang", at least until someone comes along with a more credible interpretation of the meaning of the red shift.

We can skip quickly also through those early periods when, first, electrons and protons were being used to build bigger and better atoms, and then the atoms to make bigger and finer molecules, and then these in turn were being compounded into giant and replicating molecules and self organizing molecular complexes and eventually that elaborate unit, the living cell.

We note for future reference that evolution keeps complicating the Universe by adding new phenomena that have new properties and new forces and that are regulated by new scientific principles and new scientific laws – all for future scientists in their respective disciplines to discover and formulate. Note also that the old simple laws and primeval forces of the hydrogen age never get lost or cancelled in the process of compounding the compounds. They do, however, get superseded, overwhelmed, and outclassed by the higher-level forces as these successively appear at the atomic, the molecular, and the cellular and higher levels.

We can turn now to what is probably the "most unanswered" problem in brain evolution. We encounter it a bit later on, presumably after organisms with nerve nets and brains have entered the picture. I refer to the first appearance of that most important of all brain properties and certainly the most precious, conscious awareness.

In any case, the fossil record notwithstanding, there seems to be good reason to regard the evolutionary debut of consciousness as very possibly the most critical step in the whole of evolution. Before this, the entire cosmic process, we are told, was only, as someone has phrased it, "a play before empty benches" – colorless and silent at that, because according to our best physics, before brains there was no color and no sound in the Universe, nor was there any flavor or aroma and probably rather little sense and no feeling or emotion.

All of these can now be generated by the surgeon's electrode tip applied to the proper region of the exposed conscious brain. They can be triggered also, of course by the proper external stimuli, but also, more interestingly, by centrally initiated dream states, illusionogenic and hallucinogenic agents, but always and only within and by a brain. There probably is no more important quest in all science than the attempt to understand those very particular events in evolution by which brains worked out that special trick that has enabled them to add to the cosmic scheme of things: color, sound, pain, pleasure, and all the other facets of mental experience.

In searching brains for clues to the critical features that might be responsible, I have never myself been inclined to focus on the electrons, protons, or neutrons of the brain, or on its atoms. And, with all due respect to biochemistry and the N.R.P., I have not been inclined to look particularly at the little molecules of the brain or even at its big macro-molecules in this connection. It has always seemed rather improbable that even a whole brain cell has what it takes to sense, to perceive, to feel, or to think on its own. The "search for psyche", in our own case at least, has been directed mainly at higher-level configurations of the brain, such as specialized circuit systems, and not just any juicy central nerve network that happens to be complex and teeming with electrical excitations. I have been inclined to look rather at circuits specifically designed for the express job of producing effects like pain, or High C or blue-yellow – circuits of the kind that one finds above a high transection of the spinal cord but not below, circuits with something that may well be present in the tiny pinhead dimensions of the midbrain of the color-perceiving goldfish but lacking in the massive spinal-cord tissue of the ox, circuits that are profoundly affected by certain lesions of the midbrain and thalamus but little altered by complete absence of the entire human cerebellum. Were it actually to come to laying our money on the line, I should probably bet, first choice, on still larger cerebral configurations, configurations that include the combined effect of both (a) the specialized circuit systems such as the foregoing plus (b) a background of cerebral activity of the alert, waking type. Take away either the specific circuit, or the background, or the orderly activity from either one, and the conscious effect is gone.

In this day of information explosion, these matters are not so much of the "ivory tower" as they used to be: To the engineer who comes around from Industrial Associates with dollars and cents in his eye and company competition in his heart the possibility is of more than theoretical interest that conscious awareness may be something that is not necessarily tied to living hardware, that it could prove to be an emergent, over-all circuit property that might, in theory, be borrowed and, given sufficient acreage, perhaps copied some day in order to incorporate pain and pleasure, sensations and percepts, into the rapidly evolving circuitry of computer intellect. When the aim is to build into your circuit systems some kind of negative and positive reinforcement, then pain and pleasure are about the best kind. And eager young theoreticians from the NASA committee or from radio astronomy already want a more educated guess about the possibility of encountering on other globes other minds with perhaps totally different dimensions of conscious awareness, and if not, why not? Then there are more imminent, practical matters such as the need, in view of certain other explosions we face, to be able to pinpoint the first appearance of consciousness in embryonic development and chart its subsequent growth and maturation.

Unless you believe that value judgments lie outside the realm of science, you may probably agree that a few reliable answers in these general areas and their implications could shake considerably the going value systems of our whole culture.

To shift now to certain lesser and subsidiary problems, but problems more approachable in research. Unlike the situation 25 years ago, most of us today are quite ready to talk about the evolution and inheritance not only of brain morphology but also of brain function, including general behavior and specific behavior traits. Earlier renunciation of the whole instinct concept in the animal kingdom generally stemmed in large part from our inability to imagine any growth mechanisms sufficiently precise and elaborate even to begin the fabrication of the complex nerve networks of behavior. This outlook was supported in the analytic studies of nerve growth all through the 1920s and 1930s which indicated that nerve fibers grow and connect in a random, diffuse, and non-selective manner governed almost entirely by indifferent mechanical factors.

Today the situation is entirely changed. The supposed limitations in the machinery of nerve growth are largely removed in the new insight that we have obtained in recent years into the way in which the complicated nerve-fiber circuits of the brain grow, assemble, and organize themselves in a most detailed fashion through the use of intricate chemical codes under genetic control (Sperry, 1950a, 1950b, 1951, 1958, 1961, 1962, 1963). The new outlook holds that the cells of the brain are labeled early in development with individual identification tags, chemical in nature, whereby the billions of brain cells can thereafter be recognized and distinguished, one from another. These chemical differentials are extended into the fibers of the maturing brain cells as these begin to grow outward, in some cases over rather long distances, to lay down the complicated central communication lines. It appears from our latest evidence that the growing fibers select and follow specific prescribed pathways, all well marked by chemical guideposts that direct the fiber tips to their proper connection sites. After reaching their correct synaptic zones, the fibers then link up selectively among the local population with only those neurons to which they find themselves specifically attracted and constitutionally matched by inherent chemical affinities.

The current scheme now gives us a general working picture of how it is possible, in principle at least, for behavioral nerve nets of the most complex and precise sorts to be built into the brain in advance without benefit of experience. Being under genetic control, these growth mechanisms are of course inheritable and subject to evolutionary development. The same is true of the differential endogenous physiological properties of the individual cell units in these networks which, along with the morphological interconnections are of critical importance in the shaping of behavior patterns. We have at present only the general outlines and general principles of the developmental picture; much of the detail has yet to be worked out. Also, the underlying chemistry of the demonstrated selectivity in nerve growth, as well as the molecular basis of the morphogenetic gradients involved, and of all the rest of the chemical "I.D. Card" concept remains a wide-open field that so far has been virtually untouched.

In connection with this emphasis on the "inherited" in brain organization, one may well question the extent to which the observed inbuilt order in the anatomical structure necessarily conditions functional performance and behavior. Some years ago, when we subscribed to the doctrine of an almost omnipotent adaptation capacity in the central nervous system and to functional equipotentiality of cortical areas and to the functional interchangeability among nerve connections in general (Sperry, 1958), Karl Lashley surmised that if it were feasible, a surgical rotation through 180 degrees of the cortical brain center for vision would probably not much disturb visual perception. Rotation of the brain center was not feasible, but it was possible to rotate the eyes surgically through 180 degrees in a number of the lower vertebrates and also to invert the eyeball, by transplantation from one to the other orbit, on the up-down or on the front-back axis and also to cross-connect the right and left eyes to the wrong side of the brain (Sperry, 1950a). All these and different combinations thereof were found to produce very profound disturbances of visual perception that were correlated directly in each case with the geometry of the sensory disarrangement. The animals, after recovery from the surgery, responded thereafter as if everything were to them upside down and backward, or reversed from left to right, and so on. Contrary to earlier suppositions regarding the dynamics of perception and cortical organization, it appeared that visual perception was very closely tied indeed to the underlying inherited structure of the neural machinery.

We inferred further from nerve-lesion experiments (Sperry, 1950b) on the illusory spinning effects produced by these visual inversions that the inbuilt machinery of perception must include also certain additional central mechanisms by which an animal is able to distinguish those sensory changes produced by its own movement from those originating outside. The perceptual constancy of an environment in which an animal is moving, for example, or of an environment that it is exploring by eye, head, or hand movements, would seem to require that, for every movement made, the brain must fire "corollary discharges" into the perceptual centers involved. These anticipate the displacement effect and act as a kind of correction or stabilizing factor. These centrally launched discharges must be differentially gauged for the direction, speed, and distance of each move. Along with the dynamic schema for body position which the brain must carry at all times, these postulated discharges conditioning perceptual expectancy at every move would appear to be a very important feature of the unknown brain code for perception. The consistent appearance of the spontaneous optokinetic reaction of inverted vision in fishes, salamanders, and toads would indicate that the underlying mechanism is basic and must have evolved very early.

Since the representation of movement at higher cortical levels generally seems to be more in terms of the perceptual expectancy of the end effect of the movement than in terms of the actual motor patterns required to mediate the movement, the postulated "corollary discharges" of perceptual constancy may not involve so much of an additional load, in terms of data processing, as might at first appear.

We are ready now for that old question: How much of brain organization and behavior should we blame or credit to inheritance and how much to learning and experience? As far as we can see now, it seems fair to say that all that central nervous organization that is illustrated and described in the voluminous textbooks, treatises, and professional journals of neuro-anatomy, that is, all the species-constant patterning of brain structure, the micro-architecture as well as the gross morphology that has so far been demonstrated anatomically, seems to be attributable to inheritance. Another way of saying the same thing is that no one has yet succeeded in demonstrating anatomically a single fiber or fiber connection that could be said with assurance to have been implanted by learning. In this same connection, it is entirely conceivable (though not particularly indicated) that the remodeling effects left in the brain by learning and experience do not involve the addition or subtraction of any actual fibers or fiber connections but involve only physiological, perhaps membrane, changes that effect conductance or resistance to impulse transmission, or both, all within the existing ontogenetically determined networks.

The foregoing picture leaves plenty of room for learning and for the combined effects of learning plus maturation during that prolonged period in human childhood when these two factors overlap. Nevertheless the present picture represents a very considerable shift of opinion over the past two decades in the direction of inheritance.

Some may find certain aspects here a bit difficult to reconcile with other inferences drawn in recent years from a series of sensory deprivation studies on mammals in which cats, monkeys, chimpanzees, and other animals have been raised in the dark or with translucent eye caps or in harnesses or holders of various sorts and in which, as a result of the various kinds of deprivation of experience in their early development the animals came to show subsequent deficits, moderate to severe, in their perceptual or motor capacities. The tendency to interpret these findings, along with those from human cataract cases, as evidence of the importance of early learning and experience in shaping the integrative organization of the brain we have long felt to have been overdone (Sperry, 1950a, 1962). In nearly all cases the findings could be equally well explained on the assumption that the effect of function is simply to maintain, or to prevent the loss of, neural organization already taken care of by growth. What the results have come to show in many of these studies is that certain of the newly formed neuronal elements, if abnormally deprived of adequate stimulation, undergo an atrophy of disuse. In much the same way cells of the skeletal muscles differentiate in development to the point at which they are contractile and ready to function, but then they too atrophy and degenerate if not activated. This basic developmental "use-dependent" property in maturing neurons, or even some evolutionary derivative of it, applied farther centrally beyond the sensory paths amid more diffuse growth pressures, especially among cortical association units, could, however, have true patterning effects and become a definitely positive factor in learning and imprinting.

We have been approaching very closely here the general problem of memory. Among brain functions, memory certainly rates as one of the prime "problems outstanding". Whatever the nature of the neural mechanism underlying memory, it seems to have appeared quite early in evolution. (Some writers say that even flatworms have memory!) We are frequently impressed in our work with learning and memory in cats, and even in fishes, with the fact that their simple memories, once implanted, seem to be strong and lasting. With respect to memory, then, what separates the men from the animals is very likely not so much the nature of the neural trace mechanism as the volume and the kind of information handled. The problems that relate to the translation and coding of mental experience, first into the dynamics of the brain process, and then into the static, frozen, permanent trace or engram system, pose the more formidable aspects of the memory problem.

Fundamental to these memory questions, as also to the problems of perception, volition, learning, motivation, and most of the higher activities of the nervous system, is that big central unknown that most of us working on the higher properties of the brain keep tangling with and coming back to. You may find it referred to variously as: the "brain code", or the "cerebral correlates of mental experience", or the "unknown dynamics of cerebral organization", or the "intermediary language of the cerebral hemispheres", or in some contexts, just the "black box". Thus far we lack even a reasonable hypothesis regarding the key variables in the brain events that correlate with even the simplest of mental activities, such as the elementary sensations or the simple volitional twitch of one's little finger.

In our own efforts to help to chip away at this central problem of the language of the hemispheres, we have been trying for some 10 years first to divide the problem in half by splitting the brain down the middle before we start to study it. (Many times we wonder if the end effect of this split brain approach is not so much to halve our problems as it is to double them.) At any rate, the brain-bisection studies leave us with a strong suspicion that evolution may have saddled us all with a great deal of unnecessary duplication, both in structure and in the function of the higher brain centers.

Space in the intracranial regions is tight, and one wonders if this premium item could not have been utilized for better things than the kind of right-left duplication that now prevails. Evolution, of course, has made notable errors in the past, and one suspects that in the elaboration of the higher brain centers evolutionary progress is more encumbered than aided by the bilateralized scheme which, of course, is very deeply entrenched in the mechanisms of development and also in the basic wiring plan of the lower nerve centers.

Do we really need two brain centers, for example, to tell us that our blood sugar is down or our blood pressure is up, or that we are too hot or too cold, and so on? Is it necessary to have a right and also a left brain center to let us know that we are sleepy or angry, sad or exuberant, or that what we smell is Arpege or what we taste is salty or that what we hear is voices, and so on and on and on? Surely most of us could manage to get along very well with only one cerebral anxiety mechanism, preferably in the minor hemisphere.

Emotion, personality, intellect, and language, among other brain business, would seem by nature to be quite manageable through a single unified set of brain controls. Indeed, the early loss of one entire hemisphere in the cat, monkey, and even in man causes amazingly little deficit in the higher cerebral activities in general.

With the existing cerebral system, most memories as well have to be laid down twice – one engram for the left hemisphere and another engram copy for the right hemisphere. The amount of information stored in memory in a mammalian brain is a remarkable thing in itself; to have to double it all for the second hemisphere would seem in many ways a bit wasteful. It is doubtful that all this redundancy has had any direct survival value (unless evolution could have foreseen that neurologists would be opening and closing the cranium to produce brain lesions under careful aseptic conditions that permit survival).

In the human brain, of course, we begin to see definite evidence of a belated tendency in evolution to try to circumvent some of the duplication difficulties. A de-duplication trend is seen particularly in the lateralization of speech and writing within the single dominant hemisphere in the majority of persons. Speech, incidentally, is another essentially symmetrical activity for which a double right and left control is quite unnecessary, even at the lower levels of the motor hierarchy. When the brain does try in some individuals to set up two central administrations for speech, one in each hemisphere, the result tends to make for trouble, like stammering and a variety of other language difficulties.

The fact that the corpus callosum interconnecting the two cerebral hemispheres is so very large and the functional damage produced by its surgical section is so very minor in most ordinary activities seems to be explainable in part by the fact that the great cerebral commissure is a system for cross communication between two entities that to a large extent are each completely equipped and functionally self-sufficient. The corpus callosum appears late in evolution, being essentially a mammalian structure, and its development is closely correlated with the evolutionary elaboration of the neocortex of the mammalian cerebral hemispheres.

Accordingly it is not surprising that it is in the human brain, and particularly in connection with speech, that the functional effects produced by surgical disconnection of the two cerebral hemispheres become most conspicuous. During the past two years we have had an opportunity to test and to study two patients, formerly unmanageable epileptics, who have had their right and left hemispheres disconnected by complete section of the corpus callosum, plus the anterior commissure, plus the hippocampal commissure, plus the massa intermedia, in what is perhaps the most radical surgical approach to epilepsy thus far undertaken. The surgery was done by Drs. Philip J. Vogel and Joseph E. Bogen (see Bogen and Vogel, 1962) of Los Angeles.†

It seemed a reasonable hope, in advance, that such surgery might help to restrict the seizures to one hemisphere and hence to one side of the body, and possibly to the distal portions of arm and leg, since voluntary control of both sides of the head, neck, and trunk tends to be represented in both hemispheres. In our colony of split-brain monkeys that have had similar surgery, we not uncommonly see epileptic-like seizures, especially during the early weeks after brain operations, and these seizures show a definite tendency to center in the distal extremities of the arm and leg and to be restricted to one side. It also seemed reasonable that this surgery might help the patients to retain consciousness in one hemisphere during an attack, if not throughout, at least during the early stages, and thereby give them a chance to do things that might help to break or control the seizures, or at a minimum to allow time in which to undertake protective measures before the second side became involved. It was further hoped that such surgery might reduce the severity of the attacks by the elimination of a very powerful avenue for the right-left mutual reinforcement of the seizures during the generalized phase, especially during *status epilepticus*, which was of major concern in both the cases cited above.

Judged from earlier reports of the cutting of the corpus callosum and from the behavior of

† The surgical treatment of these cases was undertaken at the suggestion of Dr. Bogen after extensive consultations on all aspects. The surgery was performed by Dr. Vogel, assisted by Dr. Bogen and other staff members at the Loma Linda Neurosurgical Unit, White Memorial Hospital. Most of the tests reported here were planned and administered by Michael S. Gazzaniga of our laboratory, with the writer collaborating on a general advisory basis.

The original work included here was supported by grants from the National Institutes of Health (MH-03372, S-F1-MH-18,080) and by the Frank P. Hixon Fund of the California Institute of Technology.

dozens of monkeys that we have observed in the laboratory with exactly the same surgery (Sperry, 1961, 1964), it seemed unlikely that this kind of surgery would produce any severe handicap, or surely none so bad as certain other forms of psychosurgery that have been used on a much more extensive scale. That the surgery might decrease the incidence of the seizures to the point of virtually eliminating them (as it seems to have done so far in both cases) was unexpected; our fingers remain very much crossed on this latter point.

Everything that we have seen so far indicates that the surgery has left each of these people with two separate minds, i.e. with two separate spheres of consciousness (Gazzaniga *et al.*, 1962, 1963). What is experienced in the right hemisphere seems to be entirely outside the realm of awareness of the left hemisphere. This mental duplicity has been demonstrated in regard to perception, cognition, learning, and memory. One of the hemispheres, the left, dominant, or major hemisphere, has speech and is normally talkative and conversant. The other mind of the minor hemisphere, however, is mute or dumb, being able to express itself only through non-verbal reactions; hence mental duplicity in these people following the surgery, but no double talk.

Fortunately, from the patients' standpoint, the functional separation of the two hemispheres is counteracted by a large number of unifying factors that tend to keep the disconnected hemispheres doing very much the same thing from one part of the day to the next. Ordinarily, a large common denominator of similar activity is going in each. When we deliberately induce different activities in right and left hemispheres in our testing procedures, however, it then appears that each hemisphere is quite oblivious to the experiences of the other, regardless of whether the going activities match or not.

This is illustrated in many ways: For example, the subject may be blindfolded, and a familiar object such as a pencil, a cigaret, a comb, or a coin is placed in the left hand. Under these conditions, the mute hemisphere connected to the left hand, feeling the object, perceives and appears to know quite well what the object is. It can manipulate it correctly; it can demonstrate how the object is supposed to be used; and it can remember the object and go out and retrieve it with the same hand from among an array of other objects. While all this is going on, the other hemisphere has no conception of what the object is and says so. If pressed for an answer, the speech hemisphere can only resort to the wildest of guesses. So the situation remains just so long as the blindfold is kept in place and other avenues of sensory input from the object to the talking hemisphere are blocked. But let the right hand cross over and touch the test object in the left hand; or let the object itself touch the face or head as in the use of a comb, a cigaret, or glasses; or let the object make some give-away sound, such as the jingle of a key case, then immediately the speech hemisphere produces the correct answer.

The same kind of right-left mental separation is seen in tests involving vision. Recall that the right half of the visual field and the right hand are represented together in the left hemisphere and vice versa. Visual stimuli such as pictures, words, numbers, and geometric forms flashed on a screen directly in front of the subject and to the right side of a central fixation point are all described and reported correctly with no special difficulty. On the other hand, similar material flashed to the left half of the visual field is completely lost to the talking hemisphere. Stimuli flashed to one half field seem to have no influence whatever, in tests to date, on the perception and interpretation of stimuli presented to the other half field.

Note in passing that these disconnection effects do not show up readily in ordinary behavior. They must be demonstrated by the flashing of the visual material fast enough so that eye movements cannot be used to sneak the answers into the wrong hemisphere, or in the testing of right and left hands vision must be excluded with a blindfold, auditory cues eliminated, and the hands kept from crossing, and so on. One of the patients, a 30-year-old housewife with two children, goes to market, runs the house, cooks the meals, watches television, and goes out to complete, 3-hour shows at the drive-in theater, all without complaining of any particular splitting or doubling in her perceptual experience. Her family believes that she still does not have so much initiative as formerly in her housecleaning, in which she was meticulous, and that her orientation is not so good, for example, she does not find her way back to the car at the drive-in theater as readily as she formerly could. In the early months after surgery there appeared to be definite difficulty with memory. By now, some 8 months later, there seems to be much improvement in this regard, though not complete recovery. Involvement of the fornix would have to be ruled out before effects like the latter can be ascribed to the commissurotomy.

In the visual tests again, one finds plenty of evidence that the minor, dumb, or mute hemisphere really does perceive and comprehend, even though it cannot express verbally what it sees and thinks. It can point out with the left hand a matching picture from among many others that have been flashed to the left field, or it can point to a corresponding object that was pictured in the left-field screen. It can also pick out the correct written name of an object that it has seen flashed on the screen, or vice versa. In other words, Gazzaniga's more recent results show that the dumb left hemisphere in the second patient is not exactly stupid or illiterate; it reads a word such as "cup", "fork", or "apple" flashed to the left field and then picks out the corresponding object with the left hand. While the left hand and its hemisphere are thus performing correctly, however, the other hemisphere, again, has no idea at all which object or which picture or which name is the correct one and makes this clear through its verbal as well as other responses. You regularly have to convince the talking hemisphere to keep quiet and to let the left hand go ahead on its own, in which case it will usually pick out the correct answer.

These minor differences of opinion between the right and left hemispheres are seen rather commonly in testing situations. For example, the left hand is allowed to feel and to manipulate, say, a toothbrush under the table or out of sight behind a screen. Then a series of five to ten cards are laid out with names on them such as "ring", "key", "fork". When asked, the subject may tell you that what she felt in the left hand was a "ring". However, when instructed to point with the left hand, the speechless hemisphere deliberately ignores the erroneous opinions of its better half and goes ahead independently to point out the correct answer, in this case the card with the word "toothbrush".

As far as we can see, about the only avenue remaining for direct communication between mind-right and mind-left is that of extrasensory perception. If any two minds should be able to tune in on each other, one might expect these two to be able to do so, but thus far no evidence of such effects is apparent in the test performances.

The conscious awareness of the minor hemisphere produced by this vertical splitting of the brain often seems so remote to the conversant hemisphere as to be comparable perhaps to that produced by a spinal transection. To go back here to some of the issues on which we started, one wonders if we can really rule out, as I implied above, the alternative contention of those who maintain that spinal cords, loaves of bread, and even single molecules have a kind of consciousness. Either way, the inferences to be drawn regarding the evolution and elaboration of consciousness for most practical purposes remain much the same.

We are often asked if each of the disconnected hemispheres must not also have a will of its own and if the two do not then get into conflict with each other. In the first half year after surgery, particularly with the first patient, we got reports suggesting something of the kind. For example, while the patient was dressing and trying to pull on his trousers, the left hand started to work against the right, pulling them off again. Or, the left hand, after just helping to tie the belt of his robe, went ahead on its own to untie the completed knot, whereupon the right hand would have to supervene again to get it retied. The patient and his wife used to refer to the "sinister left hand" which sometimes tried to push the wife away aggressively at the same time that the hemisphere of the right hand was trying to get her to come and help him with something. These antagonistic movements of right and left hands are fairly well restricted to situations in which the reactions of left and right hand are easily made from the same common supporting posture of body and shoulders. Generally speaking, there are so many unifying factors in the situation and functional harmony is so strongly built into the undivided brain stem and spinal networks, by express design, that one sees little overt expression or overflow into action, at least, of conflicting will power.

This matter of having two free wills packed together inside the same cranial vault reminds us that, after consciousness, free will is probably the next most treasured property of the human brain. Questions and information relating to the evolution of free will have practical impact rating right at the top, along with those of consciousness. As such it probably deserves at least a closing comment. Some maintain that free will is an evolved, emergent property of the brain that appeared between man and the higher apes, or, depending on whom you read, maybe somewhere after bacteria perhaps, but before houseflies.

Unlike "mind", "consciousness", and "instinct", "free will" has made no comeback in behavioral science in recent years. Most behavioral scientists would refuse to list free will among our problems outstanding, or at least as an unanswered problem. (To agree that behavior is unlawful in this respect might put them out of work as scientists, you see, and oblige them perhaps to sign up with the astrologers' union.) Every advance in the science of behavior, whether it has come from the psychiatrist's couch, from microelectrode recording, from brain-splitting, or from the running of cannibalistic flatworms, seems only to reinforce that old suspicion that free will is just an illusion. The more we learn about the brain and behavior, the more deterministic, lawful, and causal it appears.

In other words, behavioral science tells us that there is no reason to think that any of us had any real choice to be anywhere else, or even to believe in principle that our presence was not already "in the cards", so to speak, 5, 10, or 15 years ago. I do not feel comfortable with this kind of thinking any more than you do, but so far I have not found any satisfactory way around it. Alternatives to the rule of causal determinism in behavior proposed so far, like the inferred unlawfulness in the dance of subatomic particles, seem decidedly more to be deplored as a solution than desired.

The above statements are not to say that, in the practice of behavioral sciences, we must regard the brain as just a pawn of the physical and chemical forces that play in and around it. Far from it. To go back to the beginning, recall that a molecule in many respects is the master of its inner atoms and electrons. The latter are hauled and forced about in chemical interactions by the over-all configurational properties of the whole molecule. At the same time, if our given molecule is itself part of a single-celled organism such as paramecium, it in turn is obliged, with all its parts and its partners, to follow along a trail of events in time and space determined largely by the extrinsic over-all dynamics of *Paramecium caudatum.* When it comes to brains, remember that the simpler electric, atomic, molecular, and cellular forces and laws, though still present and operating, have been superseded by the configurational forces of higher-level mechanisms. At the top, in the human brain, these include the powers of perception, cognition, reason, judgment, and the like, the operational, causal effects and forces of which are equally or more potent in brain dynamics than are the outclassed inner chemical forces.

The underlying policy here: "If you can't lick 'em, join 'em," or, as Confucius might say, "If fate inevitable, relax and enjoy", or, "There may be worse fates than causal determinism". Maybe, after all, it is better to be embedded firmly in the causal flow of cosmic forces, as an integral part thereof, than to be on the loose and out of contact with these forces, "free floating" as it were and with behavioral possibilities than have no antecedent cause and hence no reason, nor any reliability when it comes to future plans, predictions, or promises.

And on this same theme, just one final point: If you were assigned the task of trying to design and build the perfect free-will model (let us say the perfect, all-wise, decision-making machine to top all competitors' decision-making machines), consider the possibility that your aim might not be so much to free the machinery from causal contact, as the opposite, that is, to try to incorporate into your model the potential value of universal causal contact; in other words, contact with all related information in proper proportion – past, present, and future.

It is clear that the human brain has come a long way in evolution in exactly this direction when you consider the amount and the kind of causal factors that this multidimensional intracranial vortex draws into itself, scans, and brings to bear on the process of turning out one of its "preordained decisions". Potentially included, thanks to memory, are the events and collected wisdom of most of a human lifetime. We can also include, given a trip to the library, the accumulated knowledge of all recorded history. And we must add to all the foregoing, thanks to reason and logic, much of the future forecast and predictive value extractable from all these data. Maybe the total falls a bit short of universal causal contact; maybe it is not even quite up to the kind of thing that evolution has going for itself over on Galaxy Nine; and maybe, in spite of all, any decision that comes out is still predetermined. Nevertheless, it still represents a very long jump in the direction of freedom from the primeval slime mold, the Jurassic sand dollar, or even the latest model orangutan.

ACKNOWLEDGEMENT

Reprinted with minor editorial changes from a publication of the American Museum of Natural History.

REFERENCES

Bogen, J. E. and Vogel, P. J. (1962) Cerebral commissurotomy in man. *Bull. Los Angeles Neurol. Soc.* **27**, 169.

Gazzaniga, M. S., Bogen, J. E. and Sperry, R. W. (1962) Some functional effects of sectioning the cerebral commissures in man. *Proc. Nat. Acad. Sci.* **48**, pt. 2, 1765.

Gazzaniga, M. S., Bogen, J. E. and Sperry, R. W. (1963) Laterality effects in somesthesis following cerebral commissurotomy in man. *Neuropsychologia,* **1**, 209.

Sperry, R. W. (1950a) Mechanisms of neural maturation. *In*: Stevens, S. S. (Ed.), *Handbook of Experimental Psychology*, New York, John Wiley and Sons, 236.

Sperry, R. W. (1950b) Neural basis of the spontaneous optokinetic response produced by visual inversion. *J. Comp. Physchol.* **43**, 482.

Sperry, R. W. (1951) Regulative factors in the orderly growth of neural circuits. *Growth Symp.* **10**, 63.

Sperry, R. W. (1958) Physiological plasticity and brain circuit theory. *In*: Harlow, H. F. and Woolsey, C. N. (Eds.), *Biological and Biochemical Bases of Behavior*, Madison, University of Wisconsin Press.

Sperry, R. W. (1961) Some developments in brain lesion studies of learning. *Fed. Proc.* **20**, 609.

Sperry, R. W. (1962) How a random array of cells can learn to tell whether a straight line is straight – discussion on. *In*: Foerster, H. von, and Zopf, G. W. Jr. (Eds.), *Principles of Self-organization*, New York, Pergamon Press, 323.

Sperry, R. W. (1963) Chemoaffinity in the orderly growth of nerve fiber patterns and connections. *Proc. Nat. Acad. Sci.* **50**, 703.

Sperry, R. W. (1964) The great cerebral commissure. *Sci. Amer.* **210**, 42.

INDEX